The General Circulation of the Tropical Atmosphere

and Interactions with Extratropical Latitudes

The MIT Press
Cambridge, Massachusetts, and London, England

The General Circulation of the Tropical Atmosphere

and Interactions with Extratropical Latitudes

Volume 1

Reginald E. Newell, John W. Kidson, Dayton G. Vincent, and George J. Boer

With contributions by Eugene M. Rasmusson and Zen-Kay Jao

Second printing, August 1973

This book was set in Linotype Baskerville
by The Colonial Press Inc.
printed on P&S offset
by Halliday Lithograph Corp.
and bound in GSB S/535 (silver black)
by The Colonial Press Inc.
in the United States of America.

Library of Congress Cataloging in Publication Data
Main entry under title:

The General circulation of the tropical atmosphere
    and interactions with extratropical latitudes.

    Includes bibliographies.
    1. Atmospheric circulation—Tropics.
2. Meteorology—Tropics. I. Newell, Reginald E.,
1931–
QC880.G44   551.5′17′0913   72–6834
ISBN 0–262–14012–8

ERRATA

p. 15   Plate 2.1. OWS-E should be plotted at 35°N in this and subsequent maps.
p. 20   Fig. 3.1. On these and subsequent cross sections the altitude scale refers to an average tropical sounding.
p. 78–81   Results for station Trindade in the South Atlantic are thought to be influenced by postvolcanic data (see Volume 2).

The work reported in this volume was done while the authors, with the exception of Eugene M. Rasmusson of the Geophysical Fluid Dynamics Laboratory, ESSA, Princeton, were members of the Department of Meteorology, Massachusetts Institute of Technology. The authors' present affiliations, where different, are

John W. Kidson
NEW ZEALAND METEOROLOGICAL SERVICE
WELLINGTON, NEW ZEALAND

Dayton G. Vincent
PURDUE UNIVERSITY
LAFAYETTE, INDIANA

Eugene M. Rasmusson
BOMAP OFFICE
NATIONAL OCEANIC AND ATMOSPHERIC ADMINISTRATION
ROCKVILLE, MARYLAND

Zen-Kay Jao
GODDARD INSTITUTE FOR SPACE STUDIES
NEW YORK

Unless otherwise indicated the text was written by the four principal authors.

Contents

4

**The Angular Momentum Budget and the Maintenance
of the Zonal Wind   131**

5

Seasonal Variation of Tropical Humidity Parameters   193
by E. M. Rasmusson

Foreword

A person entering meteorological research has a choice, speaking broadly, of two courses of action. He may either undertake theoretical work, or he may devote himself to the study of observations in any of the various aspects of the subject. Upon completion of my first year of graduate study, I had in my possession an accumulated set of analyzed weather charts which I had used in making practice forecasts. Since during the 1930s support for our research was miniscule by current standards, as a matter of economy I proceeded to turn these maps over and on their blank reverse sides to write lengthy mathematical equations pertaining to certain theoretical problems which interested me. Observing this seemingly sudden reversal in my preoccupations, our mentor, Professor C.-G. Rossby, inquired with his characteristic air of mild amusement whether I indeed preferred theory to other branches of research. Receiving an affirmative reply, he said that that was fine, but offered the admonishment, "Do not forget that everything in our subject comes from the observations." This declaration I immediately accepted as a genial platitude and returned to my manipulation of symbols.

Although physical science is a domain of human endeavor which has widened to our view most fundamentally through observational exploration and experiment, anyone with a modicum of proper experience must sense the fallacy that lurks in an attempt to draw too fine a line between theoretical and observational research. In meteorology, theoretical work which lacks enough actual or potential contact with observational fact, let us say either in its assumptions or its conclusions, is sterile. Likewise observational study which does not impinge meaningfully on important theoretical questions tends in the main also to be sterile. As to absolute value, work done in each area has often been excellent—and on occasion execrable. Choices then depend upon individual inclinations; nor can it be forgotten that the proportions of the two ingredients may be varied in an essentially continuous fashion. With variations of some latitude, my own impulses have been to employ such a mixture, and to investigate by the use of data the manner in which the atmosphere fulfills the most primitive and reliable theoretical requirements. This in essence is the philosophy put into practice also by the authors of the present volume. No further special explanation

ix

need therefore be given for my interest in and sympathy with the objectives sought in the studies that are herein contained.

Fashions and centers of interest in our studies change continuously as a result of changing practical needs and due to seemingly accidental and other factors. If one looks back, the sands of science are dotted with abandoned hulks of stranded research projects which now appear unworthy of continuance, or which were bypassed through other efforts. At the moment especially, the winds of change are striking up an unexampled gale. As a consequence it is now particularly difficult to look ahead and foresee the coming situation for even a year or two in advance, to say nothing of the future for five or ten. Since a deep-seated craving of the research worker in basic science is to achieve results endowed with permanence if he can, we find ourselves beset upon by insecure feelings. We know not whether the enterprises upon which we bestow so much care and attention are presently to assume the aspect of a supportive, service type activity, important only for the time being in the context of the current combination of circumstances, but doomed to be forgotten as the state of knowledge changes its character. There exists no absolute guarantee against this and other like dangers. Luck is a notorious factor in science. Some precautionary measures are nonetheless possible. These relate in good part to the selection of topics for investigation. If they deal with the main mechanisms of the systems studied, be the treatment theoretical or chiefly observational, and if later in the final estimate it turns out that the findings correspond to fact, then prospects for survival and longevity are much enhanced.

The significance of the belt of the atmosphere flanking both sides of the equator at low latitudes, in its relation to the general circulation over the rest of the globe, has been something of a *bête noire* in meteorological investigations. Its vast length, traversing both extensive land and ocean areas, the sparsity of proper observing stations, and the peculiar character of meteorological conditions in its vicinity have caused no end of difficulty in securing a handle on the important effects it has for the overall planetary air flow. Here then is a research topic whose study, if it can be substantially advanced, fits the requirement we have specified. It seems that the authors of this book have succeeded in dealing with the problem in a superior way, through the most complete assembly of data so far brought to bear upon the larger specific questions pertaining to the subject.

I am not here in a proper position to make free with the various actual materials and results to be found on the following pages, however interesting they may be to me. Thus unlicensed, I shall limit my closing comments to one rather general point concerning the contents; a point which is not here at issue, nor is it elsewhere among those who have given the question proper (and warranted) consideration. The text presents once again evidence, were it needed, of the actuality of a negative eddy viscosity in the atmosphere. This action tends to maintain the mean differential rotation of the atmosphere. With such a well-documented instance of the process on record, every other use of the concept of eddy viscosity becomes subject to a need for renewed scrutiny to see whether upon reexamination its action is indeed positive or negative. In many well-known applications this may be a perfunctory open-and-shut procedure, but in other cases where less explored systems are considered, crucial issues are involved.

Victor P. Starr
Cambridge, Massachusetts

Preface

This monograph grew from our efforts to answer a single question: What is the physical origin of the biennial oscillation in the tropical stratosphere? We decided that the best procedure was to undertake a comprehensive study of the tropical general circulation, which would include all levels for which data were available. There was no central agency which collected meteorological data from the tropics, and much of the data had to be gathered piecemeal by writing to each country concerned. We were fortunate that a parallel study of 5 years of data from the Northern Hemisphere was already underway. This study was sponsored by the National Science Foundation under the supervision of Professor V. P. Starr, who readily agreed to allow us to incorporate his Northern Hemisphere data. We aimed for a relatively long period of record (1957–1964) to ensure sampling several cycles of the biennial oscillation, and we tried to include all available data for the region 40°N to 40°S. It soon became clear, as we processed the data, that other items of interest could be investigated as well as the biennial oscillation. The semiannual oscillation in both troposphere and stratosphere; changes thought to be due to the Bali volcanic eruption of March 1963; the basic climatology of the tropics in terms of vertical motion and rainfall and their year-to-year changes; the genesis of hurricanes; interhemispheric interactions; the detailed time evolution of the Hadley cell circulation, particularly its longitudinal variations; the basic heat, momentum, and energy budgets; and many other items all proved of interest per se. Of course all the items are interrelated, and the biennial oscillation is another manifestation of a complex of feedback interactions operating among various atmospheric regions.

The work was initiated by J. M. Wallace and the senior author in 1963. On graduation in 1966, J. M. Wallace moved to the University of Washington, and J. W. Kidson of the New Zealand Meteorological Service joined our group. Sufficient data had accumulated by then to enable him to extend the coverage from the stratospheric region, studied by Wallace, down to the surface. It became clear that the many maps portraying the mean circulation and eddy fluxes in the tropics could best be presented to others through some form of publication like the present volumes, and we began assembling this monograph. In the fall of 1967 we were joined by D. G. Vincent.

All our water vapor results were supplied to E. M. Rasmusson of the Geophysical Fluid Dynamics Laboratory of ESSA at Princeton, and he performed the analysis of the tropical moisture budget. He drew extensively on V. P. Starr's Northern Hemisphere results as well. This analysis, contained in Chapter 5, was completed in December 1968; revisions were made in the spring of 1969.

Two points of view concerning the biennial oscillation emerged in 1967–1968. The school of thought at the University of Washington favors a wave-interaction mechanism. The chapter outlining this view, written by J. R. Holton and J. M. Wallace, was received in January 1969 and is presented in Volume 2.

The first drafts of the chapters in Volume 1 and those concerning the heat and energy budgets and the biennial oscillation were prepared in the fall of 1968. Dr. Kidson returned to New Zealand in January 1969. We planned to go to press in the summer of 1969, but early in the year we were invited to review the global energy budget at the joint Royal Meteorological Society–American Meteorological Society Conference on the Global Circulation of the Atmosphere held in London in August 1969. This opportunity to set the heat and energy budget in global perspective naturally delayed our publication plans. Much of the 1969–1970 academic year was spent by the senior author with a flood tide of twelve graduate students—supervising and discussing their thesis work. The work on radiation by T. G. Dopplick was completed during this period. We also discovered that our results were influenced by the volcanic eruption of Mt. Agung on Bali in March 1963, and we extended our period of study and added stations thought to be helpful in understanding this event. The eruption brought home to us very clearly the dangers of applying terms like "climatological normal" to the stratosphere. Mt. Agung seemed to interrupt the biennial oscillation, the focal point of our study; from studies of the oscillation we have learned about the general circulation, while studies of the eruption have taught us much about the oscillation.

In the summer of 1970, Dr. D. G. Vincent departed for Purdue University, and I was joined by Dr. G. J. Boer. He has written several chapters, rewritten others, reviewed the entire text, made substantial revisions and additions, and has introduced fresh ways of presenting the data. He has packed several years of accomplishment into the past two years.

By a fortunate coincidence, Dr. Kidson was able to visit us in April 1971, and we spent a week reviewing our final drafts of Volume 1 and clearing up many questions. Dr. Boer and I then spent three months in final checking and correction of each of the maps and cross sections.

The purpose of these two volumes is to present our results with some commentary; there is no attempt to provide a comprehensive review of the rapidly expanding literature of tropical meteorology.

Work on the monograph has constituted part of the educational process for us all over the past several years. It includes some information from the theses of T. G. Dopplick, J. W. Kidson, D. G. Vincent, and J. M. Wallace. It has been fun to explore the subject this way.

Reginald E. Newell
Cambridge, Massachusetts

August 1971

Acknowledgments

Our thanks go to the sponsor of the work, the Fallout Studies Branch of the U.S. Atomic Energy Commission, represented by Drs. J. Z. Holland, E. W. Bierly, and R. J. Engelmann. They recognized that the most comprehensive solution to the applied problem of the redistribution of trace elements by the atmosphere would be obtained by an attack on the basic problem of the atmospheric balance of momentum, mass, and energy.

In the early years of collecting tropical data we were greatly encouraged by Professor V. P. Starr, who saw the advantages to be gained by focusing on this part of the globe, and he generously shared with us the fruits of his collection of data for the Northern Hemisphere. H. M. Frazier and S. F. Seroussi, formerly of the Travelers Research Corporation, were of great help in providing this data in convenient form. We are grateful to the directors and staffs of the meteorological services of many countries for their attendance to our inquiries and for provision of data. This has been a most gratifying experience in an age marked by bitterness among governments. We make some other acknowledgments for data at appropriate points in the text.

Our staff deserve a special word of thanks. Mrs. Dorothy Berry and Mrs. Susan Ary have plotted the maps and performed the numerous hand calculations that arise in such a study. Miss Isabelle Kole has helped with the map analysis and drafted the maps and most of the figures. Mr. S. Ricci has cheerfully contributed to the figures also. Robert Crosby has helped with much of the programming work, and Mrs. Mary Andresen has punched many thousands of cards. Miss Susan Davies and Miss Mary Ferris helped us gather together the reference material and turned our chaos into order.

The General Circulation of the Tropical Atmosphere

and Interactions with Extratropical Latitudes

1
Introduction

The tropical and subtropical atmosphere may be loosely defined as that part of the atmosphere between the thirtieth parallels of latitude. This region accounts for one-half of the earth's surface area, but unfortunately it does not contain one-half of the upper-air observing stations. It clearly merits equal attention with middle latitudes in any study of the general circulation as a whole. The general circulation of the tropics differs from that of middle latitudes in several basic respects, so much so that our picture of the workings of the atmosphere must be quite different depending on whether we focus attention on one region or the other, as will become apparent.

A survey of the problems of tropical meteorology has been given by Alaka (1964) in terms of twelve general categories: the delineation of the meteorological tropics, general circulation in the tropics, the Asian summer monsoon, convection, disturbances in the tropics, hurricanes and typhoons, rainfall in the tropics, observations in the tropics, synoptic analysis, forecasting, numerical methods in the tropics, and weather modification. The study presented here gives basic attention to the general circulation of the tropical atmosphere, but some aspects of most of the remaining topics are also investigated. Questions about the region include those pertaining to the origin of deserts (where the lack of rainfall is due to areas of large-scale sinking motion in the atmosphere), to the time variations of desert fringe vegetation, to the distribution of rainfall with latitude, longitude, and season (essentially a problem of determining areas of large-scale rising motion in the atmosphere), to the yearly variations in monsoon rains, and to the frequency and distribution of hurricanes. Such problems involve not only the tropical circulation per se, but also interactions between the circulation of the Northern and Southern Hemispheres and between the tropics and middle latitudes, the latter so aptly termed "teleconnections" by Bjerknes (1966, 1969).

Meteorological differences between the tropics and middle latitudes lie not so much in the absolute magnitudes of the basic parameters as in their relative importance and in their time variability. The day-to-day variability of the wind, for example, as measured by the temporal standard deviation, shows a minimum in the tropical lower troposphere which is matched only by that of the quiet summertime stratospheric

1

circulation at middle and high latitudes. The maximum middle-latitude standard deviation is two to four times greater than the tropical values at the same altitude.

The stability of the tropical low-level winds enabled Halley to construct mean wind maps in 1686; one of them is reproduced in Shaw's book (1926). Halley suggested that in the region where the North East trades of the Northern Hemisphere and South East trades of the Southern Hemisphere come together, now called the Intertropical Convergence Zone (ITCZ), there exists rising motion due basically to the air being heated more by the sun near the equator than at higher latitudes. These ideas were extended by Hadley (1735), who put forward a hypothesis for the direction of the prevailing winds which suggested that the air flowed equatorward at low levels, then rose, and returned poleward at higher levels. He maintained that equatorward flow exhibited a relative velocity toward the west because of a tendency for air parcels to conserve their angular velocity. Similarly, at upper levels, parcels moving poleward showed a tendency to retain the higher angular velocity of the low latitudes; they moved faster than the earth's surface at the higher latitudes and produced westerly winds. As Lorenz (1967) has emphasized in a more recent discussion of these early theories, the Hadley hypothesis would have been quite acceptable if angular momentum had been considered instead of angular velocity. Hadley's general idea of air with rising motion at low latitudes, poleward motion at upper levels, sinking motion at higher latitudes (which he left unspecified) and equatorward motion at low levels fits very well with the observations now available in the tropics and subtropics.

As will be discussed in the body of the monograph, it is these "Hadley cells" and the relative importance of mean meridional motions in the tropical atmosphere that show one of the most marked differences between the workings of this region and those of middle latitudes. The Hadley cells are important not only in the mass circulation of the atmosphere but also in the heat, momentum, and overall energy budgets. An important component in the heat balance of the subtropical highs, for example, is the heating due to adiabatic subsidence in the descending motion region of the Hadley circulation (Volume 2, Chapter 7). The poleward transport of angular momentum in the tropical region is associated, in large part, with the mean meridional motions of the Hadley circulation, while at middle latitudes transport due to large-scale eddy processes dominates completely (Chapter 4). The meridional circulation also plays an important part in the energy budget; a large energy conversion rate is associated with kinetic energy production by mean meridional overturning (Volume 2, Chapter 8).

The differences between middle latitudes and low latitudes may also be seen in the patterns of other meteorological variables. Rainfall at some stations near the equator shows a semiannual maximum as the sun passes overhead twice each year; near 30°N most stations show a single maximum. At some equatorial stations the prevailing surface wind also changes direction twice each year, again related to the sun's passage, while most middle-latitude stations show essentially no change in the direction of the prevailing wind. At higher levels, say 25–35 km, the prevailing wind direction over the equator changes from one year to the next, being westerly for roughly twelve months then easterly for roughly twelve months; in middle and high latitudes the prevailing direction is westerly in winter and easterly in summer. Higher still at 50 km, the west-to-east change occurs twice each year near the equator, while at middle latitudes the situation is the same as at 30 km. The temperature field at middle latitudes in the troposphere shows quite large seasonal changes with lower temperatures in winter, but in the tropical middle troposphere there are seasonal changes of only one or two degrees.

Certain derived parameters used in studying planetary circulations are also seen to differ in value between low and middle latitudes. An approximation to the basic equations of motion enables us to derive the so-called *geostrophic wind*, for which horizontal pressure gradients and Coriolis forces are in balance (the definitive paper on this topic is by Phillips (1963). Specifically, the conditions are that the frictional forces and horizontal relative accelerations must be small compared with the Coriolis force and pressure gradients. The ratio of the horizontal relative acceleration to the Coriolis force is given as $CL^{-1}/2\Omega \cdot \mathbf{k}$, where C and L are characteristic speeds and lengths associated with the motion, Ω is the earth's rotational velocity, and $\mathbf{k}$ is the local vertical unit vector. The denominator is termed the *Coriolis parameter* and is denoted henceforth by f (a list of symbols appears in Table 1.1). Taking representative values for C of 10 m sec^{-1} and 30 m sec^{-1} and an assumed value of L of 2000 km and 3000 km (for the tropics and middle latitudes, respectively), this ratio, termed the *Rossby number,* has values of 1.0, 0.4, and 0.14 at 2, 5, and 30 degrees latitude. It thus diminishes toward middle latitudes, with the consequence that the geostrophic approximation is a better one there than in the tropics. Indeed, horizontal relative accelerations clearly cannot be ignored near the equator. The variation of the Coriolis parameter plays an important part in the dynamics of the motions in the region and in theories of the ITCZ (Charney 1969; Holton et al. 1971).

Model experiments with rotating fluids have been made which attempt to simulate the main features of the atmospheric circulation (Fultz et al. 1959 give a comprehensive account of these experiments). When a parameter analogous to the Rossby number, the so-called *thermal Rossby number,* is large, greater than about 0.3, the experiments show a direct-cell Hadley type circulation similar to that in the tropical atmosphere. At higher rotation rates where the thermal Rossby number is less than 0.3, the steady circulation breaks down into a wavelike pattern that often resembles a map of the middle-latitude upper-troposphere pressure pattern. This wavelike circulation in the rotation experiment has been termed the *Rossby regime* (a historical review of the theory of Rossby waves in the atmosphere and oceans has been presented by Platzman 1968). Of course the critical value of the Rossby number separating the two flow regimes cannot be translated literally to conditions in the real atmosphere; nevertheless by analogy with the experiments there is reason to expect two atmospheric regimes, due essentially to the latitudinal variation of the vertical component of the earth's rotation.

Further derived properties of the zonal mean wind field are the vertical component of the absolute vorticity $f - (a \cos \varphi)^{-1} \partial [\bar{u}] \cos \varphi / \partial \varphi$ and the divergence $(a \cos \varphi)^{-1} \partial [\bar{v}] \cos \varphi / \partial \varphi$. They can both be simply derived from the data presented later. In middle latitudes the absolute vorticity is much larger than the divergence and is dominated by the term due to the earth's rotation. At low latitudes, the absolute vorticity has decreased by more than an order of magnitude from its mid-latitude value; its magnitude is comparable to that of the divergence, and it is dominated by the term due to the curl of the mean wind vector.

The potential temperature $\theta = T(p/p_0)^\kappa$ is a derived property of the temperature field, and is customarily defined as the temperature an air parcel would attain if moved adiabatically to a pressure of 1000 mb. Meridional cross sections of this term, based on our observations, are shown in Volume 2, Chapter 8. There is a marked difference between the configuration of the isentropes at low and middle latitudes. In the latter region the isentropes slope upward across pressure surfaces toward the pole. The slope represents a large store of available potential energy, as will be discussed later, which can be readily converted to kinetic energy by the wavelike disturbances in the region. Eady (1949) has presented a classic discussion of this process, showing that maximum energy conversion occurs when the slope of the motions in the disturbances is about one-half the slope of the isentropes. At low latitudes the isentropes practically parallel the isobars, and the available potential energy is very small. This does not imply that no kinetic energy is being generated; indeed to maintain this temperature field mean meridional motions are necessary (as will be seen in Volume 2, Chapter 7) together with the effects due to latent heat release and radiation. In this case, air moves down the pressure gradients and kinetic energy is released (Volume 2, Chapter 8). The contrast between the regions is evident from the potential temperature cross sections; this topic will be dealt with again later. As many of the earlier studies of the atmospheric energy budget (e.g., Oort 1964) have taken the region between 20 and 80°N as a base, tropical studies clearly do more than refine the values so obtained.

It is possible to combine the derived parameters of the temperature and the wind field into the vertical component of potential vorticity $PV = -(\zeta + f)(\partial \theta / \partial p)$ (e.g., Danielsen 1968). There is a range of three orders of magnitude between the values in the lower stratosphere of middle latitudes and the tropical troposphere. Clearly exchange of air between the regions will have profound effects on both.

Theoretical analyses of the atmosphere and of the model experiments support the idea of different flow regimes in tropical and higher latitude regions. The analyses indicate that the most efficient conversion of potential to kinetic energy is accomplished by meridional motions at low latitudes and by wave motions at higher latitudes (e.g., Kuo 1956a).

It will become clear, as we present the data, that meridional motions and wave or "eddy" motions are a persistent feature of all regions of the atmosphere, even though their relative importance differs greatly. The low-latitude upper troposphere, where the maximum of the meridional winds of the Hadley circulation are found, is also a region of active eddy motions. Similarly at middle latitudes, the eddy motions are accompanied by cellular mean motions, the Ferrel circulations. Again, these circumstances were anticipated in theoretical studies (e.g., Eliassen 1951; Phillips 1954; Kuo 1956b).

The existence of several more or less distinct energetic regimes in the atmosphere, as discussed above, and the possibility of interactions among them has led to the suggestion that a particular time change in a certain region may be explained on the basis of a prime physical cause in another region. For example, we have suggested that the biennial modulation of wind in the equatorial lower stratosphere can be brought about by a variation of quasihorizontal eddy momentum fluxes from the middle-latitude lower stratosphere—these in turn depending on energy supplied from the middle-latitude lower troposphere (Newell 1964a). This hypothesis was prompted by our earlier findings that the lower stratosphere was dynamically driven from below. It provided an alternative to the then prevailing views that radiative changes in situ were responsible for the oscillation. The problem then became to account for the time changes in the stratosphere in terms of its connection to the middle-latitude troposphere. The amount of energy supplied to the lower stratosphere from the troposphere was estimated from the theoretical formulation for large-scale standing waves developed by Eliassen and Palm (1961). The same theory showed that these waves produce an equatorward flux of energy in the troposphere. This provides a source of energy for the tropical troposphere which, like the lower stratosphere, exhibits some of the characteristics of a refrigerator insofar as the eddy flux of heat is up the temperature gradient (Peixoto

1960; Starr and Wallace 1964). Viewing this countergradient flux as a conversion from eddy-available to zonal-available potential energy (see Volume 2, Chapter 8), we found that the amounts involved were almost the same in the two refrigerators (Newell 1964b; Newell and Miller 1965). Furthermore, just as the standing wave flux upward provides energy to the lower stratosphere (Newell 1964b), the standing wave flux equatorward can provide an amount to the tropical troposphere almost equal to the energy conversion occurring there (Newell 1966). This raised the interesting possibility that the biennial oscillation is the result of a redistribution of energy from the middle-latitude troposphere between the upward and equatorward fluxes, and that no external source such as solar radiation changes need be invoked. We therefore set as one of our long-range goals the measurement from observation of the equatorward and upward energy fluxes to see if their year-to-year variations were in phase or out of phase. This question led us to reexamine the entire energy budget of the tropical atmosphere; it was not clear how the magnitude of the equatorward energy flux compared with the energy conversions continually operating in the tropics. We now have a much better appreciation of the problem (Volume 2, Chapter 8). In the process of studying it we have unearthed other possibilities by which the stratospheric-tropospheric interaction may act to bring about a biennial oscillation.

The main purpose of this work is to establish background values of the momentum, energy, and mass budgets of the tropical circulation, to examine the interrelationships with middle latitudes, and to seek a physical cause for the biennial oscillation in the equatorial lower stratosphere. The data consist mainly of the free air balloon observations available from stations within about 40 degrees of the equator for an 8-year period. We display our results as functions of latitude, longitude, height, and time. Other data which are just becoming available, such as constant-level balloon data and satellite cloud, temperature, and radiation observations, have not been incorporated herein except in passing references. On the other hand, we have drawn freely on the classical surface observations, many of which are available for very long periods.

1.1 Basic Equations

The basic equations of motion in spherical and pressure coordinates are

$$\frac{du}{dt} = \frac{\tan\varphi}{a} uv + fv - \frac{1}{a\cos\varphi}\frac{\partial\phi}{\partial\lambda} + F_\lambda \tag{1.1}$$

$$\frac{dv}{dt} = -\frac{\tan\varphi}{a} u^2 - fu - \frac{1}{a}\frac{\partial\phi}{\partial\varphi} + F_\varphi \tag{1.2}$$

$$\frac{\partial\omega}{\partial p} + \frac{1}{a\cos\varphi}\left(\frac{\partial u}{\partial\lambda} + \frac{\partial}{\partial\varphi} v\cos\varphi\right) = 0 \tag{1.3}$$

$$\frac{dT}{dt} = \frac{\kappa T\omega}{p} + \frac{\dot{Q}}{C_p} \tag{1.4}$$

$$\frac{\partial\phi}{\partial p} = -\frac{RT}{p} \tag{1.5}$$

$$p = \rho RT. \tag{1.6}$$

Here

$$\frac{d(\ )}{dt} = \frac{\partial(\ )}{\partial t} + \frac{u}{a\cos\varphi}\frac{\partial(\ )}{\partial\lambda} + \frac{v}{a}\frac{\partial(\ )}{\partial\varphi} + \omega\frac{\partial(\ )}{\partial p},$$

and the symbols have their usual meteorological meanings as given in Table 1.1. The coordinates of these equations are longitude (λ), measured positive in the Eastern Hemisphere, latitude (φ), measured positive in the Northern Hemisphere, and pressure (p), which is always positive and which is zero at the "top" of the atmosphere, that is, at an infinite distance from the earth.

The velocity components of motions relative to the surface of the earth are given by

$$u = a\cos\varphi\frac{d\lambda}{dt}$$
$$v = a\frac{d\varphi}{dt}, \tag{1.7}$$

which are positive for horizontal motions, respectively, toward the east and toward the north. The wind is said to be a west wind when it blows from the west to the east. Thus the terms *westerlies* and *easterlies* im-ply winds blowing from (not toward) the west and the east; similar usage applies for winds from the north and south. The term $\omega = dp/dt$ occupies the position of the vertical velocity in this coordinate system. The relation between ω and w—the geometric vertical velocity—is

$$\omega = \rho\left(\frac{\partial\phi}{\partial t} + \frac{u}{a\cos\varphi}\frac{\partial\phi}{\partial\lambda} + \frac{v}{a}\frac{\partial\phi}{\partial\varphi} - gw\right). \tag{1.8}$$

As mentioned previously, under the appropriate assumptions the motions obey the geostrophic equations

$$u_g = -\frac{1}{fa}\frac{\partial\phi}{\partial\varphi}$$
$$v_g = \frac{1}{fa\cos\varphi}\frac{\partial\phi}{\partial\lambda}. \tag{1.9}$$

Under these assumptions the vertical gradient of the wind is related to the horizontal gradient of temperature through the thermal wind equations

$$\frac{\partial u_g}{\partial z} = -\frac{g}{fa}\frac{1}{T}\frac{\partial T}{\partial\varphi}$$
$$\frac{\partial v_g}{\partial z} = \frac{g}{fa\cos\varphi}\frac{1}{T}\frac{\partial T}{\partial\lambda}. \tag{1.10}$$

Thus, as a first approximation away from the equator, the winds are related to the geopotential through the geostrophic relations, and the variation of the velocity in the vertical is related to the horizontal temperature gradient.

The study of the general circulation is a study of the properties of the statistics of various meteorological parameters. We define here the various averaging processes used throughout the monograph. The time-mean of a quantity, say u, is defined as

$$\bar{u} = \frac{1}{\Delta t}\int_t^{t+\Delta t} u(\lambda, \varphi, p, t)\, dt. \tag{1.11}$$

Commonly, the average will be over a month or a 3-month season. The actual values used in the monograph were obtained primarily from daily upper air measurements. These were simply summed over all days, for a particular month or season, for which data were available, and divided by the number of such values.

A further grand mean or "long-term" average is obtained by averaging over corresponding months or seasons for a number of years of data. If no data were missing, this would be simply the average of the individual monthly or seasonal means of the years in question. In the face of missing data, the long-term average is formed by summing over the data for all days for all years for the particular month or season and dividing by the number of such values. The assumption is that the atmosphere has, at least in a restricted sense, a set of long-term statistical properties, and that such averaging of the data will give stable estimates of these statistics. When applied to a time derivative

$$\overline{\frac{\partial u}{\partial t}} = \frac{u(t + \Delta t) - u(t)}{\Delta t} \tag{1.12}$$

with the corresponding meaning for the long-term average.

Zonal or longitudinal averages are defined as an average around a latitude circle, that is,

$$[u] = \frac{1}{2\pi} \int_0^{2\pi} u(\lambda, \varphi, p, t)\, d\lambda. \tag{1.13}$$

Deviations from the temporal and zonal averages are written as

$$u' = u - \bar{u} \tag{1.14}$$

$$u^* = u - [u]. \tag{1.15}$$

In numerous cases the long-term zonally averaged value of a term of the form uv, uw, qv, etc., is required. Consider the term $[\overline{uv}]$, which may be interpreted as a momentum flux. Using the definitions given above, the term may be decomposed as follows:

$$u = \bar{u} + u', \quad v = \bar{v} + v', \qquad \text{whence}$$

$$\overline{uv} = \bar{u}\bar{v} + \overline{u'v'} \qquad \text{and}$$

$$[\overline{uv}] = [\bar{u}\bar{v}] + [\overline{u'v'}]. \tag{1.16}$$

Further, $\bar{u} = [\bar{u}] + \bar{u}^*, \quad \bar{v} = [\bar{v}] + \bar{v}^*, \qquad$ whence

$$[\bar{u}\bar{v}] = [\bar{u}][\bar{v}] + [\bar{u}^*\bar{v}^*]. \tag{1.17}$$

Substituting equation (1.17) into equation (1.16) gives

$$[\overline{uv}] = [\bar{u}][\bar{v}] + [\bar{u}^*\bar{v}^*] + [\overline{u'v'}]. \tag{1.18}$$

The zonally and time-averaged momentum flux may be thought of as composed of these three components, called, respectively, the mean motion flux, the standing eddy flux, and the transient eddy flux components. The terms are so named because the mean motion flux depends on the zonally and temporally averaged motions; the standing eddy flux depends on the spatial correlation of the time-averaged motions, that is, on the nontransient or standing motions; and the transient eddy flux depends on the temporal correlation at a point of the deviations from the time-mean motions, that is on the transient component of the flow. The appellation "eddy" is applied to both a deviation from a time or a zonal mean in this context.

The notation used here is relatively standard in general circulation studies, and further definitions and derived equations are introduced where appropriate in the text. A list of symbols appears in Table 1.1. A good general discussion of the equations used in general circulation studies together with a discussion of the pressure coordinate system is given by Lorenz (1967).

Table 1.1. List of Symbols

a	Radius of the earth taken as a sphere of equivalent volume (6.371×10^6 m)
C	Characteristic speed of synoptic scale motions
C_D	Surface drag coefficient
C_p	Specific heat of air at constant pressure, 996 $m^2 \sec^{-2} \deg^{-1}$ or approximately $7R/2$
C_v	Specific heat of air at constant volume, 709 $m^2 \sec^{-2} \deg^{-1}$ or approximately $5R/2$
d	Separation distance in objective analysis scheme
D	Influence radius in objective analysis scheme
E	Rate of evaporation from surface, per unit area
f	Coriolis parameter
F_λ, F_φ	The components of the frictional force in the zonal and meridional directions
g	Magnitude of the acceleration of gravity, 9.81 $m \sec^{-2}$
G	The convergence of the eddy flux of relative angular momentum
$\boldsymbol{i}$	Unit vector directed eastward
$\boldsymbol{j}$	Unit vector directed northward
J_λ, J_φ	Zonal and meridional components of the angular-momentum flux density
J_p	Quasivertical angular-momentum flux density
$\boldsymbol{k}$	Unit vector directed along the local vertical
K_z	Coefficient of turbulent viscosity
L	Latent heat of condensation
L	Characteristic length scale of synoptic motions
M	Absolute angular momentum taken about the earth's axis
p	Pressure
p_o	Standard pressure of 1000 mb
p_s	Surface pressure
p_u	Pressure at which water vapor amounts become negligible (taken as 300 mb)
P	Rate of precipitation per unit area
$P(u, v)$	Two-dimensional normal probability density
p_w, p_e	Surface pressure on the western and eastern sides of a mountain range
PV	Vertical component of the zonally averaged potential vorticity
q	Specific humidity
q_s	Specific humidity of saturated air at given temperature and pressure
$\dot{Q}$	Rate of diabatic heating
Q_λ, Q_φ	Zonal and meridional components of the water-vapor flux density
r	Radial distance from the center of the earth
R	Gas constant for dry air, 287 $m^2 \sec^{-2} \deg^{-1}$
t	Time
T	Temperature
u	Eastward (zonal) component of $\mathbf{V}$
u_g	Eastward component of the geostrophic winds
v	Northward (meridional) component of $\mathbf{V}$
v_g	Northward component of the geostrophic wind
$\mathbf{V}$	Horizontal wind velocity
$|\mathbf{V}|$	Magnitude of $\mathbf{V}$
w	Vertical velocity
W	Precipitable water; the water vapor content of a unit column of air
W_i	Weights used in the calculation of latitude band means
z	Geopotential height above sea level
ζ	Vertical component of the relative vorticity
θ	Potential temperature
κ	R/C_p
λ	Longitude, positive eastward
ρ	Density
ρ_0	Sea-level density
τ_λ	Frictional stress in the zonal direction
τ_0	Net surface stress in the zonal direction
τ_F	Net surface stress in the zonal direction calculated from the angular momentum balance
τ_M	Mountain torque contribution to the angular momentum balance
φ	Latitude, positive northward
ϕ	Geopotential, potential energy per unit mass
Ψ	Stream function for zonally averaged meridional flow, mass flux
Ω	Angular velocity of the earth ($7.292 \times 10^{-5} \sec^{-1}$)
$\omega = \dfrac{dp}{dt}$	Individual pressure change, the "vertical velocity" in pressure coordinates
χ	Stream function for zonally averaged absolute angular momentum
x	Arbitrary dependent variable
$\bar{x}$	Time average of x
x'	Deviation from time average
$[x]$	Zonal average of x
x^*	Deviation from zonal average
$\sigma(x)$	$\sqrt{\overline{x'^2}}$ Standard deviation of x

References

Alaka, M. A. 1964. *Problems of Tropical Meteorology*. Technical Note no. 62. Geneva: WMO. 36 pp.

Bjerknes, J. 1966. A possible response of the Hadley circulation to variations of the heat supply from the equatorial Pacific. *Tellus* 18:820–829.

———. 1969. Atmospheric teleconnections from the equatorial Pacific. *Mon. Weath. Rev.* 97:163–172.

Charney, J. G. 1969. On the intertropical convergence zone and the Hadley circulation of the atmosphere. Tokyo: *Proceedings of the WMO–IUGG Symposium on Numerical Weather Prediction,* part III, pp. 73–79.

Danielsen, E. F. 1968. Stratospheric-tropospheric exchange based on radioactivity, ozone, and potential vorticity. *J. Atmos. Sci.* 25:502–518.

Eady, E. T. 1949. Long waves and cyclone waves. *Tellus* 13:33–52.

Eliassen, A. 1951. Slow thermally or frictionally controlled meridional circulations in a circular vortex. *Astrophys. Norvegia.* 5:19–60.

———, and E. Palm. 1961. On the transfer of energy in stationary mountain waves. *Geof. Pub.* vol. 22. 23 pp.

Fultz, D., R. R. Long, G. V. Owens, W. Bowan, R. Kaylor, and J. Weil. 1959. Studies of thermal convection in a rotating cylinder with some implications for large-scale atmospheric motions. *Meteorol. Mon.* 4. Boston: American Meteorological Society.

Hadley, G. 1735. Concerning the cause of the general trade winds. *Phil. Trans.* 39:58–62.

Halley, E. 1686. An historical account of the trade winds and monsoons. *Phil. Trans.* 26:153–168.

Holton, J. R., J. M. Wallace, and J. A. Young. 1971. On boundary layer dynamics and the ITCZ. *J. Atmos. Sci.* 28:275–280.

Kuo, H.-L. 1956a. Energy-releasing processes and stability of thermally driven motions in a rotating fluid. *J. Meteorol.* 13:82–101.

———. 1956b. Forced and free meridional circulations in the atmosphere. *J. Meteorol.* 13:561–586.

Lorenz, Edward N. 1967. *The Nature and Theory of the General Circulation of the Atmosphere.* Geneva: WMO. 161 pp.

Newell, R. E. 1964a. A note on the 26-month oscillation. *J. Atmos. Sci.* 21:320–321.

———. 1964b. Stratospheric energetics and mass transport. *Pure and Appl. Geophys.* 58:145–156.

———. 1966. The energy budget and momentum budget of the atmosphere above the tropopause. In *Problems of Atmospheric Circulation. A Session of the Sixth International Space Science Symposium,* R. V. Garcia and T. F. Malone, eds. New York: Spartan Books, pp. 106–126.

———, and A. J. Miller. 1965. Some aspects of the general circulation of the lower stratosphere. In *Radioactive Fallout from Nuclear Weapons Tests,* Proceedings of the Second Conference, November 3–6, 1964, A. W. Klement, Jr., ed. Washington, D.C.: United States Atomic Energy Commission, Division of Technical Information, pp. 392–404.

Oort, A. H. 1964. On estimates of the atmospheric energy cycle. *Mon. Weath. Rev.* 92:483–493.

Peixoto, J. P. 1960. *Hemispheric Temperature Conditions during the Year 1950.* Report no. 4. Cambridge: Massachusetts Institute of Technology, Department of Meteorology, Planetary Circulations Project. 211 pp.

Phillips, N. A. 1954. Energy transformations and meridional circulations associated with simple baroclinic waves in a two-level, quasi-geostrophic model. *Tellus* 6:273–286.

———. 1963. Geostrophic motion. *Rev. of Geophys.* 1:123–173.

Platzman, G. W. 1968. The Rossby wave. *Quart. J. Roy. Meteorol. Soc.* 94:225–248.

Shaw, Sir Napier. 1926. *Manual of Meteorology.* Vol. I. Cambridge: Cambridge University Press. 342 pp.

Starr, V. P., and J. M. Wallace. 1964. Mechanics of eddy processes in the tropical troposphere. *Pure and Appl. Geophys.* 58:138–144.

2

Data: Sources, and Processing and Analysis Techniques

This chapter outlines the sources of data and the processing and analysis techniques used to obtain global values of the general circulation parameters discussed in the remainder of these volumes.

Tropical Data

The basic data consist of the daily station values of the meteorological parameters u, v, T, z, q, which are measured routinely by radiosonde and radiowind techniques. These daily values are first combined into time-averaged values of

$$\bar{u}, \bar{v}, \bar{T}, \bar{z}, \bar{q}, \quad \sigma(u), \sigma(v), \sigma(T), \sigma(z), \sigma(q), \quad \overline{u'v'}, \overline{u'T'}, \overline{v'T'},$$
$$\overline{u'z'}, \overline{v'z'}, \overline{u'q'}, \overline{v'q'},$$

where the bar denotes a time average and the prime the deviation from it. Next, the spatial fields of these time-averaged values at constant pressure levels are inferred from the values at the stations. This may be done either objectively (that is, numerically) or subjectively by drawing contour lines by hand on maps on which the station values have been plotted. In either case, values of the parameters are extracted at points on a two-dimensional grid at each pressure level for which the analysis has been done. These grid point values are obtained numerically in the case of objectively analyzed fields or subjectively by reading points from the hand analyzed charts.

Zonally averaged statistics are obtained from these grid point values by simply summing over all grid points around a latitude circle and dividing by the number (N) of such values. For instance:

$$[\bar{u}] = \frac{1}{N} \sum_n \bar{u}_n, \quad [\bar{v}] = \frac{1}{N} \sum_n \bar{v}_n, \quad [\overline{u'v'}] = \frac{1}{N} \sum_n (\overline{u'v'})_n,$$

where n is an index indicating the number of the grid point around the latitude circle. The standing eddy components of the various parameters are obtained from the deviations from these zonal averages, for instance

$$[\bar{u}^*\bar{v}^*] = \frac{N \sum\limits_n \bar{u}_n \bar{v}_n - \sum\limits_n \bar{u}_n \sum\limits_n \bar{v}_n}{N^2}.$$

9

In some instances where sufficient data are not available to permit accurate grid point analysis, so-called "latitude band means" are calculated. In this case all time-averaged station values within a 10-degree latitude band are averaged (with appropriate weighting to reflect the relative positions of the stations) to obtain zonally averaged values of the parameters. In this case it is obviously impossible to derive standing eddy terms.

2.1 The Observational Network

The results presented for the tropics have been based on the data from 330 radiosonde and radar wind stations for the overall period July 1957 to December 1964. Data from more recent periods are also included in the Southern Hemisphere because of our interest in volcanic effects. Virtually all stations between latitudes 35°N and 45°S were included, together with data from a sufficient number of regularly reporting stations to extend the area of coverage to about 45°N. Data from stations making only pilot balloon observations were not used since the coverage is poor at higher levels in the troposphere, and resulting statistics tend to be biased in favor of light winds and fair weather. The quality and quantity of the data varied considerably among stations. Many stations did not report regularly throughout the entire period, and 15 stations had fewer than twelve months of observations. The principal reporting time was 0000 GMT but, where necessary, data at 0600 and 1200 GMT were included to improve the coverage. Plate 2.1 shows the location of the stations for which data were available, and the type of data reported.

Daily data for most of the Northern Hemisphere stations were supplied by the National Weather Records Center at Asheville, North Carolina on card decks used for the Northern Hemisphere Data Tabulations. For 212 of these stations the data for levels up to 100 mb were copied from tapes prepared by the Travelers Research Center (TRC) for V. P. Starr's 5-year Northern Hemisphere study mentioned previously. For one-third of these stations, the data were augmented by further data on cards obtained directly from Asheville. This was done to improve the coverage and to extend the period of study from the 5-year period used by Starr (May 1958–April 1963) to the period of this study.

Data from the same sources for levels above 100 mb for 250 Northern Hemisphere stations had already been processed by J. M. Wallace (1966), and for 120 of these his monthly mean statistics were used instead of the original daily values.

In the Southern Hemisphere, data from approximately one-fifth of the stations were again supplied by Asheville, but most station data were obtained on card decks, tapes, and microfilm from the Australian, Brazilian, British, French, New Zealand, and South African Meteorological Services. To fill some of the gaps, cards were punched from Portuguese and Mauritian tabulations and from the IGY–IGC microcards published by the World Meteorological Organization. A complete station list will be found in Appendix I.

2.2 Basic Processing

The daily data at each station were first used to compute monthly mean statistics at all available levels from the surface to 7 mb. The quantities

$$\bar{u}, \bar{v}, \bar{T}, \bar{z}, \bar{q}, \quad \sigma(u), \sigma(v), \sigma(T), \sigma(z), \sigma(q),$$
$$\overline{u'v'}, \overline{u'T'}, \overline{v'T'}, \overline{u'z'}, \overline{v'z'}, \overline{u'q'}, \overline{v'q'}$$

were computed. The mean specific humidities $\bar{q}$ and water vapor transports $\overline{u'q'}, \overline{v'q'}$ were supplied to Dr. E. M. Rasmusson of the Environmental Science Services Administration (ESSA), U.S. Department of Commerce, who included them in his analysis of the water vapor budget (Chapter 5).

The standard deviations of temperature, $\sigma(T)$, and geopotential height, $\sigma(z)$, were used as a check for gross errors in the data. When these values were excessive the reason could usually be traced to one or two observations with obvious coding or punching errors. Initially these observations were removed by hand and the monthly statistics recomputed, but during the course of the work the computer in use at the Massachusetts Institute of Technology was changed from the IBM 7094 to the IBM 360 and, in reprogramming, a data check was included. The machine editing procedure was also used for data which were obtained on tapes; it was extended to cover the wind components u and v. At each level, the standard deviations were computed, and if any were found to be above a reasonable upper limit a checking procedure was initiated. This checking procedure rejected observations differing from the mean by more than 2 standard deviations, or by more than a given amount if the standard deviation was particularly large. All terms were then recomputed for the checked data. Although at first sight a cut-off value of 2 standard deviations appears rather small, it must be remembered that the checking procedure was executed only in cases where the standard deviation was already unusually large at the level in question. In most cases errors could be attributed to coding or punching errors which produced artificially high standard deviations.

The IBM 360 programs were also adapted to increase flexibility and to include common data checking and output subroutines. Assembler language subroutines were used to decode the varying card formats and store the observations by level and day in standard units of m sec^{-1}, °C, geopotential meters, and g kg^{-1}. For the Australian and Indian data, the winds were interpolated between the reporting levels at 5000-ft intervals to obtain values at constant pressure surfaces, by using either the reported height of the surface, or a climatological height if it was missing.

The monthly mean station statistics consist of the 17 terms listed earlier in this section, together with the numbers of observations at all levels from the surface to 7 mb. In practice only the main synoptic reporting levels had sufficient data for analysis. The covariances computed earlier by Wallace for data at 100 mb and above did not include $\overline{u'T'}$ or the terms involving z, and these terms are not available.

2.3 Computation of Long-Term Seasonal Mean Values

The long-term seasonal mean values were calculated from the monthly mean statistics discussed in the previous section. If $\bar{u}_m$, N_m are the values of the monthly means and the number of data points entering the respective means for all months for all years in a typical season, the resultant long-term seasonal mean is simply

$$\bar{u} = \frac{\sum_m N_m \bar{u}_m}{\sum_m N_m},$$

and similarly for the other terms. The bar notation for a time average is

used both for seasonal and monthly means, but the extent of the averaging will be clear from the context. In turn, the deviation from a time mean $u' = u - \bar{u}$ is the deviation of the daily value from the long-term seasonal mean in this context, not from the monthly mean. Thus the variances and flux quantities calculated in this fashion are not simply those which would be obtained by averaging the monthly means of these quantities, and a different partitioning of the transports between standing eddies and transient eddies is obtained.

The long-term statistics were calculated for all parameters except those involving water vapor. As the measurements of winds and temperatures are not always taken simultaneously, the calculated values for the covariances $\overline{v'T'}$ and $\overline{v'z'}$ may not be strictly correct if the numbers of observations of v and T or z differ. The means of v, T, and z used to regenerate the monthly sums were obtained from all observations in the month, not just those used for the covariance computation. In individual months the same values for the covariances are obtained as before, but the results obtained from a combination of a number of months could show some bias, depending on the differing number of wind and radiosonde observations. Some possible causes of bias are (i) balloon lost during tracking but radiosonde data still taken; (ii) observations rejected during computation of the monthly covariances. The first situation occurs mainly during strong wind conditions, and it is generally recognized (e.g., Priestley and Troup 1964) that it may significantly affect the mean wind and flux quantities. This applies not only to the heat and geopotential fluxes under discussion, but also to the momentum flux. Although it is conceivable that some bias might result from the rejection of observations, as mentioned in case (ii), over the long run the errors should be randomly distributed, and they should average to zero.

Data were grouped into four seasons, December–February, March–May, June–August, and September–November. The seasonal wind and temperature statistics were obtained using data for the 6-year period July 1957 to June 1963, and separately for even-year and odd-year combinations in order to examine possible changes related to the biennial oscillation. A further, later, refinement was to terminate the period used for the odd and even summaries with February 1963 to avoid the influence of the Mt. Agung volcanic eruption in March 1963. The even-year and odd-year calculations were not made for the geopotential terms.

2.4 Analysis Techniques

It was originally intended to process all the data for this monograph by objective numerical techniques. A comparison of objectively and subjectively analyzed maps showed, however, that a considerable amount of detail was lost in the objective analysis, although the overall patterns were quite comparable. In order to preserve as much real information as possible, the maps presented in later chapters have been hand analyzed. A comparison of zonally averaged statistics for the various meteorological parameters investigated here showed generally good agreement between values obtained from objectively and subjectively analyzed data, and the zonally averaged statistics which were obtained objectively have occasionally been used in what follows.

2.4.1 Subjective Analysis

Subjective analysis proceeds from the monthly or seasonally averaged station data to the original map analysis. Station data calculated from fewer than 30 observations for means of single quantities and fewer than 90 observations for covariances were given little weight in the analysis. The zonally averaged values were obtained from grid-point values read from the subjectively analyzed maps at 10-degree increments of latitude and 10-degree increments of longitude in the region 40°N–40°S.

2.4.2 The Objective Analysis Technique

The objective analysis technique used here grew out of the screening regression approach of Eddy (1967a, b), which is appropriate for isotropic, stationary scalar fields. Although rather complex and time-consuming, this technique has several advantages over simpler schemes. The modified scheme which we used is discussed more fully in Appendix II.

2.4.3 Latitude Band Means

Latitude band means were used to obtain zonally averaged cross sections of the basic parameters on a monthly basis in the troposphere and long-term means in the stratosphere. All stations were grouped into 10-degree latitude bands centered at the eight latitudes for 40°N to 30°S. In view of the dense network of stations over the continents, par-

ticularly the United States and China, it was decided to try to reduce possible bias by giving each station a weight based on its distance to its nearest neighbors within the latitude band. Tests of different weighting schemes were carried out and the results were compared against zonal means based on grid point values from analyzed maps. Some earlier runs were also available in which the station values had been weighted by the corresponding number of observations without regard to station location. Although the calculated means did not vary much for the different weighting schemes, it appeared that the weighting by numbers of observations was the least successful. The scheme finally adopted had weights computed as follows:

$$W_i = \frac{\min (\lambda_i - \lambda_{i-1}, \Delta) + \min (\lambda_{i+1} - \lambda_i, \Delta)}{\Delta},$$

where λ_i is the longitude of station i and $\lambda_{i\pm1}$ are the longitudes of its nearest neighbors within the same band. Δ is twice the average arc length per station, which is obtained by dividing 360° by the number of stations in the band; min indicates the minimum of the two quantities in parentheses. It will be seen from the form of this function that the weight for isolated stations cannot exceed 2, while if the stations are evenly distributed each would have a weight of 1. The minimum weight was set at 0.1 so that all stations could give some contribution. Summation of the station weights over 30°-longitude intervals showed the weighting to be reasonably uniform over most sections except over the empty oceans. Although in principle the station weights could have been computed for each month and level on the basis of the stations reporting, it was decided simply to compute a single weight for each station and to use this for all calculations.

Early calculations showed the need for some data checking as some large and erroneous station values were responsible for errors which were particularly prominent in 12-month running mean values. After all station values for the month had been read in and sorted into latitude bands, checking proceeded as in the first stage of the objective analysis scheme. The zonal means and standard deviations were formed, and station values which differed from the mean by more than 3.5 standard deviations were excluded. Each station was required to have a minimum of 3 observations for the month; if fewer than 5 stations satisfied the limit no mean was computed for the band.

2.5 Elimination of the Biennial Component

In computing the long-term means, care was taken to reduce the effects of the pronounced biennial oscillation found in the stratosphere, and hence to present values which would be representative of the unperturbed seasonal cycle. Because of the nearly 2-year periodicity of the biennial component, the long-term means were obtained from the average of the seasonal means computed for odd years and even years separately. If the oscillation was a perfect sinusoid with a 2-year period, this would permit its complete elimination in the resulting averages. As it is, the amplitude is reduced to the point where it becomes indistinguishable from errors in the data.

2.6 Summary of Tropical Data Reduction Techniques

The techniques used to obtain the tropical data which are discussed in the remainder of these volumes are summarized below. The terms involving q, which were analyzed by Rasmusson and which are discussed in Chapter 5, are not included here.

(1) Maps of $\bar{u}$, $\bar{v}$, $\bar{T}$, $\overline{u'v'}$, and $\overline{v'T'}$ were obtained at the standard pressure levels from 1000 to 100 mb, for four seasons, by subjective analysis of station data.

(2) Maps of $\sigma(u)$, $\sigma(v)$ at 850, 500, and 200 mb, for December–February and June–August, were obtained by subjective analysis of station data values.

(3) Maps of $\bar{u}$ and $\bar{T}$ at 70 and 50 mb, for the four seasons, were obtained by subjective analysis of station data for the average of odd-numbered years, even-numbered years, and a combined average of both.

(4) Values of the zonally averaged parameters $[\bar{u}]$, $[\bar{T}]$, $[\bar{u}^{*2}]$, $[\bar{v}^{*2}]$, $[\bar{T}^{*2}]$, $[\overline{u'v'}]$, $[\bar{u}^*\bar{v}^*]$, $[\overline{v'T'}]$, $[\bar{v}^*\bar{T}^*]$ were obtained by appropriate averaging of grid point values read from the maps in (1) above.

(5) Zonally averaged values of $\sigma(u)$, $\sigma(v)$, and $\sigma(T)$ were obtained by objective analysis in the tropical troposphere.

(6) Values of the parameters $[\bar{u}]$, $[\bar{T}]$, $\sigma(u)$, $\sigma(v)$, $\sigma(T)$, $[\overline{u'v'}]$, and $[\overline{v'T'}]$ were obtained as latitude band means in the entire tropical re-

gion. Values were obtained on a monthly and a seasonal basis for an average of odd-numbered years, even-numbered years, and the average of both.

(7) Values of $[\bar{v}]$ were obtained by a combination of direct measurement and indirect calculations in the tropical region. This term is notoriously difficult to obtain directly but, as discussed in Appendix III, some weight can be placed in the directly measured values in the tropical regions where $[\bar{v}]$ attains its largest values. The indirect method of calculating $[\bar{v}]$ depends on satisfying the angular momentum equation under the assumption that the vertical flux of momentum by the eddies is small except in the lower boundary layer (see Gilman 1965; Vincent 1968; Holopainen 1967; and Lorenz 1967 for examples). This method breaks down in the equatorial region of the tropics and directly measured $[\bar{v}]$ were used in combination with the indirectly calculated $[\bar{v}]$ to complete the coverage in the tropics. Values of $[\bar{\omega}]$ were also obtained in this process.

(8) Values of $\bar{\omega}$ were calculated from the grid point data of $\bar{u}$ and $\bar{v}$ (from 1 above), using the continuity equation.

Extratropical Data

In order to discuss the interactions between tropical and extratropical circulations it was necessary to expand the data coverage to the entire globe. The sources of data are listed below.

(1) Crutcher (1961, 1966): seasonally averaged maps of $\bar{u}$ and $\bar{v}$. Grid point values were extracted in the region 90°N–40°N, 1000–100 mb.

(2) Station data given in the *U.S. Navy Marine Climatic Atlas of the World*, Vol. VII, Antarctic (1965): seasonally averaged station values of $\bar{u}$, $\bar{v}$, $\sigma(u)$, $\sigma(v)$, and $\overline{u'v'}$ were taken at standard levels from 1000 to 20 mb. Some station data of $\bar{u}$ and $\bar{v}$ from the *Atlas of Wind Characteristics of the Southern Hemisphere*, compiled by Guterman (1967, 1970), were used to enhance coverage. These station data were used, together with those in the tropical region data collection, to analyze a complete set of Southern Hemisphere maps of $\bar{u}$, $\bar{v}$, $\sigma(u)$, $\sigma(v)$, and $\overline{u'v'}$. Values of these parameters were then read at grid points at every 10 degrees of latitude and longitude.

(3) Newell and Richards (1969): three months averaged together to form seasonal averages of IQSY stratospheric data for $\bar{u}$, $\bar{v}$, $\bar{T}$, $\overline{u'v'}$, and $\overline{v'T'}$ in the region 100–10 mb, 90°N–20°N.

(4) Goldie et al. (1958): one-month average maps of T for mid-season months. Grid points were extracted in the region 90°N–40°N, 700–100 mb.

(5) Hann and Suring (1943): one-month averages of surface temperature for January and July and their averages used for April and October in the region 90°N–40°N, 40°S–50°S.

(6) Holopainen (1967): seasonal averages of $[\bar{u}^*\bar{v}^*]$ and $[\overline{u'v'}]$ for the region 90°N–30°N, 1000–100 mb.

(7) Oort and Rasmusson (private communication): seasonal values of $[\overline{vT}]_E = [\overline{v'T'}] + [\bar{v}^*\bar{T}^*]$ in the region 90°N–40°N, 1000–50 mb.

(8) Wiin-Nielsen (personal communication): three monthly values of $[\overline{vT}]_E = [\overline{v'T'}] + [\bar{v}^*\bar{T}^*]$ were averaged together to obtain seasonal values for the region 90°N–40°N, 1000–100 mb.

(9) Peixoto (personal communication): six-month averages of $[\overline{v'T'}]$ in the region 30°S–90°S, 1000–50 mb.

The sources of the extratropical data values are summarized below.

$[\bar{u}]$	90°N–40°N, 1000–100 mb	(1)
	90°N–20°N, 100–10 mb	(3)
	40°S–90°S, 1000–100 mb	(2)
	20°S–90°S, 100–20 mb	(2)
$[\bar{v}]$	Values calculated indirectly from the data for $[\bar{u}]$, $[\bar{u}^*\bar{v}^*]$, and $[\overline{u'v'}]$	
$[\bar{T}]$	90°N–40°N, 700–100 mb	(4)
	90°N–40°N, 40°S–50°S, 850 and 1000 mb	(4) and (5) were used to infer these values
	90°N–20°N, 100–10 mb	(3)
	40°S–90°S, 1000–100 mb	(2)
	20°S–90°S, 100–20 mb	(2)

$[\overline{u'v'}]$, $\sigma(u)$, $\sigma(v)$	90°N–40°N, 1000–100 mb	(1)
	90°N–20°N, 100–10 mb	(3)
	40°S–90°S, 1000–100 mb	(2)
$[\bar{u}^*\bar{v}^*]$, $[\bar{u}^{*2}]$, $[\bar{v}^{*2}]$	90°N–40°N, 1000–100 mb	calculated from grid point values of $\bar{u}$ and $\bar{v}$ from (1)
	90°N–20°N, 100–10 mb	(3)
	40°S–90°S, 1000–100 mb	calculated from grid point values of $\bar{u}$ and $\bar{v}$ from (2)
$[\overline{vT}]_E$	90°N–40°N, 1000–50 mb	(7)
	90°N–40°N, 1000–100 mb	(8)
$[\overline{v'T'}]$	30°S–90°S, 1000–50 mb	(9)
$[\bar{v}^*\bar{T}^*]$	30°S–90°S, 1000–100 mb	calculated from grid point values of $\bar{v}$ and $\bar{T}$ from (2)

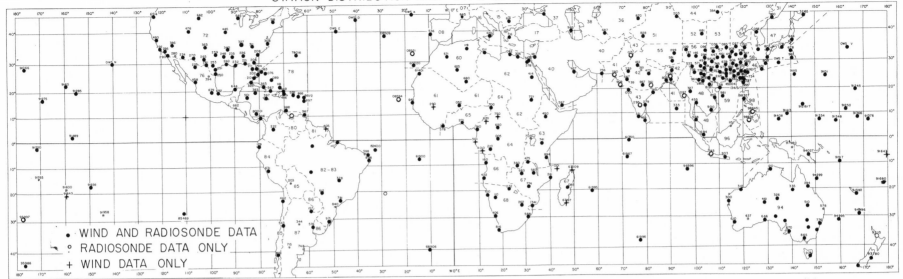

STATION DISTRIBUTION AND TYPE OF DATA

• WIND AND RADIOSONDE DATA
○ RADIOSONDE DATA ONLY
+ WIND DATA ONLY

Plate 2.1

References

Crutcher, H. L. 1961. *Meridional Cross-Sections of Upper Winds over the Northern Hemisphere*. Technical Paper no. 41. Washington, D.C.: U.S. Department of Commerce, National Weather Records Center, U.S. Weather Bureau. 307 pp.

———. 1966. *Components of the 1000 mb Winds (or Surface Winds) of the Northern Hemisphere*. Navair 50–1C–51, published by direction of the Chief of Naval Operations. Washington, D.C.: Government Printing Office.

Eddy, A. 1967a. The statistical objective analysis of scalar data fields. *J. Appl. Meteorol.* 6:597–609.

———. 1967b. *Two-Dimensional Statistical Objective Analysis of Isotropic Scalar Data Fields*. Austin: University of Texas Atmospheric Science Group, publ. no. 5. 100 pp.

Gilman, P. A. 1965. The mean meridional circulation of the Southern Hemisphere inferred from momentum and mass balance. *Tellus* 17:277–284.

Goldie, N., J. G. Moore, and E. E. Austin, 1958. *Geophys. Mem.* vol. 13, no. 101. London: HMSO. 101 pp.

Guterman, E. G. 1967, 1970. *Atlas of Wind Characteristics in the Southern Hemisphere*, vols. 1 and 2. Moscow: Institute of Aeroclimatology.

Hann, J., and R. Süring. 1943. *Lehrbuch der Meteorologie* vol. 1, no. 15. Leipzig: Willibald Keller. 480 pp.

Holopainen, E. O. 1967. On the mean meridional circulation and the flux of angular momentum over the Northern Hemisphere. *Tellus* 19:1–13.

Lorenz, Edward N. 1967. *The Nature and Theory of the General Circulation of the Atmosphere*. Geneva: WMO. 161 pp.

Newell, R. E., and M. E. Richards. 1969. Energy flux and convergence patterns in the lower and middle stratosphere during the IQSY. *Quart. J. Roy. Meteorol. Soc.* 95:310–328.

Priestley, C. H. B., and A. J. Troup. 1964. Strong winds in the global flux of momentum. *J. Atmos. Sci.* 21:459–460.

U.S. Navy Marine Climatic Atlas of the World, Vol. VII, Antarctica, 1965. NAVWEPS 50–1C–50.

Vincent, D. G. 1968. Mean meridional circulations in the Northern Hemisphere lower stratosphere during 1964 and 1965. *Quart. J. Roy. Meteorol. Soc.* 94:333–349.

Wallace, J. M. 1966. *Long-Period Fluctuations in the Tropical Atmosphere*. Report no. 19. Cambridge: Massachusetts Institute of Technology, Department of Meteorology, Planetary Circulations Project. 167 pp.

3

Mean Temperature and Wind Fields

There are at least four methods used in meteorology for the presentation of time-mean temperature and wind data. They are maps at constant pressure on which station values are plotted and isopleths drawn; meridional cross sections, with latitude and pressure as coordinates, either at a particular longitude or for an average over all longitudes; zonal cross sections with longitude and pressure as coordinates; and mean vertical soundings for particular stations or regions, with the data as a function of pressure.

3.1 The Temperature Field

The most comprehensive study of upper air temperature over the world is reported in the *Geophysical Memoir* by Goldie, Moore, and Austin (1958). The *Memoir* contains meridional cross sections, mean soundings, and a large selection of maps for the months of January, April, July, and October. The present study focuses attention on four 3-month seasons, centered on these months. Three-month periods are required to obtain reliable estimates of the flux quantities, for example the heat flux, which together with the temperature gradient is important for the energy budget. Maps of temperature for the region 45°N–45°S, at the mandatory pressure levels between 1000 mb and 100 mb, have been analyzed by hand, using station data. Above 100 mb year-to-year differences, to be discussed later, paucity of data, and instrumental inaccuracies preclude long-term mean map analysis. Also included are a selection of zonal and meridional cross sections based on map analyses, a table of zonally averaged temperatures for the globe, and a set of mean vertical temperature soundings at several latitudes obtained from the zonal averages. Taken all together, these data effectively illustrate the temperature patterns which characterize the region.

Maps of $\overline{T}$ for the four seasons are presented for the 1000, 850, 700, 500, 400, 300, 200, 150, and 100-mb pressure levels in Plates 3.2–3.10. At low latitudes the mean heights of these pressure levels are 0.1, 1.5, 3.1, 5.9, 7.6, 9.7, 12.4, 14.2, and 16.6 km, respectively. Maps showing the intersection of the 1000, 850, and 700-mb pressure surfaces with the earth's surface are shown in Plate 3.1.

The 1000-mb level is close to the surface over most of the earth. Surface temperatures from *World Weather Records 1951–1960* (ESSA

17

1967) were used to enhance the data coverage over the continents of Africa and South America, where 1000-mb data from the upper air network were relatively sparse. Only data from stations with elevations less than 300 meters were used, and temperatures were adjusted by an amount based on the mean sea-level pressure at the station. The maps clearly illustrate the influence of the underlying surface on the temperature patterns. For reference, the ocean surface temperatures have been reproduced in Plate 3.11 for January, April, July, and October (from the *World Atlas of Sea Surface Temperatures*, Hydrographic Office, United States Navy, 1944). The zone of maximum temperatures at 1000 mb moves backward and forward across the equator with the sun. The air temperature pattern over the oceans roughly parallels that of the ocean surface temperature. This is in turn greatly influenced by the ocean currents (Plate 3.12) such as the cold Benguela and California currents, which flow toward the equator, and the warm Gulf Stream and Brazil currents, which flow toward the poles. There is close coupling between wind velocity, water temperature, current direction, and air temperature. The current direction itself is related to surface wind pattern. Water temperature is partly governed by upwelling processes along the coasts, which are again controlled by surface wind patterns. In the Indian Ocean, reversals of the wind flow during the monsoon are accompanied by reversals in the Somali current.

Leaving aside cause-and-effect discussion for the present, perhaps the most important point to stress is the large east-west difference in surface air temperature across the continents of South America and South Africa. It should be noted that we have used mainly 0000 GMT data, so that the values given over the Sahara, for instance, will correspond to local nighttime observations and will be lower than mean values. We have made use primarily of the upper air network, and much more detail is available for surface temperatures. Texts on climatology such as those by Haurwitz and Austin (1944) and Kendrew (1961) should be consulted for detailed discussions of the surface temperature field.

Salient features of the 850-mb maps are the double maximum over Africa, evident in all seasons, and the extremely high temperatures over the southwestern United States in June–August. The double maximum over Africa has also been noted by Thompson (1965) for January, April, July, and October, and it moves north–south with the sun. Our values for the amplitude difference between these two maxima are 0, 4, 10, and 2 °C for the four seasons, with the Northern Hemisphere value always being the larger of the two. As may be inferred from the topographic map, high temperatures over the United States are due to the influence of the underlying surface, an effect accentuated here by the use of data obtained primarily at 0000 GMT, which corresponds to late afternoon in the region. The moderating effect of the oceans in the Southern Hemisphere can be seen in the relatively small seasonal variation of temperature as compared to the Northern Hemisphere.

The 700-mb December–February map shows temperature maxima of almost equal magnitude on either side of the equator, over most regions except the Pacific. A comparison may be made with the work of Goldie et al. (1958), for which 700 mb is the lowest level. Their January map is quite similar to the December–February map; in particular, it shows a double maximum over Africa, as does the map of Thompson (1965). By March–May, a high-temperature region stretching from Africa to Indonesia dominates the pattern, so that the zonally averaged maximum temperature is in the Northern Hemisphere; traces of a double maximum still exist. In June–August the band of highest temperatures is over the land masses of the Northern Hemisphere. In September–November the double maxima are again in evidence over Africa. The October chart by Goldie et al. does not show this double maximum. Over the equatorial western Pacific the entire annual range (based on seasonal means) is less than 1 °C at 700 mb.

The 500-mb December–February map shows maximum temperatures in the region between 10 °N and 20 °S. The field is very flat in this zone with all the station values falling between −4.5 and −6.5 °C. The −6.0 °C isotherm shows some evidence of a double maximum over the Indian Ocean region. The overall maximum occurs over the western Pacific. In March–May the pattern is essentially the same, with meridional gradients diminished slightly in the Northern Hemisphere middle latitudes. The double maximum is again evident over the Indian Ocean. In June–August there is a very warm area over the India–China monsoon region; this is presumably due to the latent heat liberated in the monsoon rains. Traces of this warm region remain on the map for September–November. The other notable feature of the September–November map is the reappearance of the double maximum over the Indian Ocean. These maxima are practically absent in the analyses of Goldie et al.

The maps at 400 mb and 300 mb show many of the same features found at 500 mb, although there are only slight traces of the double maximum over the Indian Ocean.

The maps for 200 mb display a more complicated pattern than those for lower levels, partly because of the juxtaposition of troposphere and stratosphere. Near the equator and in the Southern Hemisphere, the horizontal gradients of temperature are very weak in all seasons. In the extratropical Northern Hemisphere the temperature pattern in December–February is reversed from that at lower levels, with warm regions over eastern North America and eastern China; these regions are, of course, part of the stratosphere. This boundary of troposphere and stratosphere is marked by the larger temperature gradients in the Northern Hemisphere. In June–August, less of the Northern Hemisphere portion is in the troposphere. The occurrence of the highest temperatures over India parallels the pattern in the middle troposphere. One of the few major differences between the present work and that of Goldie et al. is that they find a warm belt over both South America and South Africa during July, whereas we find it over South Africa and Australia but not over South America. Note the reappearance of the double maximum over Africa.

Maps for 150 mb also show patterns which reflect the juxtaposition of the troposphere and the stratosphere. The pressure surface passes through the region between the upper tropopause, characteristic of equatorial regions, and the lower tropopause, characteristic of extratropical regions. The large-scale temperature gradient now shows a clear reversal from its tropospheric pattern, with a band of low temperature in the equatorial region and high temperatures in middle latitudes.

The maps for 100 mb represent either the tropopause level or the stratosphere; only the region quite close to the equator is in the troposphere. The coldest regions in all seasons are over the central and western Pacific, with traces of a double minimum found in the latter

region. Note that the lowest temperatures occur in the earlier part of the year, a finding that was also evident in the work of Goldie et al. At this level, and also to some extent at 200 mb and 150 mb, some discrepancies between adjacent stations can be noted. They are probably due to the use of different types of radiation corrections for different temperature-measuring elements. These discrepancies are quite easy to pick out and were ignored in the analyses.

The zonally averaged temperatures given in Table 3.1 were used to construct the mean profiles given in Figure 3.1, at 20-degree increments of latitude, for December–February and June–August. At low latitudes in December–February the mean tropopause is located near the 90-mb level. It is characterized by the lowest temperatures as well as by a reversal in the sign of the vertical temperature gradient. At high latitudes the tropopause is near 300 mb; the temperatures above this level are nearly isothermal, while the lapse rates below are approximately the same as those at low latitudes. The same remarks are essentially true for June–August, except that the tropical tropopause is lower (100 mb) and warmer. Note the very small meridional temperature gradients near 200 mb. Maps of the pressure and temperature at the tropopause are presented by Goldie et al.; they also reproduce a number of mean temperature soundings for typical stations which illustrate the variety of tropopause structures which are found.

To illustrate some of the salient features of the temperature patterns more clearly, two sets of vertical cross sections, with latitude and longitude as ordinates, were compiled. Cross sections at each 10 degrees of latitude from 40°N to 40°S, for the December–February and June–August seasons, are given in Figures 3.2 and 3.3. These are based on grid point data obtained from the hand-analyzed temperature charts shown in Plates 3.2–3.10. For clarity, zonal averages were subtracted from the temperature values at grid points so that it is $\bar{T}^* = \bar{T} - [\bar{T}]$ which is displayed. In what follows, the cross sections at the corresponding latitudes in the Northern Hemisphere and Southern Hemisphere, for the same season, are compared. Thus December–February at 40°N is compared with June–August at 40°S, and this set is termed the winter season at 40° latitude.

(a) For the winter season, at 40° latitude in the Northern Hemi-

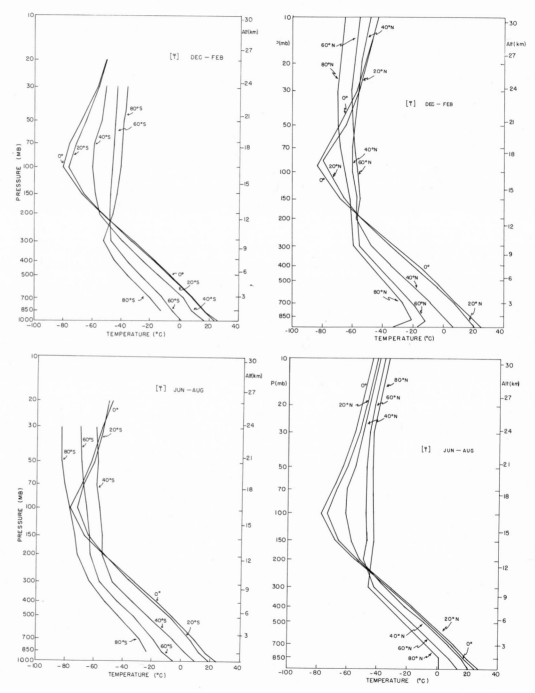

Figure 3.1

sphere, lower temperatures occur on the eastern sides of the continents and higher temperatures over the oceans. In the middle and upper troposphere, cold and warm regions slope westward with height. In the Southern Hemisphere winter the corresponding slope is eastward. The amplitudes of the temperature deviations are smaller, and the patterns are less directly tied to surface features in the absence of continent-ocean contrasts.

(b) At 30° in winter, the pattern is basically the same as that at 40°. In general in the Southern Hemisphere winter, the influence of cold ocean currents to the west of South America and Africa appears to control much of the low-level pattern. Over Australia a substantial region of lower temperature occurs in the lower troposphere.

(c) For the latitudes from 20° to the equator in winter, smaller horizontal gradients of temperature are apparent in the Northern Hemisphere. The low-level patterns, reflecting continent-ocean contrasts, are weaker, and regions of relative warmth exist over Africa and America. The pool of warm air at 100 mb is a prominent feature at these latitudes. The general pattern shows a reversal from that seen at 40°. In the Southern Hemisphere winter, at these latitudes, similar remarks apply. The effects of the cold ocean currents west of South America and Africa are prominent in the low-level patterns in the Southern Hemisphere. Again, in the free atmosphere, the pattern is roughly the reverse of that at 40°.

(d) The major differences in the winter patterns in the Northern and Southern Hemispheres are those which reflect the different continent-ocean distribution and the effects of cold ocean currents in the regions. Generally smaller mid-tropospheric temperature gradients are apparent in the Southern Hemisphere.

(e) In summer, the Northern Hemisphere pattern at 40° is roughly the reverse of the winter pattern, with low-level warm and cold regions over the continents and oceans, respectively. In the Southern Hemisphere the pattern is similar to the winter pattern, with considerably less change than the Northern Hemisphere pattern. As land masses occupy only a small fraction of the area at these latitudes, it is reasonable to expect that the ocean surface temperatures (see Plate 3.11) should be the dominant influence on the low-level temperature patterns.

(f) At 30° in summer, the Northern Hemisphere pattern is similar to that at 40°. Enhanced warmth over the monsoon region is apparent. In the Southern Hemisphere the cold ocean currents to the west of the continents show a marked effect on the temperature structure.

(g) Latitudes from 20° to the equator show summer temperature patterns in both hemispheres which are similar to those seen at higher latitudes. Continent-ocean effects are marked in the lower levels. There is a less marked winter-summer change at these latitudes than at more poleward latitudes, as would be expected.

Meridional cross sections of $\bar{T}^*$, for longitudes 150°E, 150°W, 80°W, 20°E, and 80°E, are presented in Figure 3.4 for December–February and in Figure 3.5 for June–August. These longitudes were selected because the data coverage is good over most latitudes and because they illustrate asymmetries in the zonally averaged patterns.

(a) The 150°E meridian of longitude is representative of the western Pacific. In December–February, at the higher northern latitudes, this is a region of relative coolness, as a consequence of the cold continents just to the west. Somewhat farther south, relative oceanic warmth is again apparent. Over the eastern tip of Australia the region of low temperatures shows larger amplitudes in the winter. Almost the whole region from 200 to 100 mb shows relative coolness in both seasons compared to the zonal average.

(b) For December–February at 150°W, which is characteristic of the eastern Pacific, low-level temperatures are generally warm in comparison to the zonal mean. This is likely to be a consequence of the relative warmth found over the ocean in the Northern Hemisphere winter, and of the fact that in the Southern Hemisphere summer, this region of the ocean is relatively warm compared to the other southern oceans (see Plate 3.11). In June–August the Northern Hemisphere ocean region shows relatively low temperatures, characterizing the relative coolness of the oceans as compared to the continents in summer. The region of relative warmth at 40°N in December–February has been replaced by a region of relative coolness.

(c) The 80°W meridian lies just along the east coast of the United States and to the west of Chile. The effect of the cold ocean currents in this latter region are apparent in the low-level pattern in both seasons.

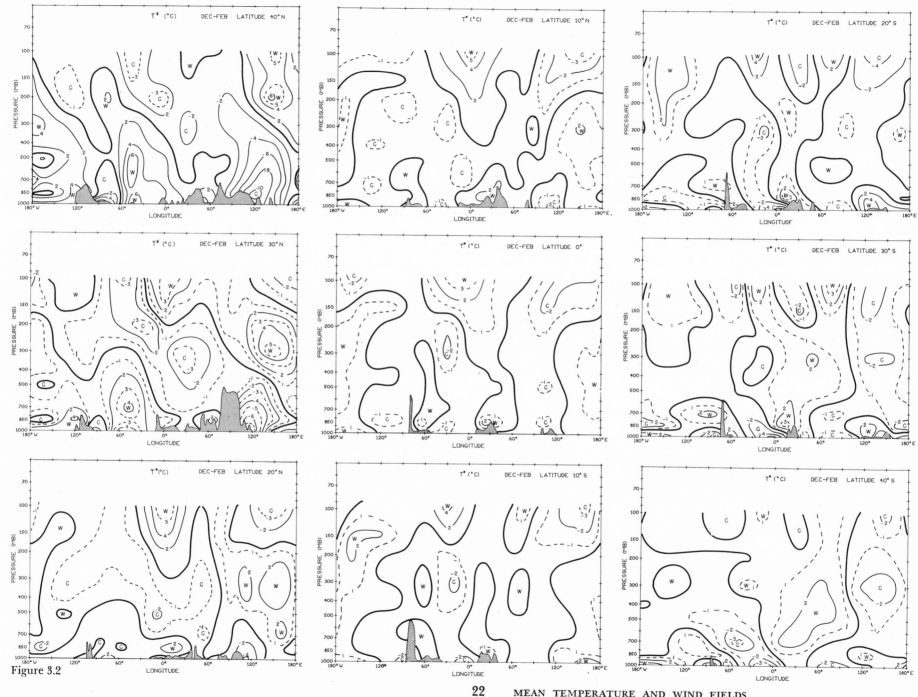

Figure 3.2

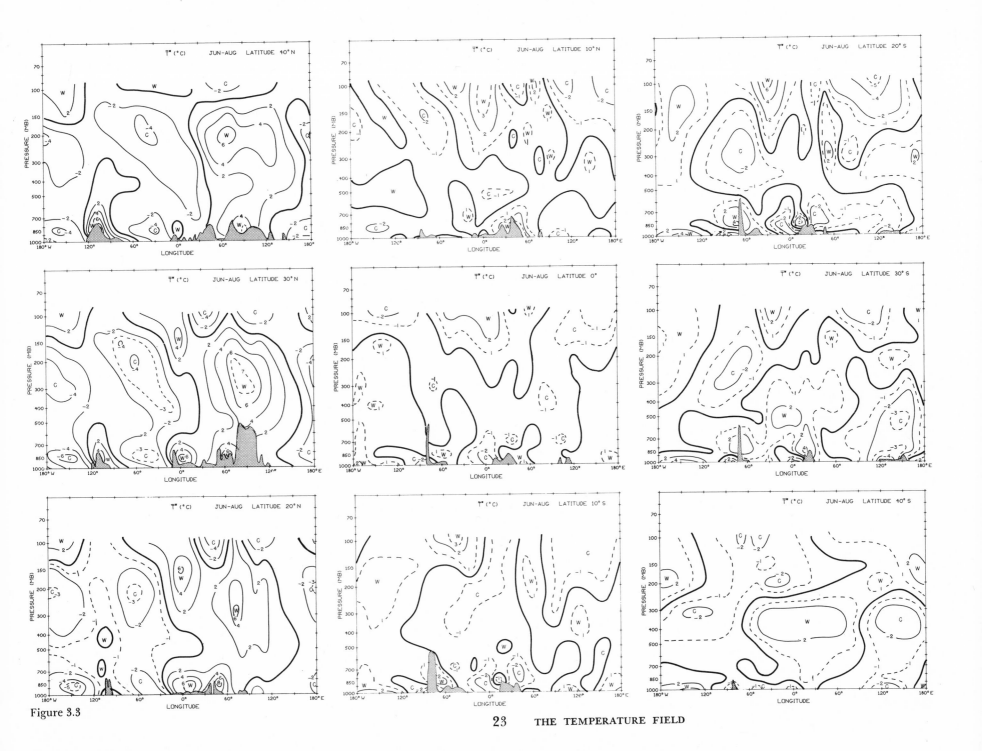

Figure 3.3

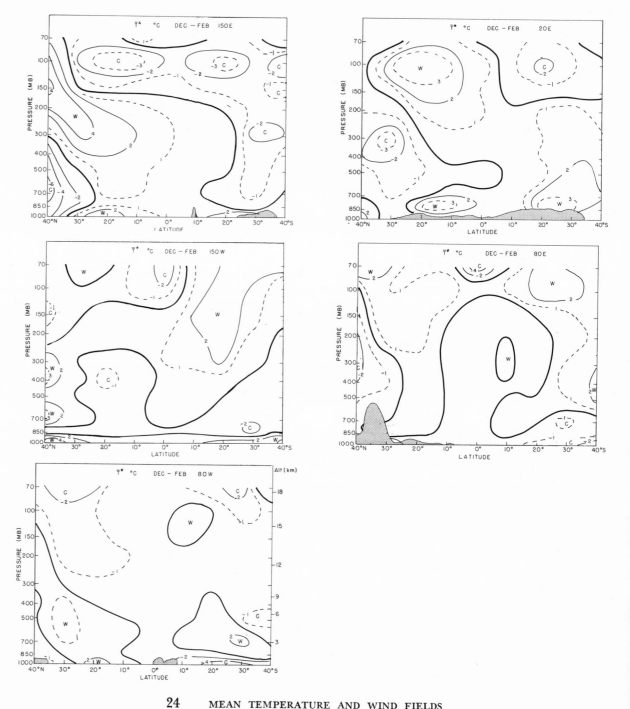

Figure 3.4

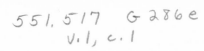

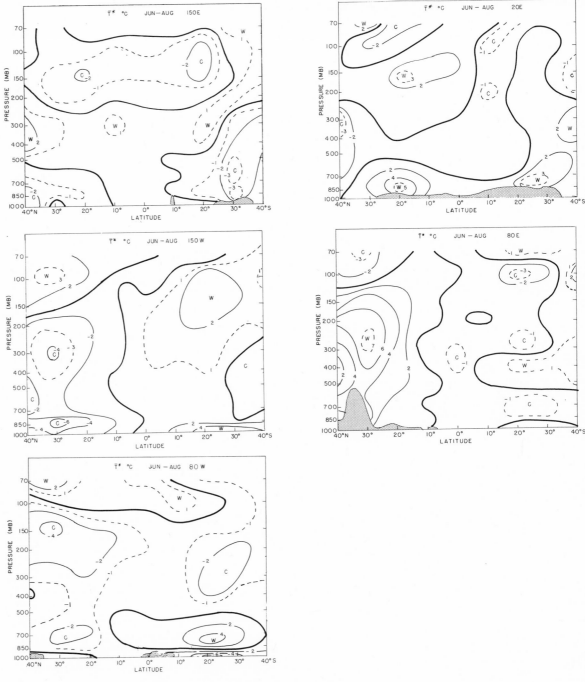

Figure 3.5

In December–February the continental effect of the United States is also apparent near 40°N. The middle and upper troposphere tends to be relatively cold in this region, compared to the zonal average.

(d) The 20°E meridian of longitude bisects Africa over most of the region. The double maximum apparent on the temperature maps over Africa is clearly seen. The temperature deviation is greater in the summer than the winter at 20°N. This effect is not as marked at 20°S. This double maximum is likely to be a consequence of the transfer of heat to the troposphere from the underlying surface. The conditions of relatively little cloud and of dry air in these regions are effective in producing high surface temperatures during the day.

(e) The 80°E meridian of longitude runs through India. The most dramatic feature of the temperature pattern is the relative warmth associated with the summer monsoon. This temperature deviation has the largest amplitude found at any longitude at this height.

(f) The zonally averaged mean temperature $[\overline{T}]$ for the four seasons are given in Table 3.1, and global cross sections are presented in Figure 3.6. The zonal averaging has eliminated all but the broad-scale features of the temperature pattern. The most noticeable features of this pattern are the seasonal displacement of the low-level equatorial temperature maximum, the seasonal change in the equatorial minimum of temperature near 100 mb, the marked temperature changes in the polar stratosphere, and the seasonal change in the Northern Hemisphere low-level temperature structure. Little evidence of the more local patterns, which have been discussed in terms of the temperature maps and cross sections, remains. The most prominent low-level features, such as the maxima on either side of the equator, have no representation in the zonally averaged temperature field.

The change in temperature between the seasons is illustrated by Figure 3.7, which was obtained by subtracting the December–February $[\overline{T}]$ values from those for June–August. The greatest change takes place in the Southern Hemisphere polar stratosphere; seasonal variations are much greater there than in the corresponding regions in the Northern Hemisphere. In the lower and middle troposphere, however, variations are generally largest in the Northern Hemisphere, particularly near the surface at higher latitudes. Below 150 mb, the line of minimum

fluctuation (zero isopleth) occurs in the Northern Hemisphere near 5°N; between 70 and 20 mb the region of minimum change is in the subtropical Southern Hemisphere. Another marked feature is that the air near the tropical tropopause is colder in December–February than in June–August, regardless of hemisphere, with the difference amounting to over 6°C at 70 mb near 10°N. This calendar year effect is also evident in the work of Goldie et al., and they also report a higher and colder tropical tropopause in January and April than in July and October.

Maps of the standard deviation of temperature were analyzed, but they are not reproduced here as a good set is available in Goldie et al. The latitudinal distribution of this quantity is portrayed in mean meridional cross sections for the region 40°N–30°S. The contribution of the trend of the temperature during the 3-month period to the standard deviation has been removed from these values. The cross sections for the four seasons (see Figure 3.8) are based on objective analysis of the observed variance. The value given is based on the observed variance less the variance due to the trend, as computed from monthly latitude band means. A relatively small time variability at low latitudes is evident.

The mean temperature at a point depends on a number of factors: the direction of the prevailing wind, underlying surface temperature, absorption of solar radiation, convergence of the vertical flux of infrared radiation, cloudiness, latent heat liberation, and mean vertical motion (producing adiabatic heating through subsidence and cooling through ascent) all play their parts. These items are not unrelated; for example, the prevailing wind direction often controls the sea surface temperature through the upwelling process. Some of these factors will be discussed in later chapters.

3.2 The Zonal Wind Field
Seasonal maps of the mean zonal wind $\bar{u}$ are presented for the surface and the 850, 700, 500, 400, 300, 200, 150, and 100-mb levels in Plates 3.13–3.21. There are several other pertinent sets of data which may be compared with these maps. Buch (1954) has presented maps of the mean wind components for the Northern Hemisphere for two 6-month

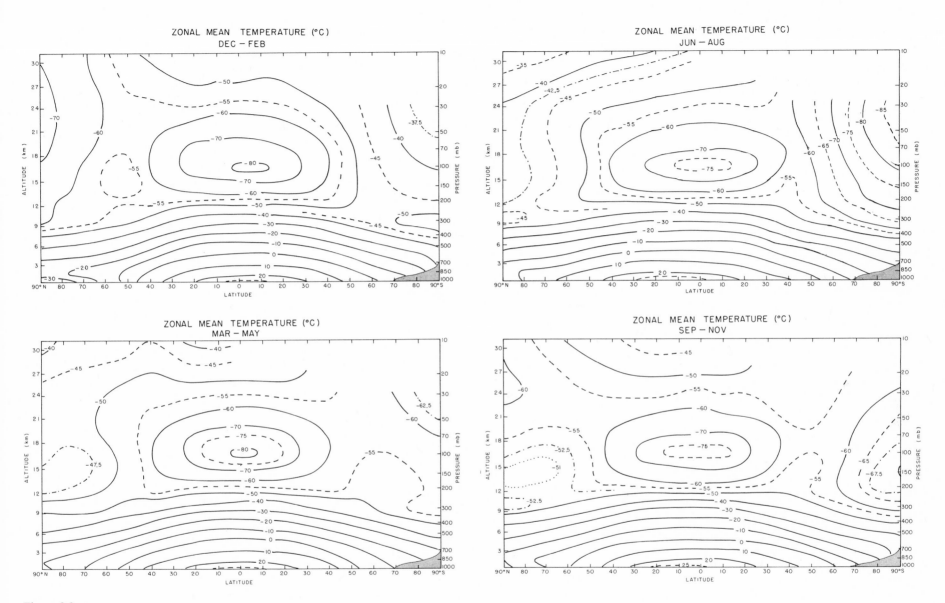

Figure 3.6

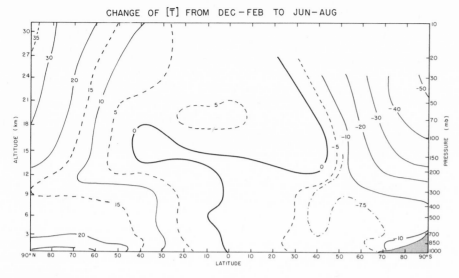

Figure 3.7

seasons at pressure levels from 850 to 100 mb. The maps are based on the observed winds for the year 1950. Heastie and Stephenson (1960) have presented maps of winds to 60°S for the months of January, April, July, and October at pressure levels from 700 to 100 mb. These winds were calculated from mean contour heights. Crutcher (1959) has prepared a set of maps of the mean wind components, and of other wind statistics, for the Northern Hemisphere for four 3-month seasons. At least 5 years of data were available for most stations. Crutcher (1961) subsequently compiled a set of meridional cross sections for each 10 degrees of longitude. His data were based mainly on the observed winds, although the gradient wind relationship (not the geostrophic approximation) was occasionally used to supplement observations in data-poor ocean areas. More recently, Crutcher (1966) has prepared monthly maps of the wind components at 1000 mb. Surface wind data were used when the 1000-mb level was below the land or sea surface. Obasi (1963), in a study analogous to that of Buch, has presented maps of wind components for the Southern Hemisphere for two 6-month seasons for pressure levels between 850 and 100 mb. The maps were based on data from the International Geophysical Year. The analysis was difficult at middle and low latitudes due to the small number of stations found there. Guterman (1967, 1970) has more recently produced two atlases containing monthly mean zonal and meridional wind components for the Southern Hemisphere, at the 850, 700, 500, 300, 200, and 100-mb levels.

The surface map for December–February shows extremely light winds, which exceed 5 m sec^{-1} in only a few regions. Crutcher's (1966) maps have been used to augment our data in the latitude region 40°N–10°N, particularly over the oceans. Between 30°N and 30°S, the flow is predominantly easterly, although patches of westerly flow exist. The magnitudes of the easterlies in the Northern Hemisphere are somewhat larger than those in the Southern Hemisphere, and the zonally averaged easterly maximum is found north of the equator. In the March–May period, the pattern is similar to that in December–February except over India, where westerlies have become more extensive. In June–August, easterlies remain over the Western Hemisphere on both sides of the equator. The westerlies over India, accompanying the

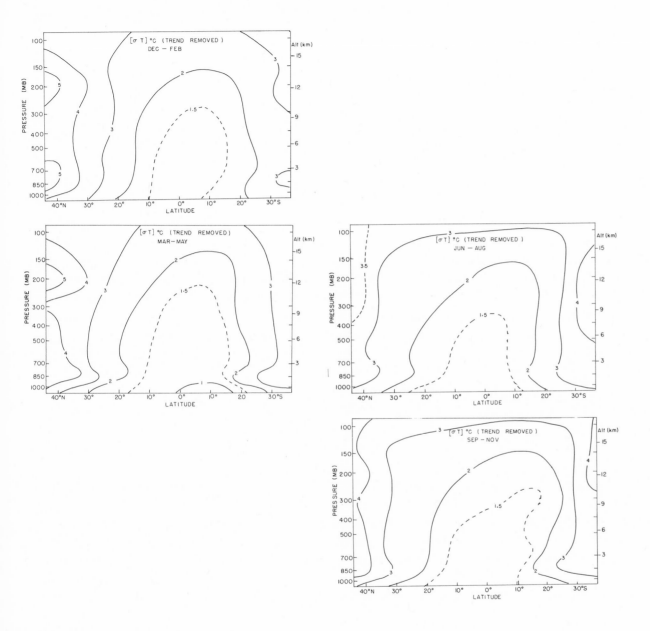

Figure 3.8

monsoon, have become stronger and more extensive and have joined with the region of westerlies over Africa. It is both because of this large region of westerlies, rather than because of weakened easterlies in the Northern Hemisphere, and because of increased easterlies in the Southern Hemisphere that the zonal averages (see Table 3.2) show the maximum easterly flow to the south of the equator. The pattern in September–November is quite similar to that of June–August but shows a weakening of the monsoon westerlies over India. The zonally averaged maximum is still south of the equator.

Superposition of the surface analyses on those for 850 mb reveals some interesting features. The general pattern for December–February is quite similar in the western Pacific, at both levels. The easterly maxima on both sides of the equator have higher speeds at 850 mb than at the surface. The most significant seasonal variation occurs in the Eastern Hemisphere, where easterlies give way to strong westerlies over India during the monsoon period, and concomitantly, Southern Hemisphere easterlies move equatorward and intensify.

At 500 mb in December–February, strong westerly winds occur within the map area. Over the eastern coast of North America and over Japan the wind maxima are to the south and west of their apparent positions at 850 mb. The peak easterlies lie to the north of the equator and are almost directly south of the peak westerlies. The easterly belt is much narrower here than at 850 mb, and there is only a trace of equatorial westerlies. A marked seasonal change occurs over southern India, where weak westerlies replace easterlies in June–August.

At 200 mb in December–February there are three cores of strong westerly winds at 30°N, each apparently associated with a maximum of easterly winds near the equator. The pattern, which is essentially the same as that at 500 mb, is also found at 400 and 300 mb. Westerly winds steadily increase in intensity between 500 and 200 mb. At 200 mb, maximum speeds of more than 60 m sec^{-1} are reached over Japan. Easterly winds are a little stronger and occur farther south at 200 mb than at 500 mb, except in the central Pacific region, where westerly winds replace easterlies above 500 mb. Very strong easterlies accompany the Indian monsoon in June–August, and persist into the following season. Easterlies even replace westerlies over the central Pacific at this

time. The seasonal variation in the amplitude of the middle-latitude westerlies is a significant feature in the middle and upper troposphere in the Southern Hemisphere. The strongest winds, greater than 45 m sec^{-1}, occur at 200 mb in June–August over Australia.

Maps of the 100-mb zonal wind components are shown in Plate 3.21. The westerly jet maxima occur below this pressure surface, and westerly winds are diminishing with altitude at this level, as can be seen by comparison with the 150-mb and 200-mb maps. This decrease accompanies a poleward increase of temperature at middle latitudes. In December–February, maximum easterly winds over Africa and South America occur a little to the south of the equator, as they do at 200 mb, while they occur over the equator in the western Pacific. Note that there are westerly winds over the equatorial Atlantic and central Pacific. In March–May the pattern near the equator is one of easterlies in the Eastern Hemisphere. Peak westerlies in middle latitudes occur between 25°N and 35°N, and westerlies cover the entire central Pacific. In June–August the peak easterly winds over India reach 30 m sec^{-1}. The center of gravity of the easterlies is now definitely in the Northern Hemisphere; even in the northern Pacific, westerlies have changed to easterlies. In September–November the main change has been the weakening of the Indian Ocean easterlies and a return to westerly flow in the central Atlantic and Pacific.

Cross sections of $\bar{u}$ are presented for December–February and June–August at each 10 degrees of latitude from 40°N to 40°S (Figures 3.9 and 3.10). Again the cross sections are discussed in pairs, corresponding to the same latitude and season in both hemispheres.

(a) The 40° and 30° latitude cross sections, in winter, are characterized by westerly flow. Maximum winds occur at the 150–200-mb level with the largest such maximum occurring at 30°N. In general, the Northern Hemisphere wind maxima are of greater magnitude than those of the Southern Hemisphere in this season. Maxima in the Eastern Hemisphere are of larger magnitude than those in the Western Hemisphere. The positions of the maxima in longitude are remarkably similar in both the Northern and Southern Hemispheres. Low-level easterlies make their appearance at 30° latitude in this period.

(b) At 10° North and South latitude, in winter, a region of strong easterly flow has replaced the region of westerly flow found at higher latitudes in the middle and upper troposphere of the Eastern Hemisphere. Westerly flow, of lesser amplitude than that found at higher latitudes, remains in the Western Hemisphere. The pattern is predominantly that of a single wave. A region of easterlies is found in the lower levels in which occasional regions of light westerly flow are embedded.

(c) The same general pattern exists at the equator with maximum easterlies occurring in the Eastern Hemisphere in June–August and maximum westerlies occurring in the Western Hemisphere in December–February.

(d) In the summer season, at 30° and 40° latitude, a prominent double maximum of westerly flow is again found. The magnitudes of the winds are considerably reduced from their winter values in the Northern Hemisphere where, at 30°N, the westerly maximum in the Eastern Hemisphere is reduced in magnitude from 60 m sec^{-1} to 15 m sec^{-1}. In the Southern Hemisphere the westerly winds at 40°S exhibit magnitudes which are somewhat less than their winter values. Easterlies at low levels penetrate to 40°N in this season.

(e) At 20°N in summer, easterlies are prominent in the Eastern Hemisphere. The maximum speed of the easterlies exceeds that of the westerlies at this time. The pattern differs greatly from the corresponding winter pattern at this latitude. The pattern at 20°S in summer resembles the corresponding winter pattern in the absence of these "monsoon" easterlies.

(f) At 10°N in summer, the monsoon easterlies reach their maximum in the Eastern Hemisphere upper troposphere. They overtop an associated lower-level westerly maximum. At 10°S in summer, the basic pattern of easterlies in the Eastern Hemisphere and westerlies in the Western Hemisphere is again apparent, but the easterly maximum is much smaller.

Cross sections of the zonal wind component along the 150°E, 150°W, 80°W, 20°E, and 80°E meridians of longitude are presented for the December–February and June–August seasons in Figures 3.11 and 3.12. These cross sections illustrate the latitudinal position of many of the features observed on the previous set of cross sections. The cross sections are characteristic of the western Pacific, the eastern Pacific, a line from the east coast of the United States to the west coast of South America, a line through Africa, and a line through India, respectively. They should be compared to the zonal means (see Figure 3.13). The cross sections show that

(a) The basic pattern is that of strong westerly winds in the Southern and Northern Hemispheres flanking an equatorial region of easterly winds, which extends in most cases through the troposphere. A notable exception to this basic pattern occurs at 150°W in December–February. Here the equatorial easterlies are confined to regions below 500 mb and above about 70 mb. The cores of the westerly winds are stronger and closer to the equator in the winter hemisphere than in the summer hemisphere. The strongest westerly winds seen on any of these cross sections occur at 150°E in December–February, with a maximum of more than 55 m sec^{-1}. This contrasts with values of 25–40 m sec^{-1} seen elsewhere.

(b) In December–February, the region of easterlies slopes southward with height, and low-level and surface easterly maxima occur predominantly in the Northern Hemisphere. In general, a secondary weak easterly maximum occurs in the low-level Southern Hemisphere wind field, and is separated from the Northern Hemisphere low-level easterly wind maximum by a region of westerlies.

(c) In June–August, the low-level easterly maximum is again found in the Northern Hemisphere at western longitudes, but appears in the Southern Hemisphere at eastern longitudes. The largest and strongest region of easterly flow is seen in the 80°E, June–August cross section over India in association with the monsoon. This section also shows an associated low-level maximum of westerly flow.

Tabular values of the mean zonal wind $[\bar{u}]$ are given in Table 3.2, and global cross sections of the winds are presented in Figure 3.13. These values were obtained from the various data sources listed in Chapter 2. The maps and cross sections of $[\bar{u}]$, which have been discussed previously, have shown that considerable longitudinal structure exists in the wind. The zonal average is not expected to characterize the flow at any particular latitude except in a general sense. The cross sections show that

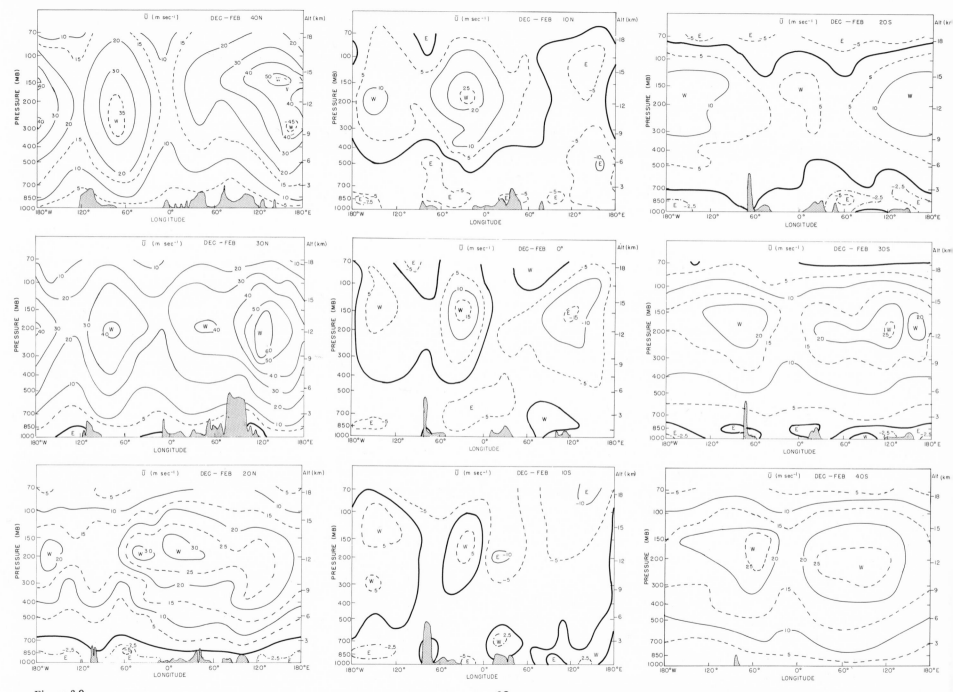

Figure 3.9

32   MEAN TEMPERATURE AND WIND FIELDS

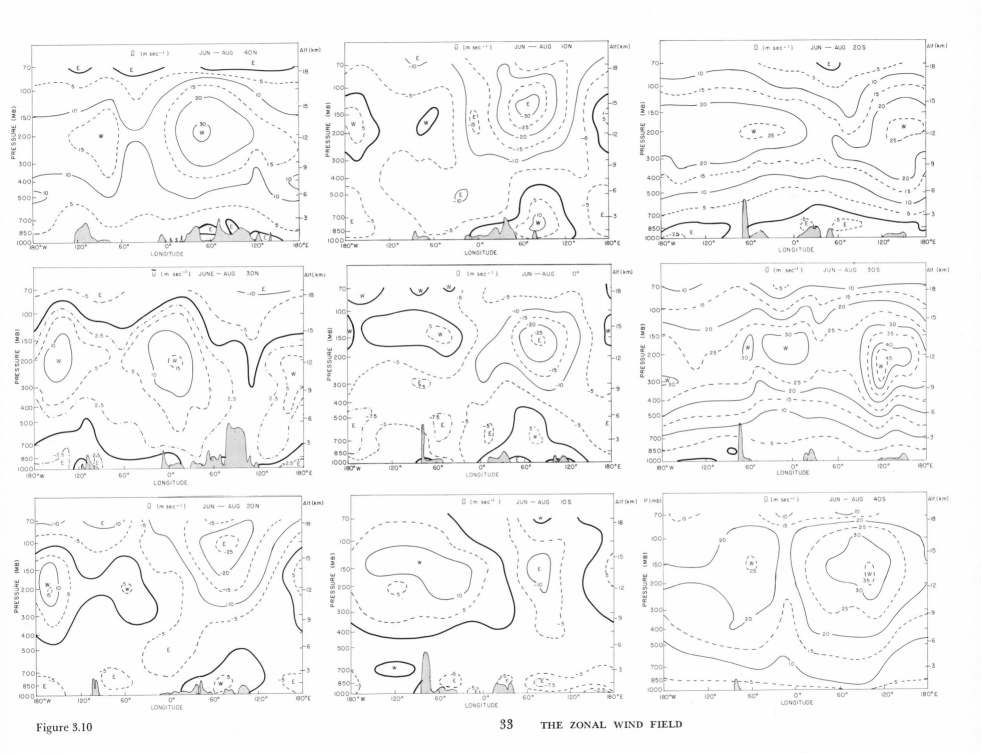

Figure 3.10

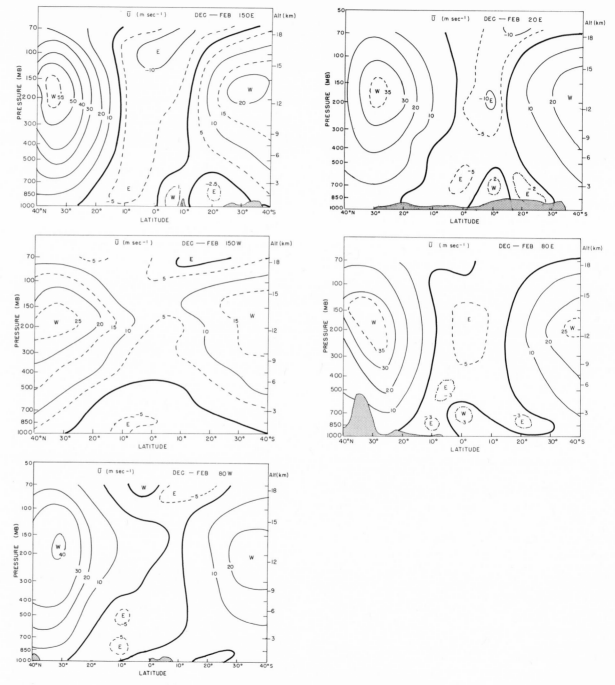

Figure 3.11

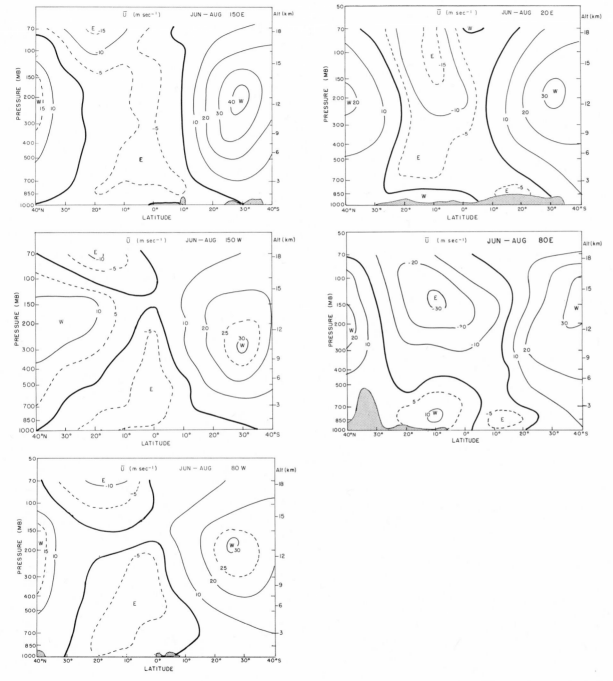

Figure 3.12

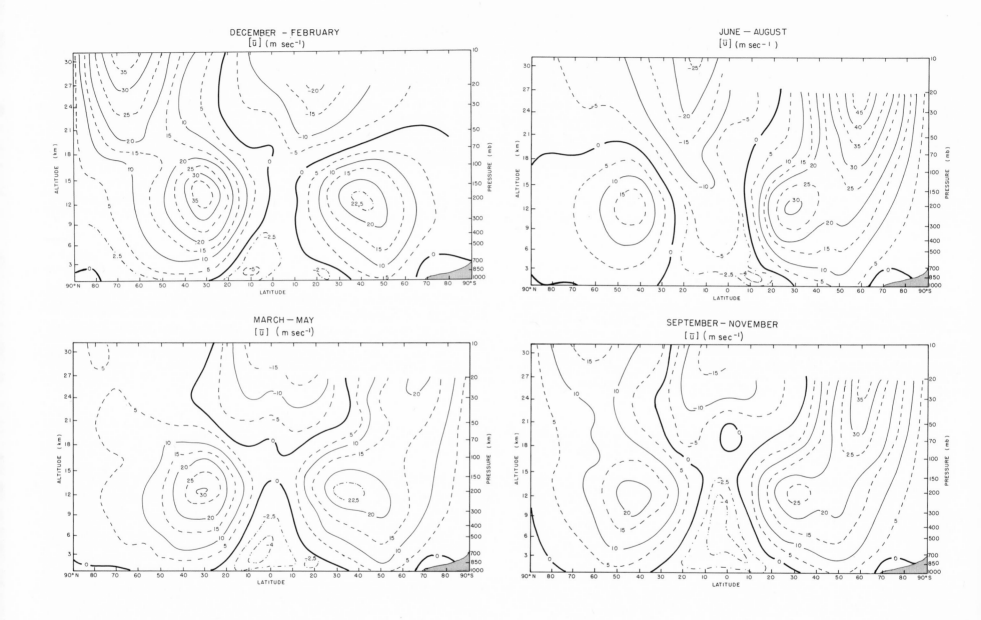

Figure 3.13

(a) The general pattern is that of two westerly wind maxima flanking an equatorial region of easterly flow. The Northern Hemisphere westerly wind maximum moves poleward and weakens considerably in summer. The Southern Hemisphere maximum also moves poleward in summer with a smaller change in speed. The region of maximum westerly wind in the polar stratosphere known as the polar night jet is a prominent feature. The Southern Hemisphere polar night jet is stronger than that of the Northern Hemisphere, and its maximum value is larger than that of the tropospheric jet. The rapid increase of zonal wind with height in this region implies, through the thermal wind equation, that a marked poleward temperature decrease exists. This gradient is clearly seen in Figure 3.6, and the difference in the gradient between the northern and southern polar regions is striking. In the Southern Hemisphere, the polar night jet appears to extend into the region occupied by the tropospheric jet, and to produce the resulting double wind maximum which is observed in the Southern Hemisphere.

(b) Easterly winds cover almost half of the globe at the surface. Their latitudinal extent diminishes up to about 200 mb before increasing again above this level. In March–May there is a band of westerly wind dividing the upper and lower easterly regimes. The easterlies occupy a larger area through the vertical in June–August than in the other seasons. The region of maximum low-level easterlies occurs in the winter hemisphere according to these cross sections. A comparison with Figures 3.11 and 3.12 shows that the zonal averaging has eliminated the region of weak equatorial westerlies often seen at low levels.

3.3 Mean Meridional Motions

The meridional component of the mean wind is a very difficult parameter to measure directly. The importance of the north–south transports of water vapor, mass, heat, and momentum by these motions is so great, however, that each independent effort to appraise them has some value. In this section we present seasonal maps of the meridional motions together with cross sections along selected meridians of longitude and around a number of latitude circles.

Maps of the 3-month long-term mean values of $[\bar{v}]$ for the surface and for the 850, 700, 500, 400, 300, 200, 150, and 100-mb pressure levels are shown in Plates 3.22–3.30. In the oceanic regions between 40°N and 10°N the data have been supplemented with those of Crutcher (1959, 1961, 1966).

The surface maps for the four seasons show patterns which are quite similar. The most marked features occur at nearly the same longitudes in each season, but shift latitude slightly with the sun. In December–February and March–May the flow in the tropics is predominantly equatorward in both hemispheres, giving rise, in general, to a zone of convergence located near the equator. In December–February the zero line shows a large excursion in latitude over eastern Africa which corresponds with what is known of the mean position of the intertropical front at this time. Air from the Northern Hemisphere penetrates southward to about 25°S (cf. Wellington 1955).

In June–August flow is generally across the equator into the Northern Hemisphere. This is particularly true in the Indian Ocean region, where southerly flow associated with the monsoon circulation extends well into the Asian continent. Note also, at the longitude of Africa, that southerly flow extends almost to the Mediterranean. In September–November the major change is the disappearance of the strong southerly flow region over southeast Asia and Ethiopia. The main convergence zone is still north of the equator, as it was in June–August.

At 850 mb, in December–February three alternating bands of northerly and southerly flow are found in the middle-latitude Northern Hemisphere. This feature was also evident on the surface maps. At low latitudes the flow is predominantly northerly in the Northern Hemisphere and southerly in the Southern Hemisphere. By June–August, the middle-latitude Northern Hemisphere pattern is dominated by the strong northerly wind maximum over the Persian Gulf. In the Southern Hemisphere, three maxima of southerly flow are observed near the equator; they are apparently associated with the continent-ocean distribution. Northerly flow predominates at higher southern latitudes. The maximum of southerly flow over eastern Africa is apparently a broadly averaged reflection of the Findlater jet which is observed in this region (Findlater 1969). A similar jet occurs over northeast Brazil, and there is some evidence of a weaker jet off northeast Australia.

The Northern Hemisphere equatorial regions show a pattern of convergence over Africa and the North Atlantic, while the Southern Hemisphere exhibits divergence over South America, southern Africa, and to a lesser extent over Australia.

The maps at 700 mb resemble those at 850 mb in general, although in June–August the large maxima of northerly and southerly flow over the Persian Gulf and eastern Africa are less apparent. Such a decrease of amplitude with height, over eastern Africa, is in accordance with what is known of the Findlater jet in this region. The region of southerly flow over South America also shows smaller magnitudes at 700 than at 850 mb. Three alternating regions of northerly and southerly flow appear in the Southern Hemisphere. The maxima of northerly flow are generally found on the western side of the continents.

At 500 mb, the December–February and March–May maps show a well-marked pattern of alternation in the direction of the flow at middle latitudes in the Northern Hemisphere. The Southern Hemisphere shows a similar pattern, although magnitudes are smaller. The magnitudes of the several maxima of the flow in the Southern Hemisphere are remarkably similar to one another, as well as to the corresponding amplitudes of the maxima at 700 mb.

At 200 mb, the magnitude of the motions has increased. In December–February, a prominent 3-wave pattern exists in both hemispheres. The Northern Hemisphere middle latitudes exhibit predominantly northerly flow, while lower latitudes exhibit predominantly southerly flow. There exists, therefore, a broad region of convergence in the Northern Hemisphere. In the Southern Hemisphere, the flow is somewhat more structured with regions of northerly and southerly flow alternating with one another around a latitude circle. Flow across the equator from the Southern Hemisphere to the Northern Hemisphere is general in this season. By June–August the pattern has changed considerably. The higher latitudes of the Southern Hemisphere exhibit southerly motion in this season, with northerly flow occurring in equatorial regions. A region of convergence similar to that found in the Northern Hemisphere in December–February occurs in the Southern Hemisphere in winter.

The flow patterns are clearly a function of season in both hemispheres. The winter-hemisphere higher latitudes are characterized by equatorward flow, while lower latitudes exhibit poleward flow. In consequence there is a general region of convergence in the winter hemisphere. The flow in the summer hemisphere is characterized by cellular patterns, with locally convergent and divergent regions contrasting with the general convergence of the winter hemisphere. In the intermediate seasons, March–May for instance, several bands of generally northerly or southerly motion occur at different latitudes. At more poleward latitudes, generally equatorward flow is bounded by poleward flow at lower latitudes in each hemisphere. Thus there is a general region of divergence near the equator and a region of convergence at more poleward latitudes in each hemisphere.

Maps for 150 mb show patterns which are essentially similar to those at 200 mb. Those at 100 mb, although again exhibiting the same general pattern, show magnitudes which are considerably smaller than at 150 and 200 mb. Above this level, increasing difficulties in measurement make the analyses unreliable.

The meridional component of the mean wind has a marked variation with longitude, as can be seen from the maps. This structure in the flow can be more clearly seen in terms of meridional cross sections. Figures 3.14 and 3.15 give meridional cross sections of $\bar{v}$ at each 10 degrees of latitude, from 40°N to 40°S, for the December–February and June–August seasons.

The patterns of meridional flow in the winter season, in the Northern and Southern Hemispheres, are quite different. The Northern Hemisphere pattern shows a regular structure of alternating maxima of northerly and southerly flow which is absent in the Southern Hemisphere. At 10° latitude, the winter flow patterns are roughly the reverse of one another in the two hemispheres, as well as of their respective 40° patterns. At 10°S, the low-level Findlater and South American jet maxima are apparent.

At the equator, the flow is predominantly from the summer to the winter hemisphere in the upper troposphere, and from the winter to the summer hemisphere at low levels.

There is a marked seasonal variation in the flow patterns, as can be seen from the cross sections for the summer hemispheres. The most notable seasonal difference in the Southern Hemisphere is in the existence of strong alternating bands of northerly and southerly flow in the upper troposphere. In the Northern Hemisphere summer, the effects of the monsoon circulation are marked.

Although relatively small values of $[\bar{v}]$ are obtained by averaging $\bar{v}$ around latitude circles, it is obvious that the values of $\bar{v}$ at a given latitude can have appreciable magnitudes. These cross sections show the marked longitudinal structure of the meridional flow, while the latitudinal structure of the flow can be more readily appreciated in terms of longitudinal cross sections. Cross sections of $\bar{v}$ along the 150°E, 150°W, 80°W, 20°E, and 80°E meridians of longitude are given in Figures 3.16 and 3.17 for the December–February and June–August seasons. Zonally averaged values are also given (Figure 3.18). As discussed in Chapter 2, the zonal averages were obtained by a combination of direct and indirect methods.

In December–February, the graph of $[\bar{v}]$ shows the modest values of meridional flow which result from averaging around latitude circles. Only the basic pattern of poleward flow in the upper tropospheric equatorial Northern Hemisphere, surmounting a compensatory equatorward flow at low levels, survives the averaging. There is a flow across the equator from the Southern Hemisphere to the Northern Hemisphere in the upper troposphere, while at low levels the cross-equatorial flow is in the opposite sense.

The individual cross sections show a pattern which is in general similar to that of the zonal mean, but which has considerably more detail and larger amplitudes. The cross sections show that two upper tropospheric maxima usually exist, one in each hemisphere. That these maxima change sign at different longitudes, and that they are by no means always in the direction implied by the zonal average, is shown by the cross section at 150°W. Regions of convergence and divergence occur in the upper troposphere as a consequence of the varying signs of these flows. The low-level equatorward maximum is a fairly constant feature of the individual cross sections.

In June–August, the $[\bar{v}]$ pattern is, crudely, the reverse of the December–February pattern. The magnitude of the flow is similar to that found in December–February.

The individual cross sections show, in general, a pattern with three

maxima in the upper troposphere. This compares to the two maxima of the December–February pattern. A marked seasonal change in flow pattern is seen at 80°E. The large-amplitude southerly flow maximum of the December–February season is replaced by a maximum of northerly flow, of even greater magnitude, which is situated just south of the equator. This "monsoon" maximum is obviously responsible for much of the amplitude of the $[\bar{v}]$ maximum in this season.

The global distribution of $[\bar{v}]$ for the four seasons is given in Table 3.3. These values were obtained by a combination of direct and indirect methods as discussed in Chapter 2.

For internal consistency, the values of $[\bar{v}]$ obtained by any method should satisfy the condition that there be no net mass flux across a given latitude; i.e., that

$$\int_0^{2\pi} \int_0^{p_0} a \cos \varphi \, v \, d\lambda \, \frac{dp}{g} = 0. \tag{3.1}$$

This condition is not strictly correct, for there are mean pressure changes in different latitude belts which imply that there is a net mass transport; but these changes may be used to show that the pressure-weighted vertical average of $[\bar{v}]$ does not normally exceed a few millimeters per second (see for example, Gordon 1953). Indirect calculations of $[\bar{v}]$ are performed so that this condition is satisfied automatically. The requirement of no mass flux has also been used to obtain a correction to measured values of $[\bar{v}]$ (see Tucker 1959 for example).

Values of $[\bar{\omega}]$ are given in Table 3.4. Approximate values of the geometric vertical velocity $[\bar{w}]$ may be obtained from the relation

$$[\bar{w}] = -\frac{R}{pg}[\bar{T}][\bar{\omega}]. \tag{3.2}$$

The form of the zonally averaged continuity equation (3.1) allows the definition of a two-dimensional stream function ψ which may be evaluated from

$$\psi = \frac{2\pi a \cos \varphi}{g} \int_p^{p_0} [\bar{v}] \, dp, \tag{3.3}$$

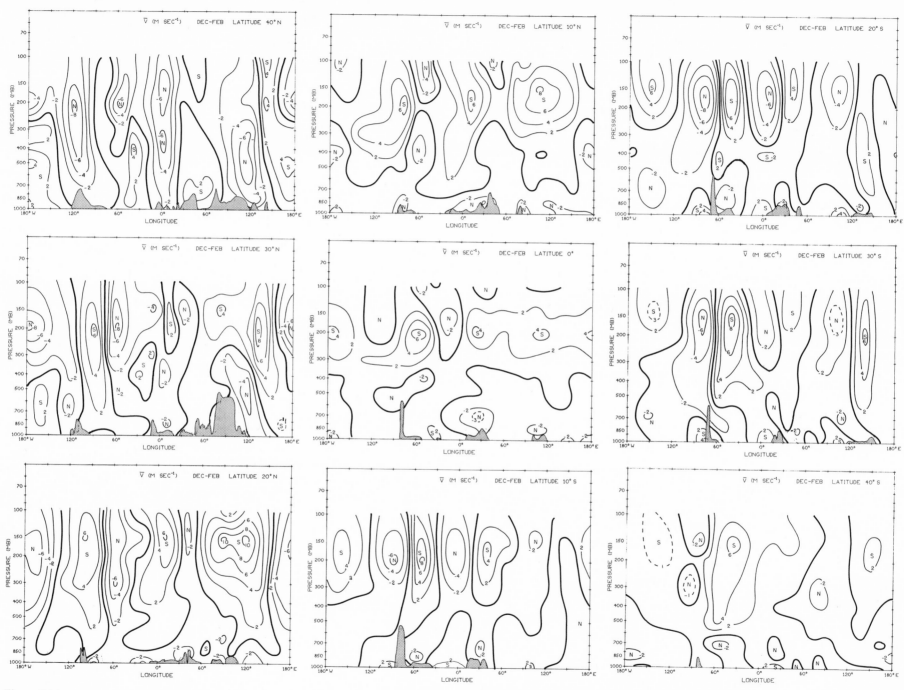

Figure 3.14

MEAN TEMPERATURE AND WIND FIELDS

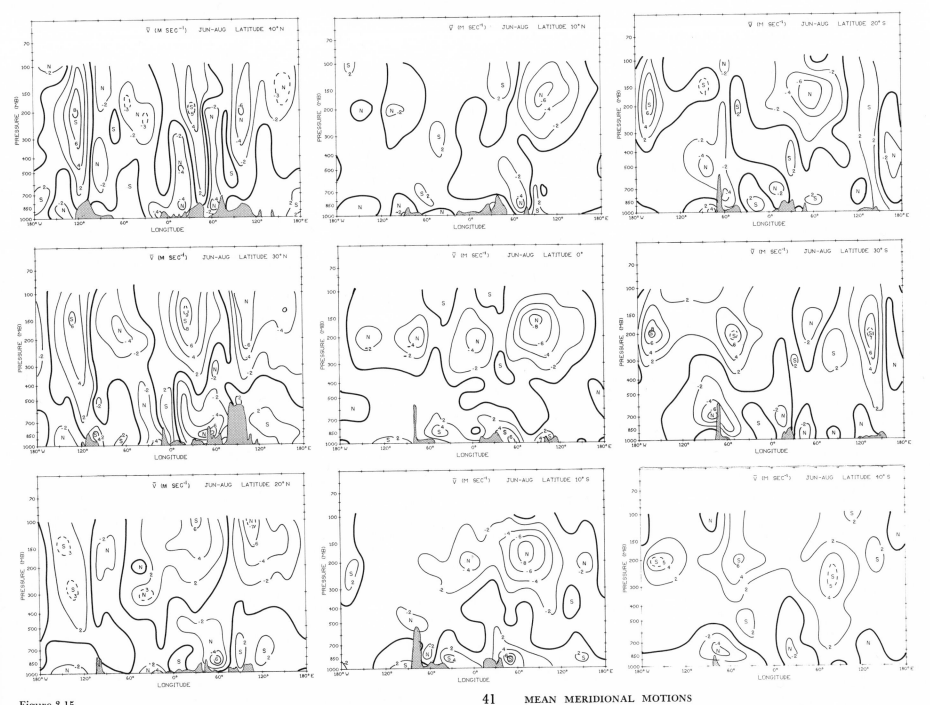

Figure 3.15

MEAN MERIDIONAL MOTIONS

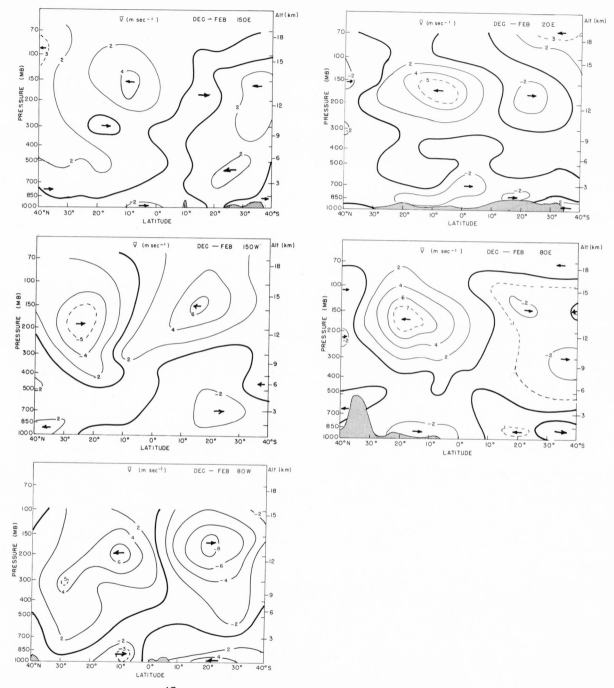

Figure 3.16

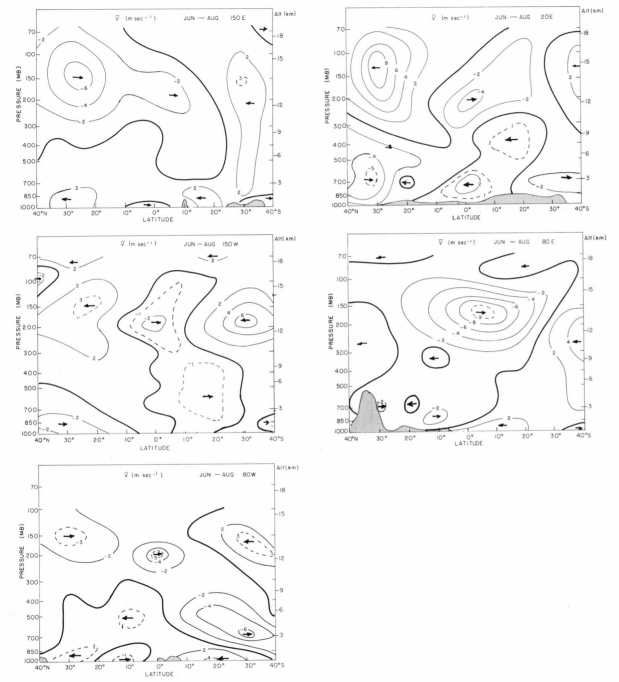

Figure 3.17

43   MEAN MERIDIONAL MOTIONS

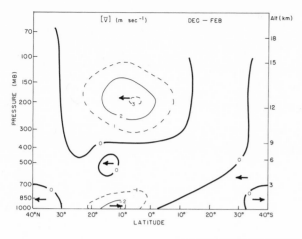

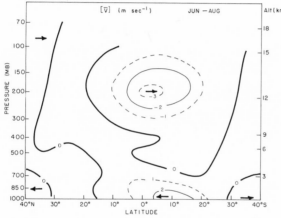

Figure 3.18

whence

$$[\bar{v}] = \frac{-g}{2\pi a \cos \varphi} \frac{\partial \psi}{\partial p},$$

$$[\bar{\omega}] = \frac{g}{2\pi a^2 \cos \varphi} \frac{\partial \psi}{\partial \varphi}. \qquad (3.4)$$

The value of $\psi(\varphi, p)$ at a point can be interpreted as the flux of mass, by the mean meridional motions $[\bar{v}]$, through the vertical surface from the 1000-mb level to that point. These values may be taken as a measure of the intensity of the circulations.

The streamline patterns for the four seasons are shown in Figure 3.19 and illustrate the following features:

(a) In December–February and June–August, a large thermally direct (Hadley) cell dominates the tropical region. A smaller, weaker direct cell is located in the subtropical summer hemisphere. In the intermediate seasons, Hadley cells are present in each hemisphere with the rising motion between them concentrated near the equator.

(b) At middle latitudes, indirect cells are evident in both hemispheres in all seasons. The possibility of indirect cells was first discussed by Ferrel (1859), and such circulations are now called *Ferrel cells*.

(c) At high latitudes, polar direct cells occur in both hemispheres in all seasons, although only a trace of such a circulation is apparent in the Northern Hemisphere summer pattern.

(d) In September–November and December–February, the tropospheric cells in the Northern Hemisphere appear to extend into the stratosphere and to tilt poleward with height, at least to the upper limit of our data at 30 km. This connection between the stratosphere and troposphere must be viewed with some caution, considering the limited resolution of the data.

(e) The maximum amplitude of the stream function (and consequently the maximum value of the integrated mass flux) for the large Hadley cell in the December–February season is approximately 1.8×10^{14} g sec^{-1}, while that for the Hadley cell in the June–August season is about 2×10^{14} g sec^{-1}. These may be compared with the values of 2.3 and 1.8×10^{14} g sec^{-1} which were obtained by Palmén and Vuorela (1963) and by Vuorela and Tuominen (1964) from Crutcher's Northern Hemisphere data. Manabe and Smagorinsky (1967) obtained

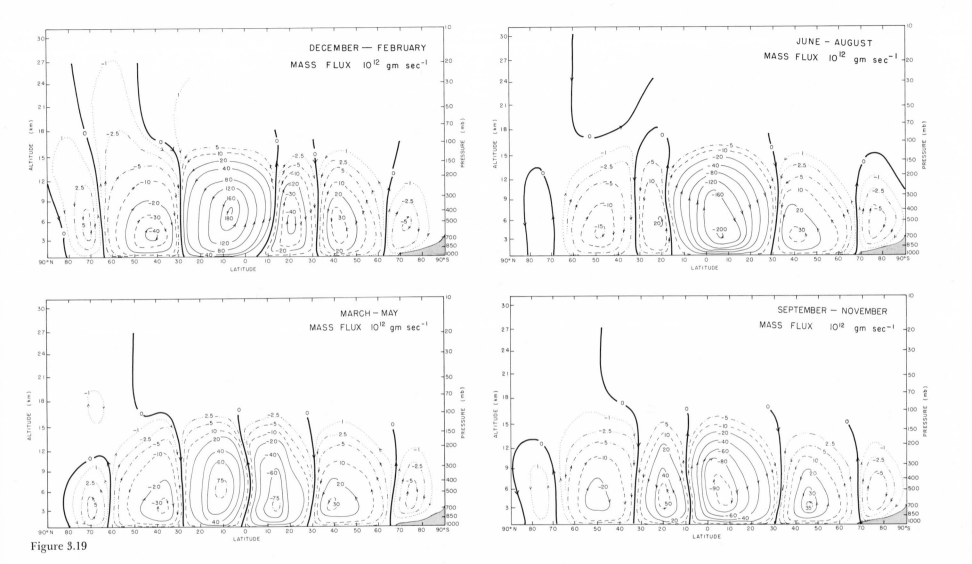

Figure 3.19

a value of 1.4×10^{14} g sec^{-1} for this quantity from their numerical modeling experiments, while Manabe et al. (1970) obtained a value of approximately 2×10^{14} g sec^{-1} from their moist tropical model.

(f) In March–May, the maximum amplitude of the mass flux for the two Hadley cells are comparable at about 0.75×10^{14} g sec^{-1}. During September–November, the Southern Hemisphere cell has a maximum value of approximately 0.9×10^{14} g sec^{-1}, which is about twice that of the cell in the Northern Hemisphere.

As will be discussed in later chapters, these meridional motions play a very important part in the momentum, heat, and energy balances of the tropical atmosphere.

3.4 The Horizontal Flow Patterns

In previous sections, a large amount of data has been presented which describes the seasonally averaged motions of the tropical atmosphere. The presentation has included maps of $\bar{u}$ and $\bar{v}$, meridional and zonal cross sections, the zonally averaged values $[\bar{u}]$ and $[\bar{v}]$, and the stream-lines of the zonally averaged meridional circulation in terms of the stream function ψ. It has been our purpose to illustrate, as well as possible, the true nature of the circulation, and this has led to considerable complexity in the presentation. One way to illustrate the overall picture of the flow is in terms of the streamlines of the flow in the horizontal. Plate 3.31 shows the streamlines of the flow at the 850 and 200-mb levels for the December–February and June–August seasons. The streamlines were constructed from the seasonally averaged values of $\bar{u}$ and $\bar{v}$ in the usual fashion (Petterssen 1956).

(a) At 850 mb in December–February, broad regions of anticyclonic flow associated with the subtropical highs are apparent. Poleward of these anticyclones, predominantly westerly flow characteristic of mid-latitudes is seen. Equatorward of the anticyclones, the northeast and southeast flows of the trade winds are apparent. The anticyclonic regions are situated somewhat farther poleward in the Southern Hemisphere than in the Northern Hemisphere in this season, and the flow across the equator is generally from the Northern Hemisphere to the Southern Hemisphere.

(b) The broad regions of divergent anticyclonic flow correspond well with the regions of descending and divergent motion indicated by the mass flux diagrams of the previous section. As well as these regions of descending and divergent flow associated with the anticyclones, there are convergent regions of cyclonic flow concentrated in the Southern Hemispheric equatorial region. These motions agree with the convergent and subsequently ascending motions shown in zonally averaged mass flux patterns.

(c) At 200 mb in December–February, the flow is seen to be much more uniform than at 850 mb. Predominantly westerly winds again occur at middle latitudes, and a broad band of easterly flow exists at low latitudes. Regions of anticyclonic divergent flow occur in more equatorial latitudes at this pressure level, above the regions of cyclonic convergent flow at 850 mb. In general, there are convergent flow patterns in each hemisphere which occur above the regions of divergence associated with the subtropical anticyclones at 850 mb. At this level the flow is predominantly from the Southern Hemisphere to the Northern Hemisphere, as opposed to that at 850 mb. The flow patterns, and the regions of convergence and divergence of the flow, agree with that required to account for the mass flux diagrams of the previous section.

(d) At 850 mb in June–August, the patterns show many of the features already noted on the December–February chart. The band of anticyclonic motion has shifted northward in this season, the cross-equatorial flow is from the Southern Hemisphere to the Northern Hemisphere, and the region of marked convergence occurs in the Northern Hemisphere. A notable change in the flow pattern has occurred over northern Africa, India, and Indochina in this monsoon season.

(e) At 200 mb in June–August, the flow patterns are again much more zonal than those found at 850 mb. Regions of convergent and divergent flow at this level overtop regions of divergence and convergence at 850 mb. The regions of marked divergence and consequent anticyclonic flow in the Northern Hemisphere, together with the region of convergence which now is apparent in the Southern Hemisphere, agree well with the Hadley circulation shown by the mass flux diagrams. The large region of divergent and anticyclonic flow associated with the monsoon is particularly marked on this chart.

(g) Taken together, the streamline patterns show the general westerly and easterly flow which is characteristic of regions poleward and equatorward of the large subtropical anticyclones. The regions of anticyclonic convergence and cyclonic convergence at the two levels agree with the patterns expected from the zonally averaged positions of the Hadley cell and the equatorward branch of the Ferrel circulation in the region. The marked latitudinal and longitudinal variation of the streamline pattern shows the importance of the standing waves in the specification of the flow. The seasonal changes in the flow patterns are those which would be expected from a consideration of the mass flux diagrams, although more local variations, such as the monsoon circulation in June–August over Asia, are of particular note.

3.5 The Variance of the Wind Components

One measure of the time variability of the wind components is given by the standard deviation. The standard deviation is an important quantity for the estimation of the diffusion rate of aerosols and for the calculation of the surface stress, and it is a measure of the eddy kinetic energy of the atmosphere.

Subjectively analyzed maps of $\sigma(u)$ and $\sigma(v)$ at 850, 500, and 200-mb levels for the December–February and June–August seasons are given in Plates 3.32–3.34. These maps show

(a) The values of $\sigma(u)$ and $\sigma(v)$ at 850 mb are remarkably similar in magnitude and in geographical distribution. The pattern is characterized by relatively small values in equatorial regions which increase rather uniformly with latitude in both hemispheres. There is a seasonal variation of the pattern, with largest values occurring in the poleward region of the Northern Hemisphere in December–February and in the poleward region of the Southern Hemisphere in June–August.

(b) The maps at 500 mb resemble those at 850 mb to a great extent, although the magnitudes of the standard deviations have increased. The patterns show somewhat less variation around latitude circles at this level than at 850 mb.

(c) The 200-mb pattern resembles those at lower levels, although the magnitudes of the standard deviations are larger.

In general it will be noted that the regions of maximum westerly flow, characteristic of middle latitudes, and the region of easterly flow in tropical regions are associated, respectively, with larger and smaller variability of the wind. This is only a general statement, however, as exceptions may be noted in a number of regions.

Zonally averaged values of the standard deviations are presented in Figures 3.20 and 3.21 and in Tables 3.5 and 3.6. The values were obtained by objective analysis of the data in the region from 30°N to 30°S and from the data sources listed in Chapter 2 for the extratropical regions. The values of $\sigma(u)$ in the region 30°N–30°S were calculated by removing a linear seasonal trend in the data before calculating the standard deviations. The variance attributable to the seasonal trend in $\bar{u}$ itself gives a sizeable contribution to the calculated value of $\sigma(u)$ in tropical regions, although it is less important in extratropical latitudes. Values of $\sigma(u)$ are not given in the tropical stratosphere, as the biennial oscillation introduces a similar complicating factor in this region. Neither of these considerations apply to the calculation of $\sigma(v)$ and this quantity was obtained in a straightforward manner. The figures show:

(a) In December–February and June–August, the values of $\sigma(u)$ and $\sigma(v)$ are very similar, as is their general distribution with latitude and height. The maximum values of $\sigma(u)$ and $\sigma(v)$ are clearly associated with the tropospheric maxima of $[\bar{u}]$ in each hemisphere (see Figure 3.13).

(b) There is a large seasonal variation in the standard deviations in the Northern Hemisphere polar stratosphere and only small seasonal changes in the troposphere.

(c) In all seasons, there is a minimum in the standard deviations in the equatorial regions where easterly winds predominate. There is also a clear association between greater easterly winds and smaller variability of the winds. This general relation was also noted in connection with the maps of the standard deviations. The relative steadiness of the easterlies has been noted by Riehl (1954, p. 4) who pointed out the inverse relation between the strength of the easterlies and the "constancy" of the flow. This effect may be at least partially explained by a restriction on the propagation of baroclinic waves from middle latitudes into regions of easterly flow (Charney 1969).

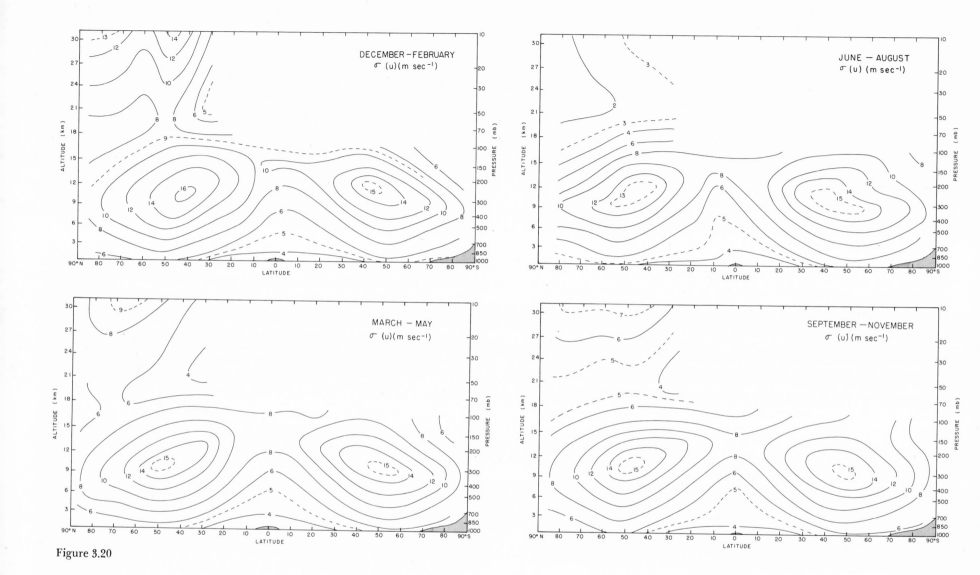

Figure 3.20

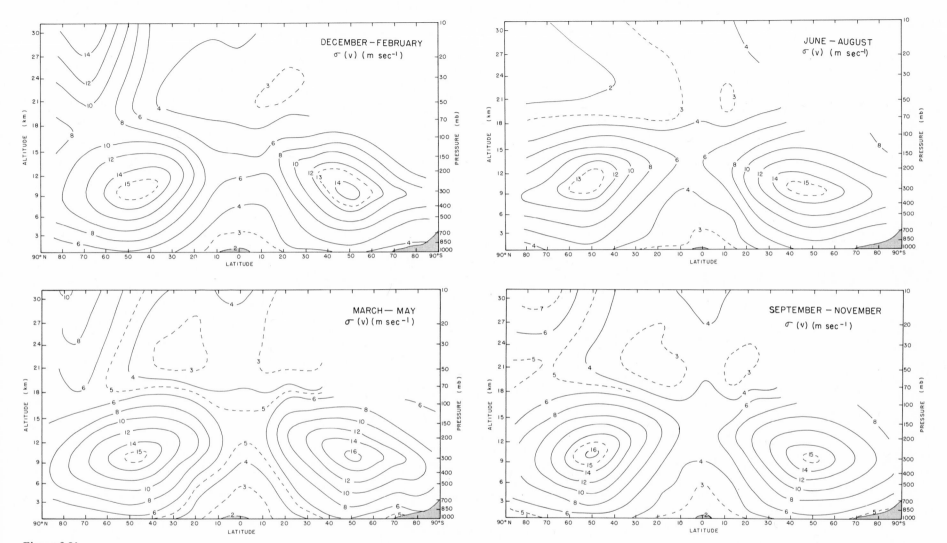

Figure 3.21

Table 3.1. Zonal Mean Temperatures $[\bar{T}]$ (°C)

(a) December–February

p (mb)	1000	850	700	500	400	300	200	150	100	70	50	30	20	10
Lat.														
90°N	−39.0	−24.0	−28.0	−41.0	−51.0	−61.0	−61.0	−65.0	−69.0	−72.0	−74.0	−74.0	−72.0	−69.0
80	−33.6	−21.0	−26.3	−39.4	−48.5	−58.9	−59.4	−61.4	−63.7	−67.2	−69.3	−69.4	−67.8	−65.0
70	−26.4	−18.0	−24.2	−37.7	−46.5	−57.2	−58.2	−58.1	−59.0	−62.2	−64.1	−64.5	−63.0	−60.7
60	−16.2	−13.0	−19.9	−34.1	−43.0	−54.4	−56.0	−54.5	−54.8	−57.4	−58.7	−59.3	−57.8	−55.1
50	−6.6	−7.9	−15.3	−29.2	−39.0	−51.3	−55.1	−52.8	−53.4	−55.5	−56.3	−56.1	−54.2	−50.7
40	7.0	−0.8	−6.6	−23.0	−33.5	−47.1	−57.1	−57.3	−59.6	−59.5	−58.7	−56.3	−53.8	−49.3
30	14.9	7.8	0.8	−14.9	−26.3	−39.9	−55.2	−61.0	−67.5	−65.4	−61.6	−55.8	−52.2	−45.7
20	21.0	14.7	7.2	−7.9	−19.4	−34.3	−53.8	−64.5	−74.3	−74.3	−64.3	−55.3	−50.5	−43.1
10	24.8	17.1	9.5	−5.8	−17.0	−32.2	−53.5	−66.5	−78.6	−75.8	−66.6	−57.4	−51.6	−43.2
0	26.3	17.5	9.3	−5.5	−16.0	−31.2	−53.6	−67.2	−80.6	−76.2	−68.1	−56.8	−50.4	
10	25.9	17.6	9.5	−5.4	−16.0	−31.2	−53.6	−67.1	−80.3	−74.8	−66.4	−58.6	−51.4	
20	23.1	17.1	9.5	−6.0	−16.8	−32.2	−53.2	−65.4	−76.5	−71.4	−64.8	−56.0	−51.9	
30	20.7	14.6	7.6	−9.0	−20.7	−35.3	−53.8	−61.7	−69.0	−67.6	−60.5	−53.4	−48.6	
40	17.5	8.8	2.8	−13.6	−24.6	−39.0	−54.9	−58.1	−60.1	−58.7	−54.0	−50.1		
50	7.8	1.1	−5.7	−20.5	−31.5	−43.8	−51.3	−51.5	−51.6	−51.5	−49.1	−46.8		
60	1.0	−4.9	−11.9	−26.0	−36.1	−47.1	−48.5	−47.3	−46.2	−45.3	−44.5	−42.9		
70	—	−9.7	−17.3	−30.7	−40.5	−50.1	−46.4	−44.6	−42.4	−46.1	−40.1	−39.3		
80	—	−12.1	−19.8	−35.3	−44.4	−52.3	−45.9	−43.1	−40.5	−39.4	−38.4	−36.5		
90°S	—	—	—	−37.7	−46.7	−53.7	−45.3	−42.3	−39.7	−38.5	−37.3	−35.7		

(b) March–May

p (mb)	1000	850	700	500	400	300	200	150	100	70	50	30	20	10
Lat.														
90°N	−23.5	−20.5	−27.0	−40.0	−47.0	−55.0	−46.5	−47.0	−48.5	−48.5	−48.5	−47.0	−44.0	−39.0
80	−16.7	−17.0	−23.5	−37.1	−44.3	−53.4	−47.6	−46.8	−46.7	−48.7	−48.8	−47.8	−45.1	−40.4
70	−9.9	−12.5	−19.6	−33.8	−42.0	−52.1	−48.6	−47.0	−46.7	−48.9	−49.0	−48.5	−46.2	−41.4
60	−1.2	−6.5	−14.4	−29.0	−38.0	−50.3	−50.2	−48.4	−49.1	−50.6	−50.7	−49.9	−47.0	−42.1
50	6.1	−1.0	−9.3	−24.5	−35.0	−47.9	−53.0	−51.2	−51.9	−52.7	−53.1	−51.5	−48.0	−42.3
40	12.9	5.7	−1.9	−18.0	−30.0	−44.3	−57.1	−57.4	−59.1	−58.6	−56.6	−53.2	−50.6	−45.0
30	18.3	12.3	4.1	−12.3	−23.9	−38.9	−55.6	−61.1	−66.5	−64.5	−59.3	−53.4	−49.1	−41.3
20	23.8	17.3	8.8	−7.5	−18.7	−33.7	−53.4	−64.5	−73.3	−72.0	−62.9	−53.4	−47.6	−38.8
10	26.3	18.6	10.1	−5.6	−16.2	−31.5	−53.0	−66.1	−78.0	−75.0	−65.6	−55.3	−48.8	−39.6
0	26.4	18.1	9.8	−5.4	−15.5	−30.7	−52.9	−67.0	−80.7	−76.4	−66.8	−55.2	−48.4	
10	25.5	17.4	9.5	−5.5	−15.9	−31.2	−53.1	−66.8	−79.5	−74.9	−65.7	−55.8	−49.4	
20	22.4	15.8	8.2	−6.6	−17.7	−33.4	−53.4	−64.6	−74.6	−71.3	−63.2	−56.1	−50.4	
30	18.2	11.3	4.8	−11.1	−22.5	−37.9	−54.7	−60.7	−67.5	−65.3	−60.0	−54.2	−50.4	
40	14.4	5.3	−0.5	−16.3	−27.6	−42.4	−56.1	−58.1	−59.3	−59.2	−56.3	−54.7		
50	4.0	−1.8	−8.2	−23.2	−34.3	−47.1	−54.4	−54.8	−56.1	−56,0	−55.2	−56.2		
60	−3.0	−8.9	−14.7	−29.0	−39.9	−51.3	−53.6	−53.5	−54.9	−55.6	−55.7	−58.0		
70	—	−15.4	−20.8	−34.9	−44.9	−54.6	−53.6	−53.5	−55.3	−56.7	−57.7	−60.3		
80	—	−19.1	−24.8	−39.8	−48.9	−57.1	−54.8	−55.1	−57.2	−59.2	−60.8	−62.7		
90°S	—	—	—	−41.0	−50.7	−59.3	−55.3	−56.3	−58.3	−60.1	−62.7	−64.0		

Table 3.1. Zonal Mean Temperatures [$\overline{T}$] (°C) (continued)

(c) June–August

p (mb)	1000	850	700	500	400	300	200	150	100	70	50	30	20	10
Lat.														
90°N	−7.0	−3.0	−9.5	−23.0	−33.0	−46.0	−42.0	−41.0	−41.0	−40.5	−40.5	−40.5	−37.5	−32.0
80	0.2	0.5	−7.0	−21.4	−31.5	−45.1	−43.9	−41.8	−41.7	−42.6	−42.4	−42.4	−38.9	−33.1
70	6.6	3.5	−4.3	−19.3	−30.0	−43.4	−46.5	−43.6	−43.7	−44.6	−44.5	−44.0	−40.2	−43.0
60	13.9	7.0	−1.4	−16.4	−27.0	−41.6	−48.9	−46.5	−46.3	−47.6	−47.5	−46.2	−42.0	−34.9
50	18.7	11.0	2.3	−12.5	−24.0	−38.6	−50.7	−50.5	−50.3	−51.8	−51.3	−48.4	−43.8	−35.9
40	21.0	15.5	7.1	−9.6	−21.0	−35.1	−52.2	−58.3	−61.6	−58.4	−54.5	−49.5	−45.3	−39.2
30	24.6	18.8	10.0	−6.3	−17.2	−31.8	−51.7	−62.4	−68.3	−62.9	−57.6	−50.8	−46.2	−39.7
20	27.5	19.5	10.6	−5.8	−16.3	−31.1	−52.2	−65.0	−72.8	−66.8	−59.9	−51.9	−47.6	−40.0
10	27.1	18.1	9.9	−5.9	−16.5	−31.6	−53.3	−66.7	−75.9	−68.9	−61.8	−53.7	−48.6	−41.8
0	25.5	16.8	9.1	−6.0	−16.5	−31.9	−53.8	−67.3	−77.3	−69.6	−62.9	−55.4	−48.6	
10	23.3	15.4	8.7	−6.2	−16.9	−32.6	−54.3	−67.1	−76.7	−70.1	−62.4	−55.0	−49.3	
20	19.5	12.6	6.5	−7.9	−19.3	−34.7	−53.6	−64.6	−72.5	−68.6	−60.8	−54.6	−50.9	
30	14.8	7.8	0.8	−14.8	−26.5	−40.2	−53.6	−59.4	−64.6	−62.9	−59.0	−54.8	−53.4	
40	10.4	2.5	−5.4	−21.8	−33.2	−47.4	−54.7	−54.4	−56.3	−59.0	−57.9	−59.5		
50	3.0	−4.7	−11.9	−27.0	−38.5	−52.0	−58.9	−59.0	−59.4	−61.5	−62.1	−63.6		
60	−4.0	−12.5	−18.4	−32.6	−43.6	−56.3	−63.1	−63.5	−65.6	−68.3	−69.7	−70.6		
70	—	−19.8	−24.7	−38.7	−48.9	−60.5	−67.9	−69.3	−72.4	−75.3	−77.7	−78.0		
80	—	−23.2	−28.2	−43.0	−52.3	−63.5	−72.1	−73.9	−77.7	−81.0	−83.0	−83.3		
90°S	—	—	—	−44.7	−54.7	−66.0	−74.0	−75.7	−80.7	−84.0	−86.7	−88.0		

(d) September–November

p (mb)	1000	850	700	500	400	300	200	150	100	70	50	30	20	10
Lat.														
90°N	−25.0	−18.0	−23.5	−37.0	−44.5	−54.5	−51.0	−50.5	−53.0	−56.0	−58.2	−62.0	−61.0	−59.5
80	−16.7	−13.0	−19.7	−33.5	−42.1	−53.0	−51.0	−50.4	−52.1	−55.0	−57.2	−60.0	−57.8	−56.4
70	−9.9	−8.0	−15.8	−29.7	−39.0	−51.1	−51.2	−50.3	−51.3	−54.2	−56.0	−57.8	−56.0	−53.2
60	−1.2	−3.5	−11.0	−25.6	−35.5	−48.5	−52.2	−51.5	−52.0	−54.2	−55.4	−55.9	−54.4	−49.7
50	6.1	2.0	−6.2	−21.1	−31.5	−45.0	−53.3	−54.3	−54.7	−55.9	−56.0	−54.8	−52.5	−46.9
40	16.1	9.0	2.2	−14.4	−25.7	−40.3	−55.2	−59.8	−62.8	−60.7	−58.1	−54.4	−51.1	−46.8
30	21.4	14.5	6.5	−9.2	−20.6	−35.5	−54.1	−63.5	−69.3	−64.3	−59.5	−53.4	−49.3	−43.6
20	25.8	17.9	9.2	−6.6	−17.5	−32.5	−53.6	−66.0	−74.0	−68.6	−61.1	−53.0	−48.1	−41.5
10	26.4	18.1	9.5	−6.0	−16.7	−31.7	−53.7	−67.4	−76.7	−70.0	−63.0	−55.0	−49.3	−41.7
0	25.7	17.3	9.2	−6.0	−16.0	−31.5	−53.8	−67.6	−78.7	−71.4	−64.4	−56.8	−49.2	
10	24.8	16.6	9.0	−6.0	−16.6	−32.1	−54.1	−67.3	−77.3	−70.2	−62.5	−54.0	−49.2	
20	21.7	15.1	7.6	−8.0	−18.3	−34.0	−53.6	−64.9	−72.9	−69.1	−61.3	−54.4	−50.2	
30	17.3	10.2	3.3	−13.4	−25.0	−38.8	−53.8	−59.4	−64.9	−63.1	−58.0	−53.4	−51.1	
40	13.1	4.4	−3.2	−18.6	−30.4	−44.1	−53.8	−54.5	−56.1	−57.3	−54.5	−52.6		
50	6.0	3.5	−9.9	−25.6	−36.2	−48.9	−55.3	−53.6	−55.0	−55.3	−54.0	−52.5		
60	−2.5	−9.9	−15.9	−31.1	−41.3	−53.5	−59.0	−56.6	−56.7	−56.5	−55.8	−54.5		
70	—	−16.4	−22.2	−36.6	−46.6	−57.9	−64.3	−63.0	−62.0	−60.4	−58.8	−56.9		
80	—	−19.4	−26.1	−40.8	−50.5	−61.5	−68.7	−68.4	−66.7	−64.6	−62.2	−59.0		
90°S	—	—	—	−41.7	−51.0	−62.7	−69.3	−70.0	−68.0	−66.5	−65.0	−60.3		

Table 3.2. Mean Zonal Winds $[\bar{u}]$ (m sec^{-1})

(a) December–February

p (mb) Lat.	1000	850	700	500	400	300	200	150	100	70	50	30	20	10
90°N	0.0	0.0	0.0	0.0	0.0	0.0	0.0	0.0	0.0	0.0	0.0	0.0	0.0	0.0
80	−0.1	−0.2	1.1	1.6	2.3	3.4	3.2	4.2	6.6	10.0	13.5	17.9	21.0	24.7
70	0.4	0.7	2.0	3.6	4.7	5.8	6.8	7.5	8.8	18.4	22.0	27.5	31.0	37.8
60	0.7	3.0	5.0	7.5	8.7	10.5	11.3	11.6	11.0	19.5	21.7	26.0	29.5	35.7
50	1.4	5.5	8.6	12.6	14.8	17.1	18.8	17.9	14.9	17.7	17.1	17.8	21.0	24.9
40	2.0	5.8	9.6	15.4	20.2	25.3	27.3	30.5	21.4	15.6	10.5	8.0	9.0	13.2
30	0.4	3.9	8.2	17.2	24.0	30.2	38.6	35.1	24.6	13.7	7.3	3.9	4.7	8.1
20	−2.3	−1.6	2.4	8.8	14.5	18.0	25.2	22.5	13.7	4.6	0.2	−3.0	−2.5	−1.7
10	−2.8	−5.6	−3.1	−2.5	0.7	3.6	7.6	6.5	2.5	−0.7	−1.3	−6.2	−8.4	−10.7
0	−1.8	−2.5	−3.1	−3.0	−1.8	−1.4	−1.1	−0.1	−0.9	0.5	−1.9	−6.9	−10.6	
10	−1.4	−0.9	−0.5	−1.0	−0.6	−0.1	−1.4	−0.5	−2.6	−7.4	−9.9	−12.4	−19.2	
20	−1.2	−1.9	−0.3	2.5	3.5	6.9	8.4	8.1	2.4	−4.2	−10.6	−16.6	−20.6	
30	−0.4	−0.3	3.7	8.3	10.6	15.0	20.0	18.6	10.3	−2.0	−6.5	−12.7	−15.8	
40	1.8	5.0	8.0	13.2	15.7	20.0	23.5	21.3	13.9	−2.0	−3.4	−9.1	−11.5	
50	4.5	9.1	11.1	16.0	18.0	21.0	20.5	17.8	12.5	4.0	−1.0	−6.6	−8.3	
60	2.0	3.1	5.0	9.5	11.0	14.5	13.9	10.7	7.5	3.8	0.4	−4.6	−5.4	
70	—	−1.5	−0.3	3.0	4.2	6.5	6.0	5.2	3.8	2.0	0.5	−2.9	−2.9	
80	—	−0.4	−1.4	1.1	1.8	2.9	3.0	2.1	1.6	0.5	−0.3	−2.0	−2.1	
90°S	—	—	—	0.0	0.0	0.0	0.0	0.0	0.0	0.0	0.0	0.0	0.0	0.0

(b) March–May

p (mb) Lat.	1000	850	700	500	400	300	200	150	100	70	50	30	20	10
90°N	0.0	0.0	0.0	0.0	0.0	0.0	0.0	0.0	0.0	0.0	0.0	0.0	0.0	0.0
80	−0.7	0.3	1.4	2.5	3.2	3.8	4.0	3.7	3.2	4.0	4.9	4.3	0.0	0.0
70	−0.3	0.7	2.3	4.8	5.9	6.8	6.5	5.5	4.3	7.2	6.6	5.4	4.5	6.9
60	0.0	2.2	3.6	6.2	7.5	8.9	9.1	7.8	5.5	7.2	5.9	3.9	3.0	3.5
50	0.9	3.3	5.8	9.2	11.4	13.7	14.3	12.7	9.3	8.0	5.0	1.9	0.8	−0.2
40	2.0	4.2	7.7	14.0	14.7	21.2	23.7	21.5	15.4	9.1	3.5	0.6	2.0	8.0
30	0.0	2.8	6.6	13.4	17.0	22.5	30.0	27.9	18.2	9.4	2.0	−1.8	−0.5	5.3
20	−1.9	−1.5	1.1	6.4	9.9	14.8	20.3	19.1	10.4	0.9	−3.8	−7.3	−6.8	−7.3
10	−2.0	−4.1	−3.8	−2.1	−0.2	2.7	7.1	5.5	1.9	−1.6	−4.6	−10.3	−12.5	−17.4
0	−1.7	−3.0	−3.7	−3.9	−3.4	−2.0	−1.1	1.1	0.3	0.1	−2.3	−9.1	−11.5	
10	−1.8	−2.9	−1.7	−1.3	−0.5	0.7	2.9	4.0	1.2	−3.3	−8.5	−11.6	−15.5	
20	−1.7	−2.4	0.6	4.8	7.6	10.9	17.8	14.8	8.1	1.4	−6.4	−11.5	−15.0	
30	−0.5	0.9	4.6	10.1	13.2	17.5	23.6	21.3	14.5	4.5	−1.0	−4.0	−3.3	
40	1.2	5.9	9.3	14.0	16.2	19.8	23.0	21.3	16.0	10.0	4.9	7.5	5.2	
50	5.0	10.5	14.0	16.9	18.0	20.5	20.5	18.6	16.5	15.0	13.5	15.5	12.5	
60	3.0	5.5	7.5	11.0	14.0	16.0	15.7	15.2	14.5	16.5	17.5	19.0	19.5	
70	—	−2.3	1.5	5.9	7.0	9.5	10.5	11.0	12.0	13.5	15.0	18.5	21.0	
80	—	−1.0	1.1	3.5	4.0	5.1	5.8	6.1	7.5	8.0	9.2	11.0	12.1	
90°S	—	—	—	0.0	0.0	0.0	0.0	0.0	0.0	0.0	0.0	0.0	0.0	

Table 3.2. Mean Zonal Winds $[\bar{u}]$ (m sec^{-1}) (continued)

(c) June–August

p (mb)	1000	850	700	500	400	300	200	150	100	70	50	30	20	10
Lat.														
90°N	0.0	0.0	0.0	0.0	0.0	0.0	0.0	0.0	0.0	0.0	0.0	0.0	0.0	0.0
80	0.0	1.9	2.5	3.5	3.3	2.9	2.7	2.2	1.3	0.0	−1.2	−2.1	−2.5	−3.6
70	−0.2	0.9	2.5	3.8	4.4	4.7	4.9	4.0	2.3	−0.3	−2.0	−4.0	−5.0	−6.2
60	0.4	1.9	3.2	5.2	6.5	7.9	8.7	7.6	4.1	1.1	−1.6	−4.5	−6.0	−7.4
50	0.9	3.7	5.8	9.0	11.0	13.7	15.7	13.9	7.7	3.2	−1.3	−5.1	−7.0	−9.1
40	1.0	2.5	4.5	8.9	10.7	14.5	18.5	14.5	8.1	0.1	−4.5	−8.7	−10.3	−12.0
30	−0.7	0.1	1.4	2.5	3.5	5.3	7.1	4.2	−1.7	−6.2	−10.3	−14.5	−15.6	−17.3
20	−1.7	−1.8	−2.6	−3.3	−2.6	−1.3	−2.1	−4.9	−10.3	−12.7	−16.8	−20.3	−21.2	−24.7
10	−1.1	−1.5	−3.3	−4.8	−5.1	−5.7	−7.6	−9.0	−10.7	−9.1	−12.2	−17.0	−19.7	−24.4
0	−1.6	−2.4	−3.2	−5.0	−5.6	−6.7	−8.0	−5.6	−4.7	−1.2	−0.1	−9.3	−9.4	
10	−2.8	−4.5	−1.8	−2.6	−1.1	2.0	3.5	5.0	0.9	−0.4	−1.3	−6.1	−6.6	
20	−1.7	−2.2	2.3	8.7	14.2	20.2	21.8	19.3	10.7	5.1	2.0	−1.6	−2.4	
30	0.2	2.9	6.8	14.2	20.2	27.7	31.2	28.0	18.3	4.8	7.5	7.0	6.5	
40	2.0	7.0	9.9	14.7	17.9	20.6	24.5	24.8	22.1	20.3	17.8	19.4	20.9	
50	6.5	9.0	11.5	15.0	17.9	20.1	21.8	23.8	25.1	29.0	29.5	34.8	36.3	
60	3.5	4.8	7.0	10.0	13.2	16.1	20.5	23.0	28.0	32.5	36.5	45.1	47.3	
70	—	−1.5	0.0	5.0	6.9	8.9	12.2	13.8	18.5	22.9	25.2	33.9	36.9	
80	—	−0.4	1.0	3.1	3.7	4.1	6.3	7.3	9.1	12.0	14.0	18.9	18.5	
90°S	—	—	—	0.0	0.0	0.0	0.0	0.0	0.0	0.0	0.0	0.0	0.0	

(d) September–November

p (mb)	1000	850	700	500	400	300	200	150	100	70	50	30	20	10
Lat.														
90°N	0.0	0.0	0.0	0.0	0.0	0.0	0.0	0.0	0.0	0.0	0.0	0.0	0.0	0.0
80	−0.2	0.0	0.5	2.1	3.1	3.9	4.6	4.8	4.4	4.0	4.6	6.1	7.5	9.7
70	0.1	0.7	2.0	4.4	5.6	7.1	8.1	8.0	6.2	7.7	8.3	9.7	12.0	15.5
60	0.7	3.7	5.2	7.9	9.0	10.7	12.2	11.9	8.7	10.5	10.0	10.7	13.5	16.2
50	1.7	5.6	8.7	13.0	15.5	18.3	20.2	18.5	13.0	12.9	10.3	9.8	11.5	14.4
40	1.0	3.8	6.7	12.1	14.0	19.1	23.6	18.8	14.1	7.2	5.0	4.0	6.0	10.0
30	−1.1	0.1	2.8	7.4	10.2	13.7	18.0	14.5	9.2	3.0	0.0	−1.1	1.3	7.0
20	−1.9	−2.7	−1.7	0.1	2.1	4.7	7.1	7.0	1.0	−5.4	−8.5	−10.7	−11.2	−8.5
10	−1.2	−3.7	−3.8	−3.5	−2.8	−1.9	−1.1	−0.7	−3.9	−2.4	−5.5	−12.5	−14.6	−16.0
0	−1.5	−2.7	−3.2	−3.9	−3.5	−3.9	−3.2	−0.8	−1.4	2.3	0.7	−9.9	−13.7	
10	−2.4	−3.8	−2.7	−1.4	−0.4	1.4	3.6	4.4	0.2	−0.4	−3.3	−10.5	−10.8	
20	−2.1	−2.0	1.6	6.1	10.8	14.1	19.6	16.0	7.6	1.2	−3.9	−6.3	−6.5	
30	−0.5	1.5	5.9	11.6	16.6	21.6	25.8	21.5	14.0	4.3	0.1	−2.5	−3.5	
40	1.4	6.2	9.3	13.5	17.2	19.5	22.1	20.5	17.4	14.5	12.5	8.3	6.8	
50	6.0	9.5	11.9	16.3	17.5	20.5	21.9	22.5	23.5	25.0	24.5	23.8	21.7	
60	3.0	5.5	8.0	11.4	12.8	15.2	17.0	18.5	23.2	26.1	32.0	34.9	35.2	
70	—	−1.1	1.0	5.0	6.8	8.5	10.7	13.1	16.0	20.3	22.7	24.2	28.6	
80	—	−0.3	0.3	2.0	2.7	4.0	5.1	6.3	8.0	9.5	10.4	12.1	14.2	
90°S	—	—	—	0.0	0.0	0.0	0.0	0.0	0.0	0.0	0.0	0.0	0.0	

Table 3.3. Mean Meridional Winds $[\bar{v}]$ (m sec^{-1})

(a) December–February

p (mb)	1000	850	700	500	400	300	200	150	100	70	50	30	20	10
Lat.														
85°N	0.29	0.25	−0.08	−0.19	−0.23	−0.23	−0.07	0.02	0.18	0.23	0.28	0.56	0.37	0.11
75	−0.11	−0.11	0.00	−0.03	0.00	0.06	0.07	0.11	0.17	0.13	0.09	0.12	−0.01	−0.31
65	−0.05	−0.02	0.02	0.03	0.03	0.04	−0.01	0.01	0.07	−0.00	−0.04	−0.20	−0.35	−0.61
55	0.33	0.33	0.01	−0.08	−0.14	−0.19	−0.15	−0.10	−0.11	−0.07	−0.08	−0.16	−0.25	−0.32
45	0.67	0.62	−0.08	−0.18	−0.19	−0.39	−0.29	−0.30	−0.16	−0.12	−0.05	0.03	0.10	0.13
35	0.45	0.42	−0.08	−0.09	−0.16	−0.21	−0.27	−0.19	−0.09	−0.03	0.02	0.03	0.07	0.12
25	−0.51	−0.53	−0.04	−0.02	0.05	0.31	0.61	0.60	0.22	0.12	0.02	0.07	0.14	0.25
15	−2.10	−1.10	−0.05	0.08	−0.10	0.51	2.25	1.90	0.51					
5	−1.15	−1.40	−0.58	−0.15	−0.33	1.10	3.15	2.00	0.27					
5	0.10	−0.49	−0.42	−0.26	−0.30	0.45	1.55	0.95	0.08					
15	0.50	0.25	−0.05	−0.08	−0.10	−0.20	−0.31	−0.02	−0.01					
25	0.36	0.34	0.03	−0.02	−0.07	−0.28	−0.43	−0.24	−0.09					
35	−0.17	−0.19	0.03	0.03	0.04	0.10	0.14	0.12	0.01					
45	−0.58	−0.50	0.03	0.11	0.17	0.38	0.36	0.23	0.11					
55	−0.16	−0.16	0.03	0.03	0.05	0.07	0.14	0.06	0.04					
65	0.05	0.07	−0.01	−0.02	−0.04	−0.05	0.01	0.01	0.01					
75	—	0.23	0.23	−0.10	−0.14	−0.19	−0.09	−0.03	−0.02					
85°S	—	—	0.05	−0.07	−0.01	0.04	−0.03	−0.04	−0.04					

(b) March–May

p (mb)	1000	850	700	500	400	300	200	150	100	70	50	30	20	10
Lat.														
85°N	0.30	0.29	−0.07	−0.09	−0.12	−0.16	−0.09	−0.04	−0.01	0.01	0.01	−0.06	−0.14	−0.27
75	−0.07	−0.08	0.01	0.04	0.06	0.09	0.03	0.02	−0.01	−0.04	−0.05	−0.12	−0.17	−0.28
65	−0.12	−0.10	0.04	0.02	0.05	0.10	0.04	0.03	0.01	−0.02	−0.05	−0.11	−0.17	−0.24
55	0.18	0.17	−0.02	−0.03	−0.08	−0.14	−0.07	−0.04	−0.01	−0.00	−0.01	−0.02	−0.04	−0.06
45	0.35	0.33	−0.04	−0.09	−0.11	−0.19	−0.22	−0.16	−0.05	0.02	0.02	0.01	0.01	0.02
35	0.47	0.43	−0.04	−0.11	−0.19	−0.22	−0.25	−0.29	−0.11	−0.03	0.01	0.05	0.08	0.08
25	−0.26	−0.23	−0.03	0.02	0.10	0.09	0.17	0.35	0.07	0.00	0.00	0.03	0.07	0.19
15	−1.20	−0.55	−0.12	0.07	0.14	0.40	1.10	0.85	0.26					
5	−0.79	−0.14	−0.39	−0.01	−0.03	0.31	1.11	0.80	−0.05					
5	0.55	0.60	−0.21	−0.18	−0.05	0.13	−0.54	−0.55	−0.29					
15	1.30	0.69	−0.02	−0.09	−0.18	−0.25	−1.15	−0.95	−0.17					
25	0.26	0.23	0.04	−0.02	−0.14	−0.18	−0.24	−0.16	−0.03					
35	−0.34	−0.34	0.06	0.10	0.16	0.16	0.17	0.16	−0.00					
45	−0.44	−0.37	0.03	0.10	0.18	0.17	0.29	0.18	0.13					
55	−0.32	−0.29	0.05	0.05	0.11	0.21	0.18	0.08	0.04					
65	0.01	−0.00	−0.02	−0.03	−0.01	0.03	0.06	0.03	−0.00					
75	—	0.37	0.32	−0.16	−0.20	−0.24	−0.11	−0.06	−0.06					
85°S	—	—	0.20	−0.12	−0.06	−0.02	−0.05	−0.09	0.03					

Table 3.3. Mean Meridional Winds [v̄] (m sec⁻¹) (continued)

(c) June–August

p (mb)	1000	850	700	500	400	300	200	150	100	70	50	30	20	10
Lat.														
85°N	0.18	0.18	0.00	−0.08	−0.10	−0.13	−0.08	−0.01	0.01	0.01	0.00	0.00	0.00	0.01
75	−0.05	−0.05	0.00	0.02	0.01	−0.01	0.02	0.07	−0.01	−0.01	−0.01	−0.01	−0.01	−0.01
65	0.02	0.02	0.00	−0.01	0.00	0.00	−0.02	−0.02	−0.02	−0.01	−0.01	−0.01	−0.01	−0.01
55	0.24	0.22	−0.03	−0.05	−0.07	−0.12	−0.17	−0.11	−0.01	−0.00	0.01	0.01	0.01	0.01
45	0.21	0.21	0.00	−0.03	−0.05	−0.15	−0.19	−0.12	−0.07	−0.01	0.01	0.01	0.01	0.00
35	0.04	0.06	−0.01	0.00	0:02	0.01	−0.05	−0.10	−0.10	−0.03	−0.01	0.01	0.01	0.01
25	−0.21	−0.16	−0.02	−0.02	0.02	0.10	0.19	0.27	0.17	0.01	−0.02	−0.02	0.00	0.03
15	−0.00	0.10	−0.03	0.10	0.20	0.10	−0.48	−0.50	0.25					
5	0.70	0.95	0.40	0.00	0.32	−0.50	−2.11	−1.74	−0.10					
5	2.10	1.98	0.44	−0.55	0.12	−0.89	−3.29	−2.25	−0.50					
15	2.10	1.60	0.10	−0.58	−0.20	−0.70	−1.95	−1.65	−0.35					
25	0.30	0.32	0.09	0.02	−0.04	−0.25	−0.51	−0.38	−0.08					
35	−0.39	−0.36	0.01	0.10	0.19	0.25	0.20	0.12	0.00					
45	−0.55	−0.46	0.10	0.15	0.18	0.25	0.27	0.15	0.08					
55	−0.33	−0.32	0.06	0.11	0.15	0.19	0.12	0.08	0.01					
65	−0.06	−0.07	0.00	0.02	0.04	0.05	0.01	0.03	0.03					
75	—	0.40	0.38	−0.18	−0.22	−0.29	−0.18	−0.08	0.01					
85°S	—	—	0.69	−0.30	−0.28	−0.27	−0.17	0.08	0.04					

(d) September–November

p (mb)	1000	850	700	500	400	300	200	150	100	70	50	30	20	10
Lat.														
85°N	0.07	0.07	−0.04	−0.09	−0.06	0.07	0.07	0.03	−0.02	−0.03	−0.04	−0.01	−0.05	−0.10
75	−0.08	−0.08	0.02	−0.00	0.04	0.09	0.06	0.03	0.01	−0.01	−0.03	−0.05	−0.11	−0.24
65	0.01	0.03	0.02	0.02	−0.00	−0.02	−0.04	−0.03	−0.00	−0.01	−0.02	−0.07	−0.14	−0.23
55	0.40	0.39	−0.02	−0.09	−0.16	−0.25	−0.24	−0.16	−0.07	−0.02	−0.02	−0.04	−0.07	−0.09
45	0.36	0.35	−0.04	−0.08	−0.12	−0.19	−0.22	−0.18	−0.07	−0.04	−0.01	−0.00	0.03	0.04
35	0.11	0.09	−0.02	−0.01	0.03	−0.04	−0.10	−0.12	−0.10	−0.03	0.00	0.04	0.05	0.10
25	−0.34	−0.33	0.00	0.05	0.10	0.21	0.30	0.18	0.11	0.02	−0.03	−0.03	0.01	0.08
15	−0.75	−0.35	−0.00	0.15	0.20	0.24	0.37	0.26	0.20					
5	0.50	0.37	−0.09	−0.01	0.31	−0.29	−0.71	−0.46	0.02					
5	1.17	0.67	0.34	−0.19	0.11	−0.37	−1.62	−1.30	−0.27					
15	1.11	0.56	0.11	−0.10	−0.20	−0.40	−1.00	−0.73	−0.03					
25	0.39	0.37	0.04	−0.07	−0.15	−0.24	−0.37	−0.26	−0.02					
35	−0.17	−0.13	0.08	0.04	0.05	−0.03	0.07	0.09	0.00					
45	−0.63	−0.56	0.03	0.18	0.27	0.42	0.28	0.14	0.09					
55	−0.38	−0.38	0.06	0.11	0.16	0.23	0.21	0.11	−0.00					
65	−0.08	−0.07	−0.02	0.06	0.03	0.02	0.07	0.03	−0.02					
75	—	0.40	0.37	−0.16	−0.20	−0.28	−0.17	−0.12	−0.06					
85°S	—	—	0.51	−0.13	−0.16	−0.19	−0.24	−0.13	−0.02					

Table 3.4. Zonal Mean "Vertical Velocities" $[\bar{\omega}]$ $(10^{-4} \text{ mb sec}^{-1})$

(a) December–February

p (mb)	1000	850	700	500	400	300	200	150	100	70	50	30	20	10
Lat.														
80°N	0.00	0.40	0.57	0.49	0.42	0.27	0.12	0.05	−0.02	−0.05	−0.06	−0.05	−0.03	0.00
70	0.00	−0.05	−0.11	−0.18	−0.23	−0.24	−0.21	−0.18	−0.16	−0.14	−0.12	−0.08	−0.04	0.00
60	0.00	−0.55	−0.81	−0.70	−0.57	−0.37	−0.18	−0.12	−0.05	−0.01	0.00	0.01	0.01	0.00
50	0.00	−0.56	−0.76	−0.56	−0.47	−0.31	−0.12	−0.02	0.05	0.07	0.08	0.06	0.03	0.00
40	0.00	0.18	0.27	0.22	0.19	0.13	0.07	0.06	0.04	0.02	0.01	0.00	0.00	0.00
30	0.00	1.29	1.91	1.82	1.70	1.37	0.73	0.35	0.10	0.03	0.02	0.02	0.01	0.00
20	0.00	1.63	2.06	1.97	1.99	1.96	1.08	0.38	0.00					
10	0.00	−0.38	0.22	0.91	1.12	0.95	0.24	−0.02	0.00					
0	0.00	−1.45	−2.16	−2.20	−2.17	−1.89	−0.87	−0.28	0.00					
10	0.00	−0.77	−1.52	−2.03	−2.21	−2.01	−0.88	−0.24	0.00					
20	0.00	0.06	−0.04	−0.17	−0.22	−0.20	−0.13	−0.06	0.00					
30	0.00	0.73	1.09	1.05	0.98	0.75	0.32	0.11	0.00					
40	0.00	0.41	0.59	0.53	0.45	0.29	0.10	0.04	0.00					
50	0.00	−0.61	−0.88	−0.79	−0.68	−0.45	−0.17	−0.06	0.00					
60	0.00	−0.31	−0.45	−0.36	−0.29	−0.20	−0.07	−0.02	0.00					
70	0.00	0.10	−0.09	−0.22	−0.17	−0.11	−0.04	−0.01	0.00					
80°S	—	0.00	0.43	0.61	0.46	0.23	0.04	0.01	0.00					

(b) March–May

p (mb)	1000	850	700	500	400	300	200	150	100	70	50	30	20	10
Lat.														
80°N	0.00	0.35	0.50	0.37	0.25	0.09	−0.04	−0.07	−0.08	−0.07	−0.06	−0.04	−0.02	0.00
70	0.00	0.11	0.12	0.08	0.08	0.04	0.01	−0.01	−0.02	−0.02	−0.02	−0.01	−0.01	0.00
60	0.00	−0.40	−0.55	−0.45	−0.37	−0.19	−0.03	0.01	0.03	0.03	0.03	0.02	0.01	0.00
50	0.00	−0.29	−0.42	−0.33	−0.28	−0.22	−0.10	−0.03	0.02	0.02	0.01	0.01	0.01	0.00
40	0.00	−0.23	−0.33	−0.30	−0.24	−0.16	−0.11	−0.06	−0.01	0.01	0.02	0.01	0.01	0.00
30	0.00	0.92	1.35	1.23	1.04	0.78	0.46	0.22	0.03	0.01	0.00	0.01	0.01	0.00
20	0.03	0.93	1.23	1.28	1.24	1.08	0.50	0.16	0.00					
10	0.00	−0.53	−0.61	−0.30	−0.19	−0.08	−0.07	−0.07	0.00					
0	0.00	−1.40	−2.02	−2.03	−1.94	−1.85	−1.03	−0.36	0.00					
10	0.00	−0.53	−0.70	−0.95	−0.94	−0.71	−0.27	−0.06	0.00					
20	0.00	1.06	1.34	1.21	1.15	1.09	0.62	0.22	0.00					
30	0.00	0.78	1.16	1.04	0.86	0.58	0.24	0.08	0.00					
40	0.00	0.02	0.02	0.06	0.07	0.08	0.05	0.03	0.00					
50	0.00	−0.24	−0.35	−0.30	−0.23	−0.19	−0.12	−0.05	0.00					
60	0.00	−0.48	−0.65	−0.50	−0.41	−0.25	−0.08	−0.03	0.00					
70	0.00	0.01	−0.36	−0.52	−0.42	−0.26	−0.09	−0.03	0.00					
80°S	—	0.00	0.62	0.79	0.59	0.30	0.08	0.04	0.00					

Table 3.4. Zonal Mean "Vertical Velocities" $[\tilde{\omega}]$ (10^{-4} mb sec^{-1}) (continued)

(c) June–August

p (mb)	1000	850	700	500	400	300	200	150	100	70	50	30	20	10
Lat.														
80°N	0.00	0.21	0.32	0.24	0.18	0.12	0.06	0.01	−0.01	−0.01	0.00	0.00	0.00	0.00
70	0.00	−0.08	−0.12	−0.09	−0.08	−0.07	−0.06	−0.03	−0.01	0.00	0.00	0.00	0.00	0.00
60	0.00	−0.33	−0.46	−0.39	−0.32	−0.22	−0.08	−0.02	0.01	0.01	0.01	0.00	0.00	0.00
50	0.00	−0.04	−0.08	−0.12	−0.13	−0.10	−0.05	−0.03	−0.01	0.00	0.00	0.00	0.00	0.00
40	0.00	0.19	0.28	0.27	0.23	0.13	0.01	−0.02	−0.01	−0.01	0.00	0.00	0.00	0.00
30	0.00	0.33	0.49	0.51	0.52	0.48	0.33	0.19	0.04	0.00	−0.01	0.00	0.00	0.00
20	0.02	−0.29	−0.46	−0.56	−0.69	−0.78	−0.48	−0.15	0.00					
10	0.00	−1.06	−1.94	−2.24	−2.26	−2.04	−1.02	−0.36	0.00					
0	0.00	−1.63	−2.35	−1.89	−1.55	−1.29	−0.58	−0.20	0.00					
10	0.00	0.34	0.86	1.19	1.34	1.38	0.65	0.18	0.00					
20	0.00	2.16	3.08	2.54	2.19	1.90	1.00	0.36	0.00					
30	0.00	0.91	1.42	1.43	1.29	0.97	0.41	0.14	0.00					
40	0.00	0.09	0.07	−0.02	−0.02	0.02	0.03	0.02	0.00					
50	0.00	−0.36	−0.47	−0.37	−0.31	−0.23	−0.10	−0.04	0.00					
60	0.00	−0.43	−0.60	−0.44	−0.33	−0.19	−0.05	−0.01	0.00					
70	0.00	−0.11	−0.56	−0.68	−0.51	−0.28	−0.09	−0.03	0.00					
80°S	—	0.00	0.61	0.69	0.55	0.33	0.12	0.04	0.00					

(d) September–November

p (mb)	1000	850	700	500	400	300	200	150	100	70	50	30	20	10
Lat.														
80°N	0.00	0.21	0.28	0.20	0.14	0.06	−0.01	−0.03	−0.04	−0.04	−0.04	−0.03	−0.02	0.00
70	0.00	−0.12	−0.19	−0.22	−0.21	−0.16	−0.07	−0.04	−0.02	−0.02	−0.02	−0.02	−0.01	0.00
60	0.00	−0.59	−0.85	−0.71	−0.58	−0.38	−0.15	−0.06	−0.00	0.01	0.01	0.01	0.01	0.00
50	0.00	−0.05	−0.05	−0.02	−0.02	−0.03	−0.02	−0.01	0.01	0.02	0.02	0.02	0.01	0.00
40	0.00	0.30	0.44	0.36	0.27	0.14	0.04	0.01	0.01	0.01	0.01	0.01	0.00	0.00
30	0.00	0.60	0.88	0.81	0.75	0.60	0.30	0.14	0.02	−0.01	−0.02	−0.01	0.00	0.00
20	0.01	0.35	0.38	0.30	0.20	0.14	0.08	0.04	0.00					
10	0.00	−1.32	−1.75	−1.53	−1.51	−1.33	−0.61	−0.20	0.00					
0	0.00	−0.66	−1.15	−1.38	−1.21	−1.09	−0.64	−0.25	0.00					
10	0.00	0.16	0.41	0.54	0.63	0.77	0.48	0.19	0.00					
20	0.00	0.66	0.85	0.89	0.85	0.75	0.37	0.11	0.00					
30	0.00	0.73	1.05	0.91	0.77	0.58	0.27	0.09	0.00					
40	0.00	0.53	0.82	0.76	0.62	0.35	0.07	0.03	0.00					
50	0.00	−0.42	−0.62	−0.55	−0.43	−0.25	−0.08	−0.04	0.00					
60	0.00	−0.51	−0.71	−0.56	−0.46	−0.28	−0.09	−0.03	0.00					
70	0.00	−0.13	−0.59	−0.69	−0.51	−0.32	−0.11	−0.03	0.00					
80°S	—	0.00	0.64	0.74	0.56	0.31	0.10	0.05	0.00					

Table 3.5. Standard Deviation of the Zonal Wind Component $\sigma(u)$ (m sec^{-1})

(a) December–February

p (mb) Lat.	1000	850	700	500	400	300	200	150	100	70	50	30	20	10
90°N	0.0	0.0	0.0	0.0	0.0	0.0	0.0	0.0	0.0	0.0	0.0	0.0	0.0	0.0
80	5.6	6.4	6.7	8.5	9.9	9.9	8.9	8.5	7.5	7.2	8.2	10.0	11.3	13.5
70	6.0	7.0	7.7	10.0	10.9	11.5	10.0	9.3	8.0	7.2	8.1	9.1	10.5	12.9
60	6.4	7.6	8.7	11.4	12.8	13.5	12.2	10.8	8.8	7.5	7.2	8.3	9.2	10.7
50	6.4	8.1	9.7	12.4	14.0	15.2	14.0	12.6	9.6	8.5	8.3	10.0	11.4	14.1
40	5.5	7.0	8.3	11.6	13.9	16.0	15.4	13.4	10.1	7.2	6.5	8.1	9.6	12.5
30	4.0	6.2	7.3	9.6	11.2	13.4	14.7	13.2	9.8	7.0	4.2	4.7	6.0	7.7
20	3.3	4.9	6.0	8.0	9.9	11.3	12.4	11.7	9.6					
10	2.3	4.2	4.9	6.3	7.6	8.5	9.7	9.9	8.7					
0	1.9	4.1	4.6	5.3	5.8	6.5	8.8	10.4	8.6					
10	2.4	4.1	4.7	5.3	5.8	6.6	9.0	10.4	8.6					
20	2.6	4.3	5.2	5.8	6.8	8.5	10.4	10.6	8.3					
30	3.2	4.7	5.7	7.5	9.1	11.2	13.0	11.4	8.6					
40	4.0	6.1	7.4	9.5	11.1	13.4	15.0	12.0	8.5					
50	5.2	6.8	8.2	10.2	12.3	14.2	14.2	11.0	7.6					
60	5.0	6.5	7.4	9.5	11.8	13.4	11.5	8.9	6.3					
70	—	5.6	6.3	8.3	10.0	11.7	8.2	6.7	5.3					
80	—	5.3	5.7	6.8	8.5	9.1	5.8	4.9	4.3					
90°S	—	—	—	0.0	0.0	0.0	0.0	0.0	0.0					

(b) March–May

p (mb) Lat.	1000	850	700	500	400	300	200	150	100	70	50	30	20	10
90°N	0.0	0.0	0.0	0.0	0.0	0.0	0.0	0.0	0.0	0.0	0.0	0.0	0.0	0.0
80	5.0	5.6	6.5	8.7	9.0	8.8	7.7	6.6	5.8	6.0	6.5	7.5	7.9	8.2
70	5.2	5.9	6.9	9.3	10.5	10.7	9.0	8.0	6.5	6.3	6.3	7.0	8.0	9.0
60	5.5	6.5	7.9	10.6	12.2	13.5	11.0	9.3	7.0	5.8	6.0	7.0	7.7	9.0
50	5.8	7.2	8.9	11.6	13.6	15.8	12.8	10.9	8.0	5.9	5.4	6.5	6.9	8.5
40	5.2	6.7	8.0	10.6	12.2	14.5	14.2	12.0	8.8	6.0	4.3	4.9	5.6	7.0
30	4.0	5.6	7.0	9.0	10.5	13.5	14.3	12.6	9.0	6.0	3.5	3.9	4.4	5.3
20	3.0	4.2	5.2	7.3	8.3	9.8	11.2	11.0	8.7					
10	2.2	3.7	4.5	5.5	6.5	7.7	9.0	9.6	8.4					
0	1.7	3.2	4.2	4.8	5.4	6.2	8.2	9.5	8.1					
10	2.2	3.7	4.4	5.5	6.0	7.0	8.9	9.6	8.4					
20	2.6	4.4	5.3	6.2	7.5	9.0	10.9	10.1	7.6					
30	3.6	5.2	6.2	8.2	9.8	12.0	13.8	11.6	8.9					
40	4.6	6.2	8.0	10.3	12.0	14.2	14.5	12.0	9.5					
50	5.5	7.5	9.2	12.0	13.4	15.5	13.9	11.6	9.2					
60	5.4	7.2	8.8	11.8	13.2	14.7	11.8	9.8	8.6					
70	—	6.4	7.4	10.2	12.0	12.8	9.5	7.8	7.4					
80	—	5.5	6.8	8.0	9.5	10.0	7.5	5.8	5.6					
90°S	—	—	—	0.0	0.0	0.0	0.0	0.0	0.0					

Table 3.5. Standard Deviation of the Zonal Wind Component $\sigma(u)$ (m sec^{-1}) (continued)

(c) June–August

p (mb)	1000	850	700	500	400	300	200	150	100	70	50	30	20	10
Lat.														
90°N	0.0	0.0	0.0	0.0	0.0	0.0	0.0	0.0	0.0	0.0	0.0	0.0	0.0	0.0
80	4.5	5.0	5.2	7.5	9.1	10.0	6.7	5.0	3.0	2.3	1.7	1.6	1.7	1.8
70	4.7	5.3	6.0	7.9	9.8	11.1	8.4	6.2	3.7	2.7	1.8	1.8	1.9	2.2
60	4.8	6.0	7.0	9.2	11.1	12.6	10.6	8.7	5.0	3.2	2.0	2.0	2.3	2.8
50	4.2	6.0	7.1	9.3	10.9	13.0	13.0	10.9	7.0	4.0	2.3	2.3	2.5	3.1
40	3.5	5.3	6.1	8.0	9.7	12.0	13.4	11.2	7.8	4.2	2.5	2.5	2.9	3.4
30	3.1	4.8	5.6	6.8	7.6	9.3	11.2	10.4	7.8	4.3	2.8	2.9	3.2	3.7
20	2.7	4.3	5.1	5.6	6.2	7.2	9.0	9.2	7.0					
10	2.2	4.0	4.6	4.8	5.0	5.5	6.0	8.5	7.7					
0	1.9	3.2	4.2	4.8	5.3	6.2	8.1	8.9	7.4					
10	2.3	3.9	4.7	5.8	6.9	8.2	9.4	9.2	7.5					
20	3.3	4.6	6.0	8.0	9.4	11.0	12.0	10.4	8.1					
30	4.0	6.0	7.2	9.7	11.5	14.0	14.5	12.0	9.2					
40	5.7	7.1	8.7	11.1	13.0	15.4	14.8	13.0	9.5					
50	6.5	8.0	9.4	11.9	14.0	15.7	13.4	12.8	9.8					
60	6.5	8.0	9.2	11.8	13.5	14.4	11.9	11.2	9.7					
70	—	7.0	8.2	10.7	12.1	12.9	10.7	9.6	8.9					
80	—	6.0	7.0	9.0	10.1	11.2	9.5	9.0	7.5					
90°S	—	—	—	0.0	0.0	0.0	0.0	0.0	0.0					

(d) September–November

p (mb)	1000	850	700	500	400	300	200	150	100	70	50	30	20	10
Lat.														
90°N	0.0	0.0	0.0	0.0	0.0	0.0	0.0	0.0	0.0	0.0	0.0	0.0	0.0	0.0
80	5.0	5.5	6.0	7.8	8.5	8.8	7.9	6.7	5.0	4.5	4.7	5.5	6.1	7.0
70	5.5	6.0	7.0	8.8	9.9	10.6	9.2	8.1	5.8	4.8	4.5	5.3	6.1	7.3
60	6.0	7.2	8.0	10.5	12.0	13.3	11.7	9.8	6.6	5.0	4.4	5.0	6.0	7.2
50	6.3	7.8	9.0	11.2	13.2	15.1	14.1	11.7	7.4	5.4	4.5	5.2	6.0	7.8
40	4.9	6.3	7.5	10.0	11.5	14.0	14.4	12.2	7.8	5.4	4.3	4.9	5.7	7.3
30	3.7	5.4	6.3	8.0	9.5	11.5	13.5	12.2	7.6	5.1	3.6	4.0	5.0	6.0
20	3.0	4.5	5.3	6.8	7.5	9.2	11.2	11.4	7.4					
10	2.3	4.1	4.8	5.4	5.7	7.0	9.5	9.4	7.0					
0	2.0	4.0	4.4	4.6	5.0	6.0	8.1	8.6	6.4					
10	2.3	3.9	4.7	5.4	6.2	7.6	9.2	9.3	7.0					
20	3.0	4.5	5.4	7.0	8.4	10.1	11.5	10.2	7.9					
30	3.9	5.4	6.8	9.2	10.8	13.0	13.5	11.7	8.5					
40	5.8	6.6	8.2	10.9	13.0	14.6	14.4	12.0	9.6					
50	6.0	7.2	8.8	11.5	14.0	15.3	14.4	12.2	10.6					
60	5.6	7.0	8.2	10.9	13.3	14.2	12.8	11.5	11.2					
70	—	6.4	7.5	9.5	11.3	12.3	10.6	10.0	10.3					
80	—	6.0	6.8	7.9	9.4	10.0	9.0	7.9	7.6					
90°S	—	—	—	0.0	0.0	0.0	0.0	0.0	0.0					

Table 3.6. Standard Deviation of the Meridional Wind Component $\sigma(v)$ (m sec^{-1})

(a) December–February

p (mb)	1000	850	700	500	400	300	200	150	100	70	50	30	20	10
Lat.														
90°N	0.0	0.0	0.0	0.0	0.0	0.0	0.0	0.0	0.0	0.0	0.0	0.0	0.0	0.0
80	4.5	5.5	6.8	9.0	9.8	10.0	9.0	8.5	7.7	8.2	8.8	10.8	12.4	14.4
70	5.0	6.2	7.6	10.0	11.2	11.9	11.0	10.3	9.0	9.0	10.4	12.4	13.8	15.5
60	5.6	7.2	8.7	11.6	13.4	14.3	13.0	11.8	9.0	9.0	9.2	10.5	11.6	13.8
50	5.8	7.6	9.2	12.4	14.4	15.8	14.0	12.4	9.2	7.2	6.4	7.5	8.5	9.9
40	5.0	6.8	8.6	11.3	13.3	15.3	14.4	12.5	9.0	6.0	4.4	5.0	5.7	6.0
30	4.3	5.5	6.7	9.2	10.7	13.0	14.0	12.0	8.0	5.1	3.5	3.5	4.4	5.0
20	3.2	3.7	4.5	6.1	7.3	8.8	9.8	8.4	6.0	4.1	3.1	3.2	3.6	4.6
10	2.0	2.7	3.0	4.0	4.8	5.7	6.6	6.4	5.2	3.8	3.2	3.5	3.6	4.3
0	1.6	2.3	2.9	3.3	4.0	5.0	6.5	6.4	5.1	3.7	3.8	3.9	3.9	4.5
10	2.5	2.8	3.2	3.6	3.9	4.8	6.1	6.0	4.8	3.6	2.9	3.3	3.6	4.0
20	3.0	4.1	4.2	5.0	5.7	6.6	8.3	8.6	6.0	4.2	3.0	2.8	3.2	3.7
30	3.3	4.9	5.6	6.8	8.0	9.9	11.4	10.0	6.0	4.2	3.5	3.0	3.2	3.6
40	4.0	6.1	6.9	9.4	11.2	13.0	13.4	11.4	7.1					
50	4.5	6.4	7.2	10.7	12.5	14.5	12.8	10.5	6.5					
60	3.8	5.1	6.2	9.1	11.2	12.9	10.2	7.9	5.5					
70	—	4.2	5.2	7.1	8.9	10.7	7.5	5.9	4.6					
80	—	3.9	4.8	6.2	8.0	9.1	5.8	4.8	4.2					
90°S	—	—	—	0.0	0.0	0.0	0.0	0.0	0.0					

(b) March–May

p (mb)	1000	850	700	500	400	300	200	150	100	70	50	30	20	10
Lat.														
90°N	0.0	0.0	0.0	0.0	0.0	0.0	0.0	0.0	0.0	0.0	0.0	0.0	0.0	0.0
80	4.5	5.2	6.2	8.9	9.2	9.0	7.7	6.4	5.4	6.0	6.5	8.0	8.8	10.4
70	4.8	5.6	7.0	9.7	11.1	11.6	10.0	8.2	5.8	6.0	6.3	7.3	8.1	9.2
60	5.0	6.2	8.1	10.9	12.9	14.0	12.0	9.7	6.0	5.0	4.9	5.5	6.2	7.2
50	5.3	6.8	8.5	11.6	13.7	15.5	13.2	11.0	7.0	4.8	3.8	4.2	4.8	5.6
40	4.8	6.5	7.8	10.4	12.4	14.4	13.4	11.2	7.5	4.9	3.2	3.0	3.8	4.5
30	4.1	5.5	6.1	8.0	9.5	11.4	12.5	10.8	7.2	4.7	3.0	2.7	3.1	3.8
20	3.2	3.5	3.8	5.1	6.5	8.2	10.0	8.5	6.0	4.0	2.9	2.9	2.9	3.6
10	2.0	2.8	3.1	3.6	4.6	5.7	6.1	5.9	4.7	3.6	3.2	3.3	3.6	4.2
0	1.7	2.5	2.7	3.1	3.6	4.1	5.5	5.6	4.8	3.9	3.8	3.8	3.8	4.2
10	2.8	2.8	3.0	3.4	4.0	4.9	5.8	5.8	4.6	3.3	2.9	2.9	3.0	3.5
20	3.1	3.8	4.4	5.4	6.4	8.2	9.9	9.1	6.3	4.2	3.0	2.9	2.6	2.8
30	4.0	5.2	6.0	7.6	9.2	10.8	12.5	10.6	6.8	3.6	2.9	2.9	2.8	3.0
40	5.5	6.8	7.7	10.0	12.0	14.0	14.0	11.4	7.4					
50	5.8	7.5	8.8	11.3	13.5	16.0	13.5	10.8	7.7					
60	5.2	6.2	8.0	10.6	12.9	14.8	11.5	9.2	7.2					
70	—	5.4	6.9	9.3	11.1	12.6	9.5	7.6	6.6					
80	—	6.0	7.2	9.0	10.2	10.2	7.7	6.8	5.8					
90°S	—	—	—	0.0	0.0	0.0	0.0	0.0	0.0					

Table 3.6. Standard Deviation of the Meridional Wind Component $\sigma(v)$ (m sec^{-1}) (continued)

(c) June–August

p (mb)	1000	850	700	500	400	300	200	150	100	70	50	30	20	10
Lat.														
90°N	0.0	0.0	0.0	0.0	0.0	0.0	0.0	0.0	0.0	0.0	0.0	0.0	0.0	0.0
80	3.0	5.0	5.7	8.2	9.7	10.0	7.6	5.7	3.6	2.8	1.8	1.6	1.7	1.8
70	4.0	5.2	6.2	8.5	10.1	11.2	9.0	7.2	4.0	3.0	2.0	1.7	1.8	2.0
60	4.6	6.0	6.9	9.2	11.1	13.1	11.2	9.1	5.2	3.2	2.0	1.8	1.9	2.2
50	4.6	6.2	7.2	9.3	11.0	13.4	13.4	10.3	5.8	3.2	2.1	1.9	2.0	2.4
40	3.8	4.8	5.6	7.2	8.4	10.5	12.0	9.8	6.0	3.0	2.2	2.0	2.2	2.7
30	2.8	4.2	4.7	5.5	6.1	7.2	9.0	8.2	5.4	2.8	2.4	2.5	2.6	3.2
20	2.4	3.4	3.6	3.8	4.4	5.4	7.0	7.0	4.4	2.8	2.6	2.8	2.8	3.5
10	2.0	3.0	3.2	3.3	3.5	3.9	5.4	5.8	4.4	3.3	3.0	3.1	3.3	3.6
0	1.7	2.5	2.9	3.3	3.8	4.7	6.0	5.9	4.8	3.8	3.7	3.6	3.7	3.9
10	2.7	3.1	3.4	3.8	4.9	6.0	7.3	7.0	4.8	3.1	3.0	3.1	3.2	3.4
20	3.4	4.2	5.0	6.9	8.3	10.2	10.4	8.5	6.2	4.0	3.3	3.6	4.0	4.6
30	4.4	6.0	6.8	9.1	11.1	13.4	12.8	10.3	7.2	4.5	3.5	4.1	4.5	5.0
40	6.0	7.2	8.4	11.3	13.0	15.2	13.5	11.3	8.5					
50	6.2	8.0	9.2	12.0	13.8	15.6	12.8	11.0	8.8					
60	6.0	7.4	8.5	11.5	13.1	14.6	11.8	10.0	9.1					
70	—	7.2	7.7	10.1	12.0	12.7	10.4	9.2	9.8					
80	—	7.0	7.5	9.2	10.4	11.4	9.3	8.2	7.6					
90°S	—	—	—	0.0	0.0	0.0	0.0	0.0	0.0					

(d) September–November

p (mb)	1000	850	700	500	400	300	200	150	100	70	50	30	20	10
Lat.														
90°N	0.0	0.0	0.0	0.0	0.0	0.0	0.0	0.0	0.0	0.0	0.0	0.0	0.0	0.0
80	4.6	5.4	6.4	8.1	8.8	9.2	8.1	7.0	5.5	5.2	4.9	5.7	6.4	7.7
70	5.0	6.0	7.0	9.3	10.8	11.7	10.2	8.6	6.2	5.4	5.1	5.7	6.4	7.6
60	5.2	7.2	8.1	11.2	13.2	14.9	12.9	10.6	7.2	5.3	4.5	4.8	5.5	6.4
50	5.0	7.8	8.8	11.9	13.9	15.9	14.7	11.9	7.8	5.1	4.0	4.0	4.5	5.2
40	4.7	5.7	7.3	9.6	11.5	13.6	13.7	11.4	7.5	4.6	3.4	3.2	3.4	3.9
30	3.6	4.8	5.5	7.0	8.0	9.6	11.0	9.8	6.6	3.7	2.8	2.8	3.2	3.8
20	3.0	3.7	3.9	5.0	5.6	6.6	8.0	7.6	4.9	3.0	2.7	3.0	3.0	3.8
10	2.3	3.0	3.2	3.4	3.8	4.5	5.6	5.9	4.2	3.1	2.9	3.2	3.4	4.1
0	1.7	2.8	2.9	3.0	3.4	4.2	5.4	5.4	4.5	4.2	3.6	3.6	4.0	4.2
10	2.7	2.8	3.3	3.5	4.1	5.6	6.8	5.8	4.3	3.0	2.9	3.2	3.6	4.0
20	3.0	4.2	4.8	6.0	6.8	8.0	9.8	8.4	6.0	3.7	2.9	3.0	3.2	3.6
30	3.9	5.5	6.4	8.2	9.8	11.2	11.8	10.0	7.0	3.5	3.4	3.6	3.8	4.0
40	4.5	6.2	7.5	10.5	12.5	14.2	13.2	11.0	8.0					
50	5.0	6.8	7.8	11.2	13.7	15.0	13.1	11.0	9.0					
60	5.0	6.3	7.5	10.1	12.0	13.7	12.0	10.4	9.0					
70	—	5.6	6.8	9.0	10.3	11.6	10.1	9.0	8.0					
80	—	5.0	6.0	7.8	9.1	9.6	8.7	7.4	7.3					
90°S	—	—	—	0.0	0.0	0.0	0.0	0.0	0.0					

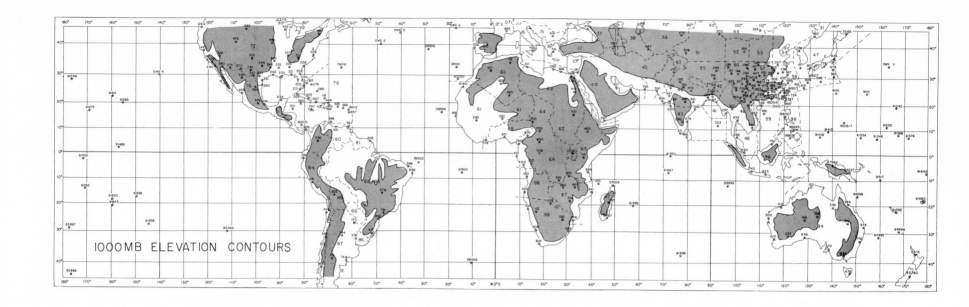

1000MB ELEVATION CONTOURS

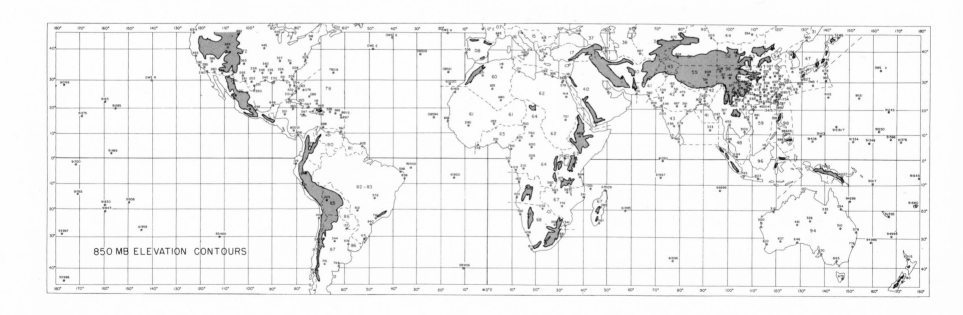

850 MB ELEVATION CONTOURS

Plate 3.1

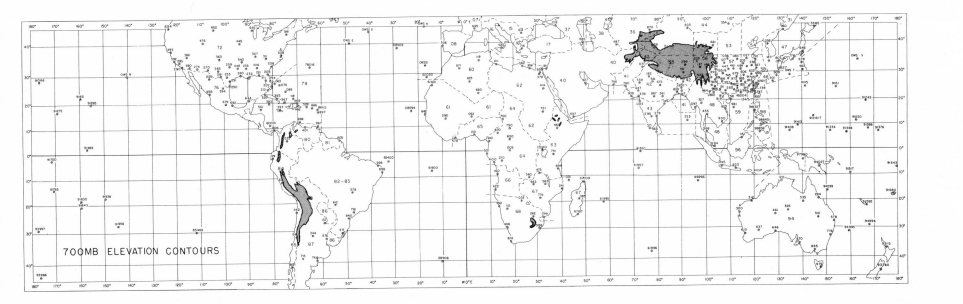

700MB ELEVATION CONTOURS

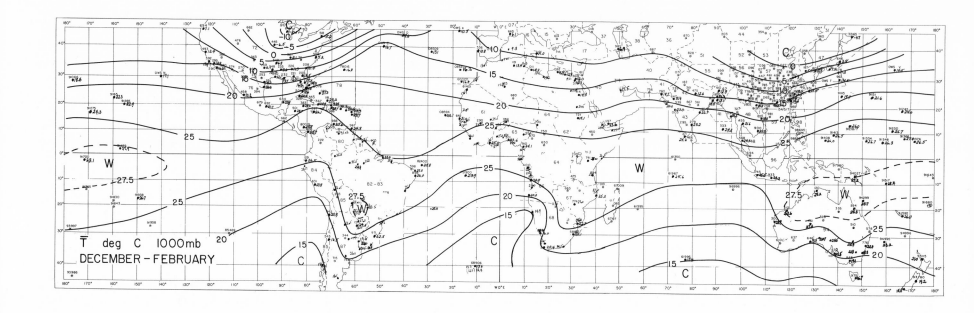

T̄ deg C 1000mb
DECEMBER – FEBRUARY

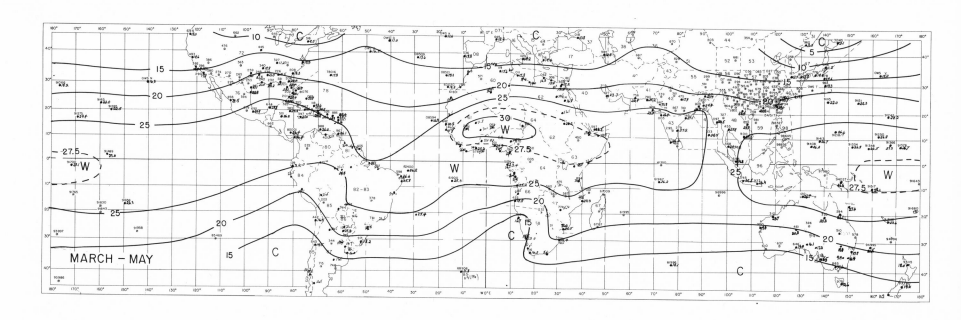

MARCH – MAY

Plate 3.2

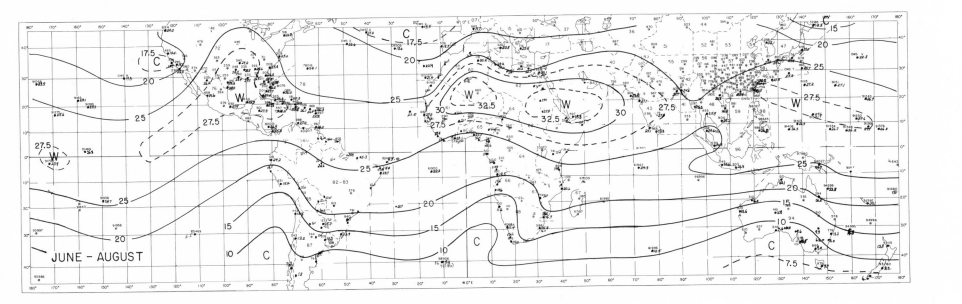

JUNE - AUGUST

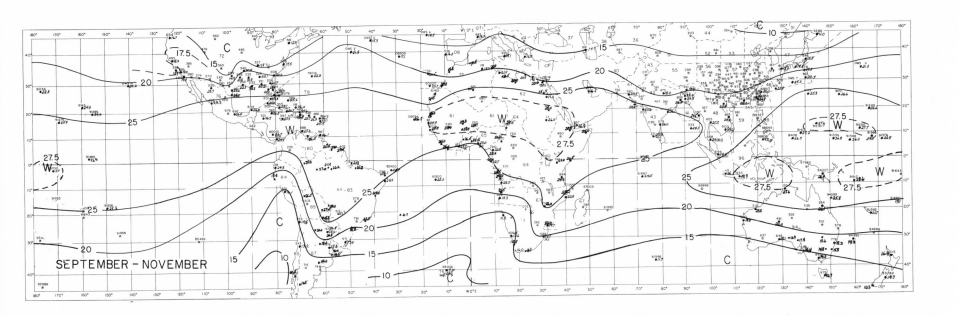

SEPTEMBER - NOVEMBER

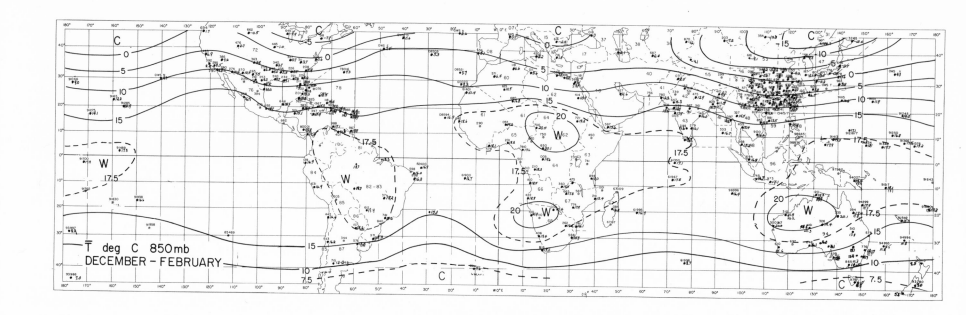

$\overline{T}$ deg C 850mb
DECEMBER - FEBRUARY

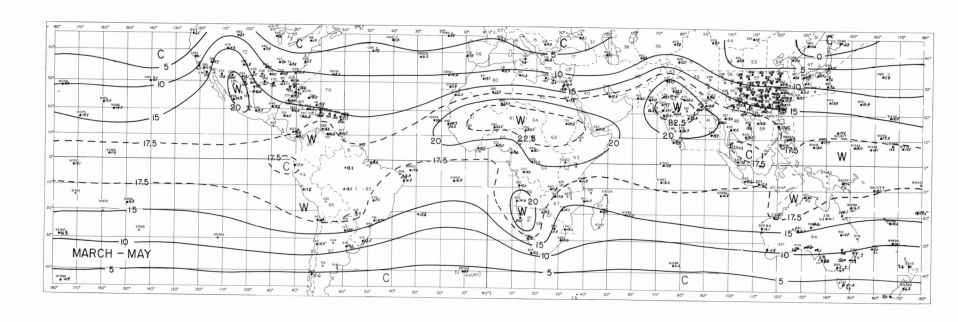

MARCH - MAY

Plate 3.3

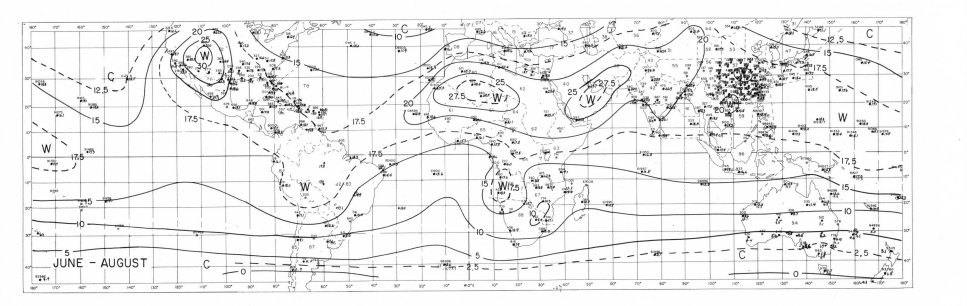

JUNE - AUGUST

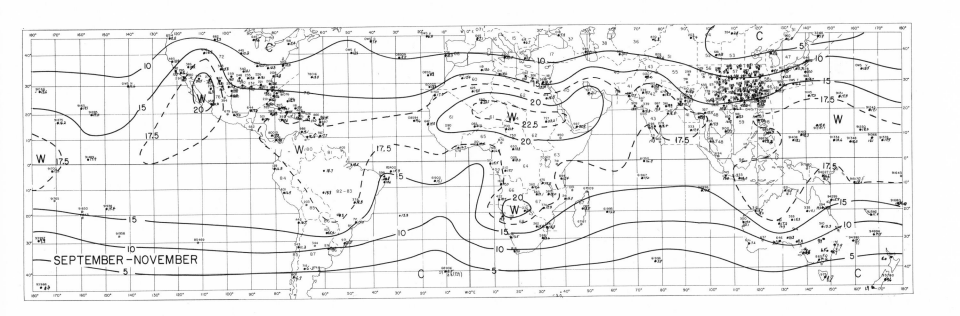

SEPTEMBER - NOVEMBER

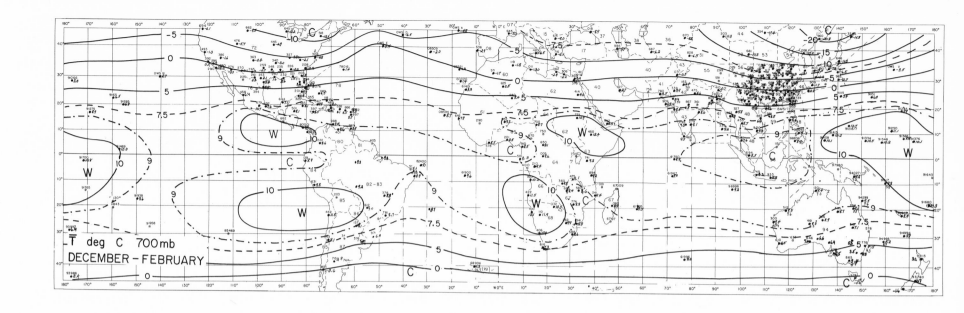

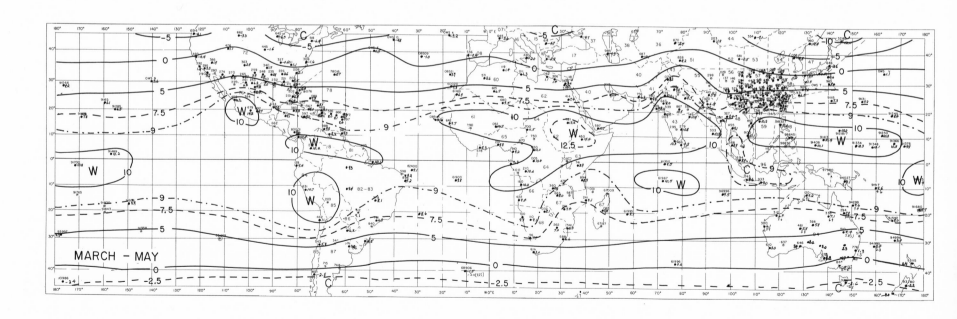

Plate 3.4

68   MEAN TEMPERATURE AND WIND FIELDS

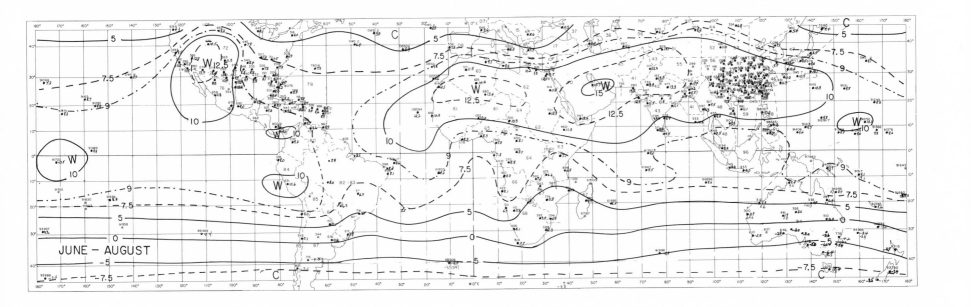

JUNE – AUGUST

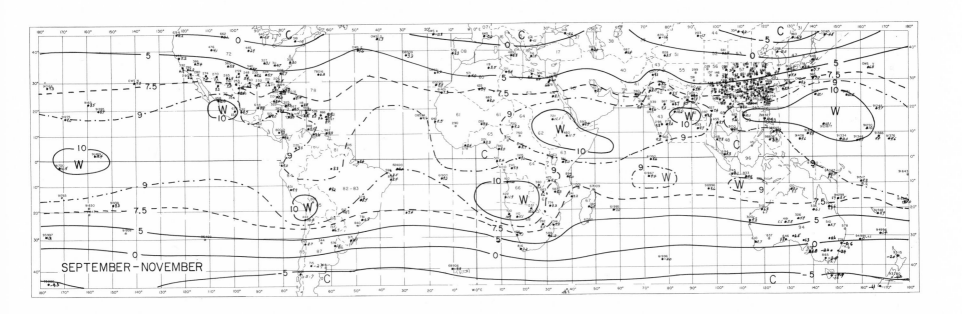

SEPTEMBER – NOVEMBER

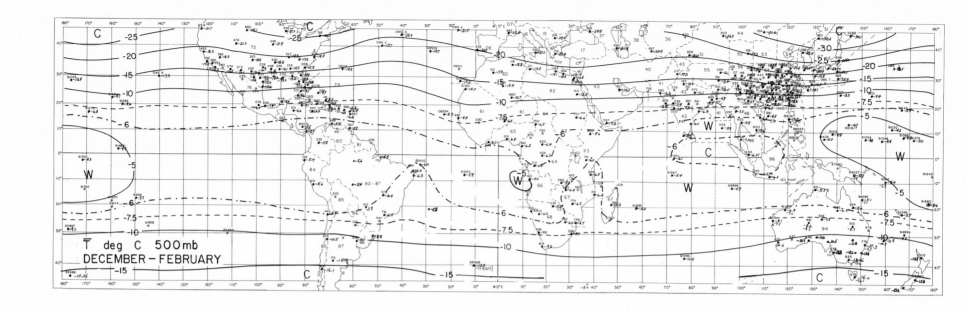

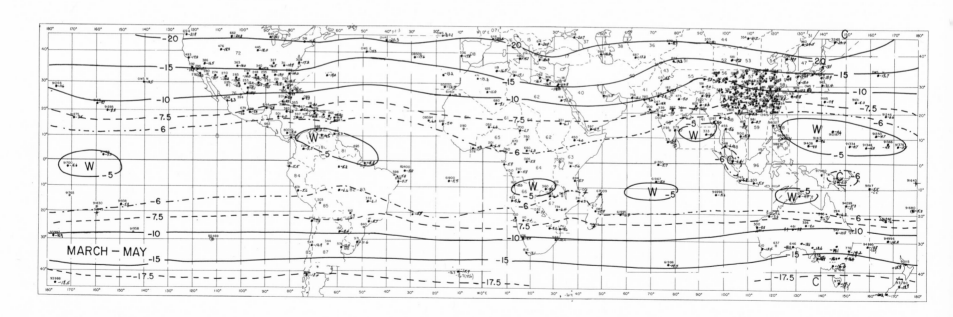

Plate 3.5

70   MEAN TEMPERATURE AND WIND FIELDS

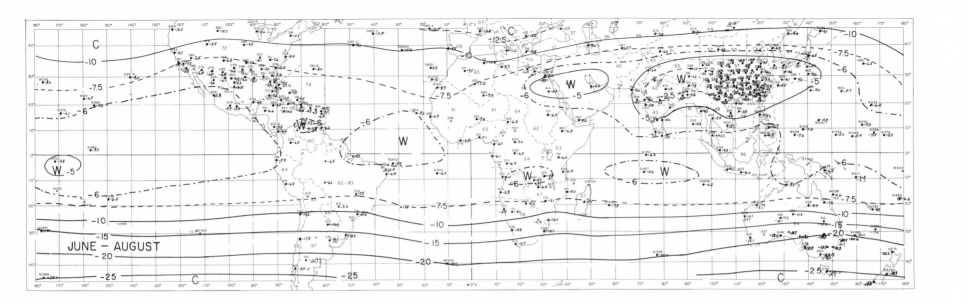

JUNE – AUGUST

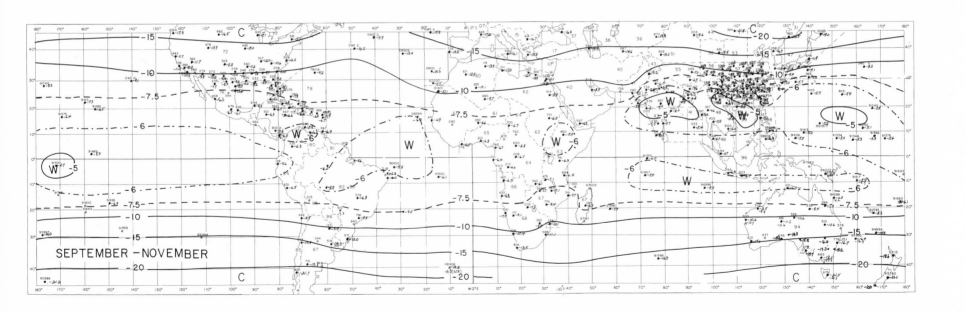

SEPTEMBER – NOVEMBER

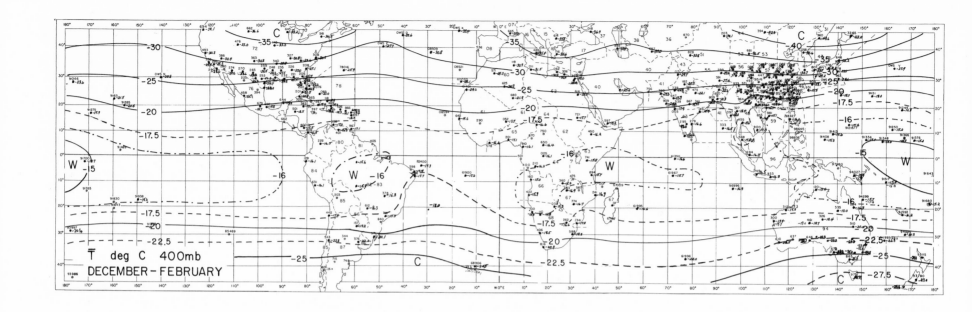

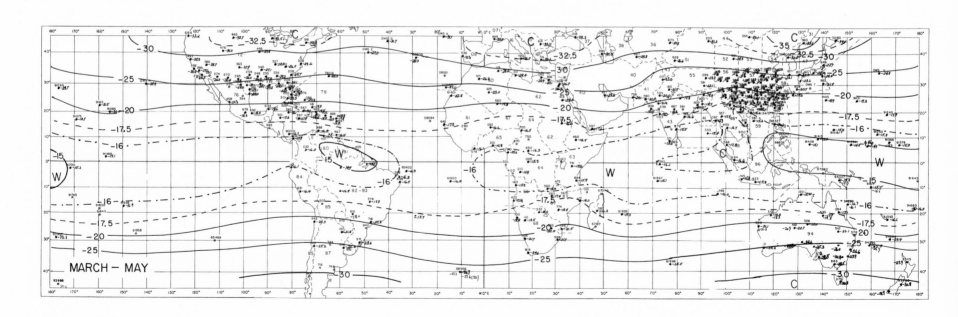

Plate 3.6

72   MEAN TEMPERATURE AND WIND FIELDS

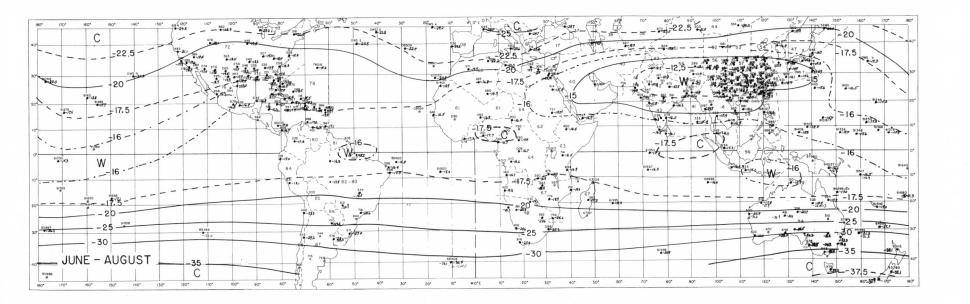

JUNE - AUGUST

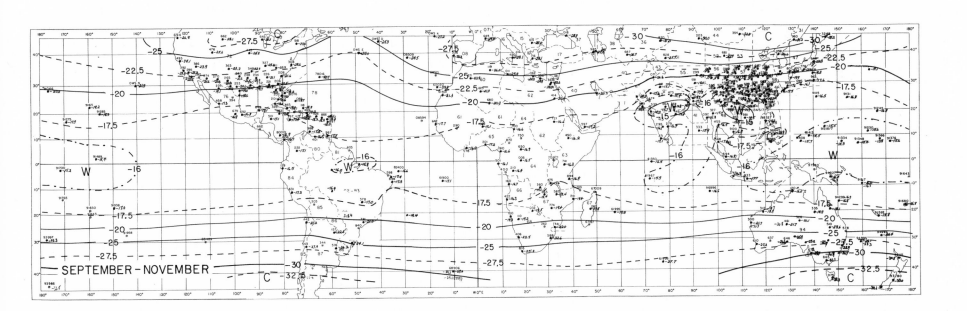

SEPTEMBER - NOVEMBER

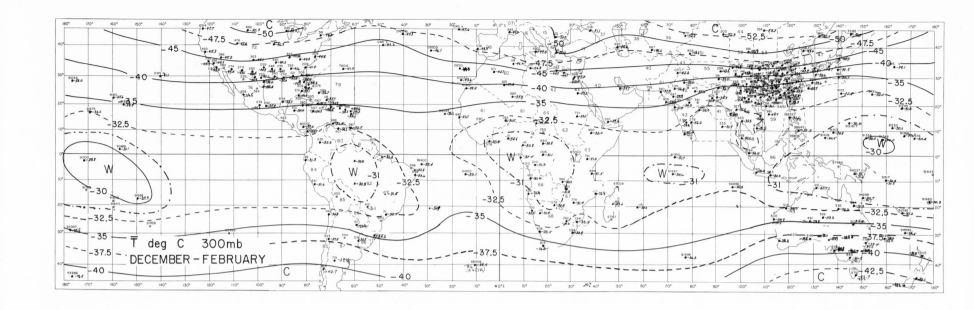

$\overline{T}$ deg C 300mb
DECEMBER – FEBRUARY

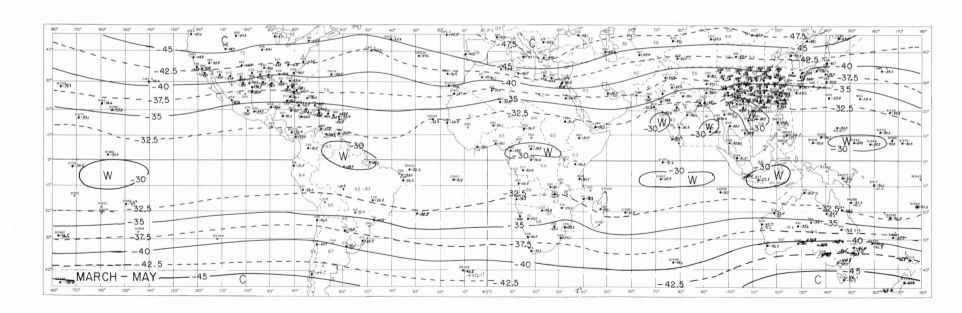

MARCH – MAY

Plate 3.7

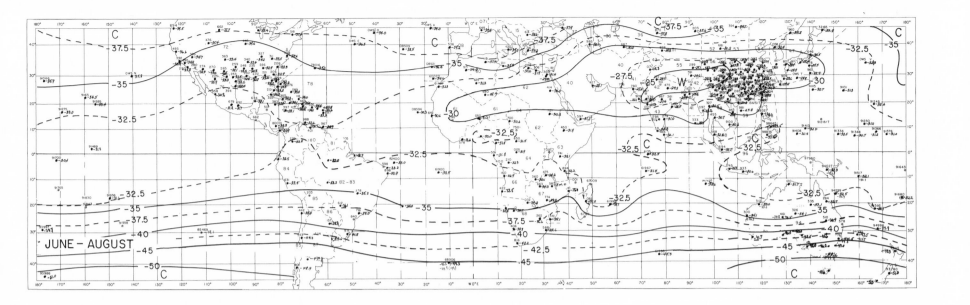

JUNE–AUGUST

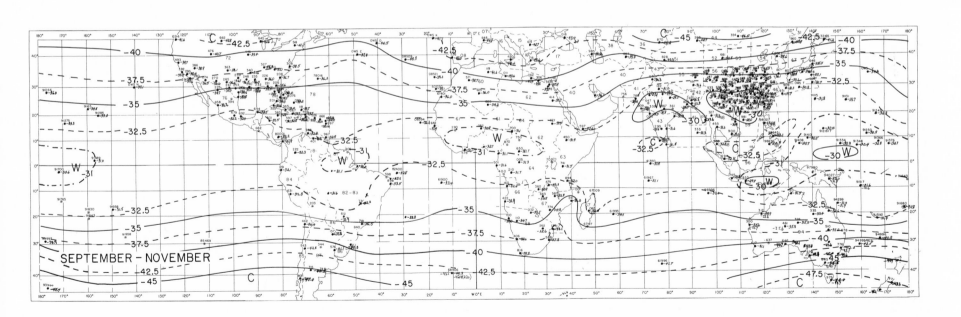

SEPTEMBER–NOVEMBER

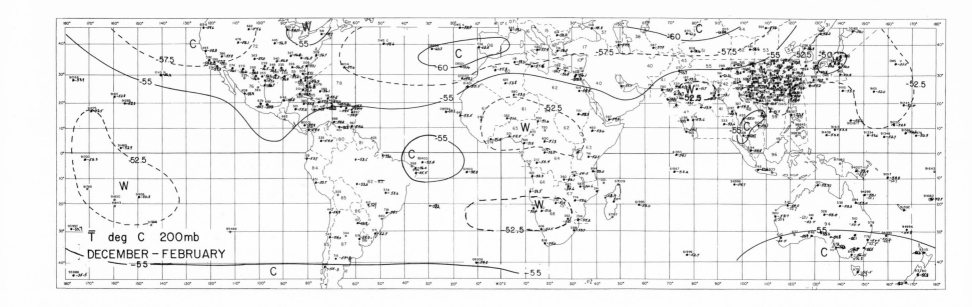

T deg C 200mb
DECEMBER – FEBRUARY

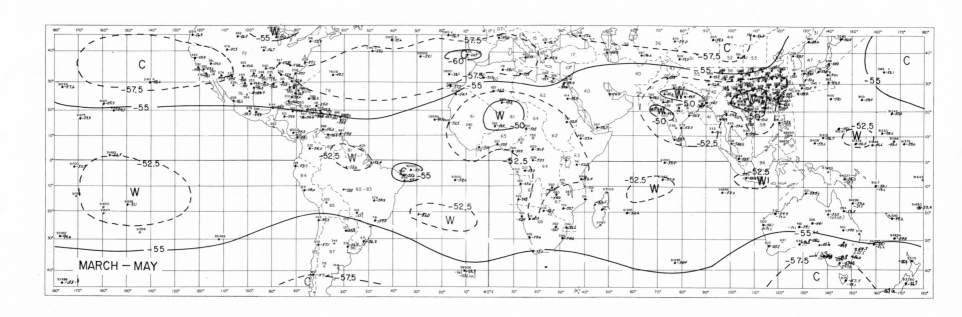

MARCH – MAY

Plate 3.8

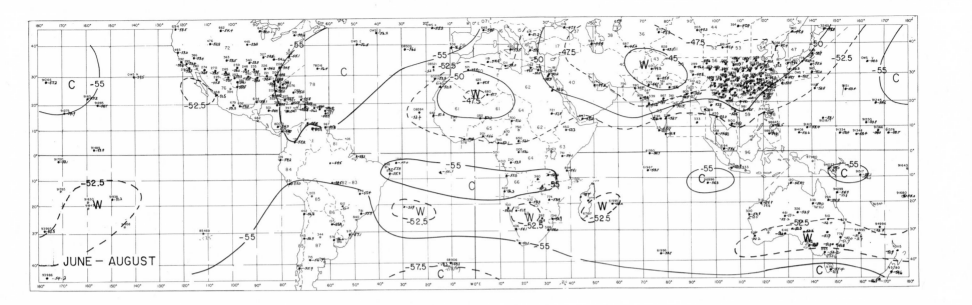

JUNE — AUGUST

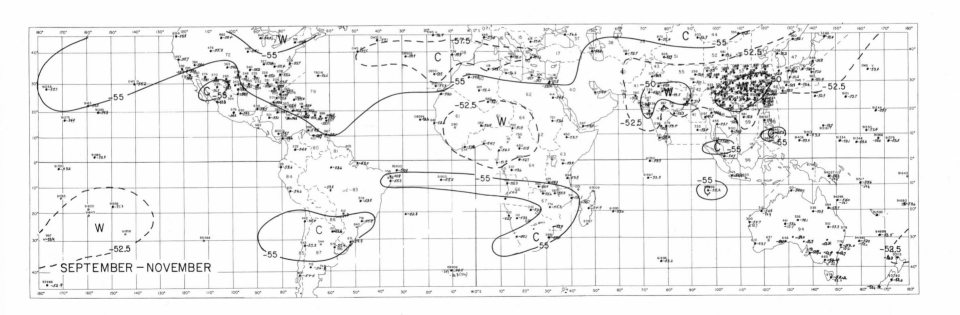

SEPTEMBER — NOVEMBER

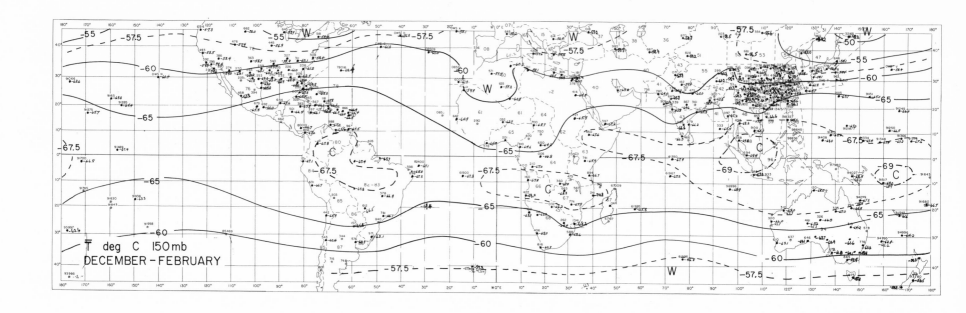

T deg C 150mb
DECEMBER - FEBRUARY

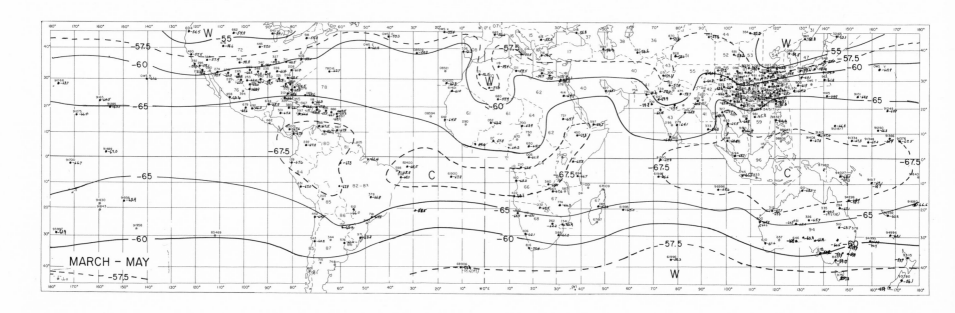

MARCH - MAY

Plate 3.9

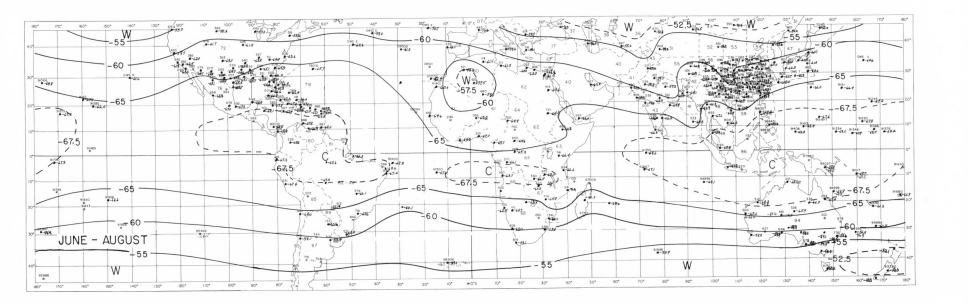

JUNE – AUGUST

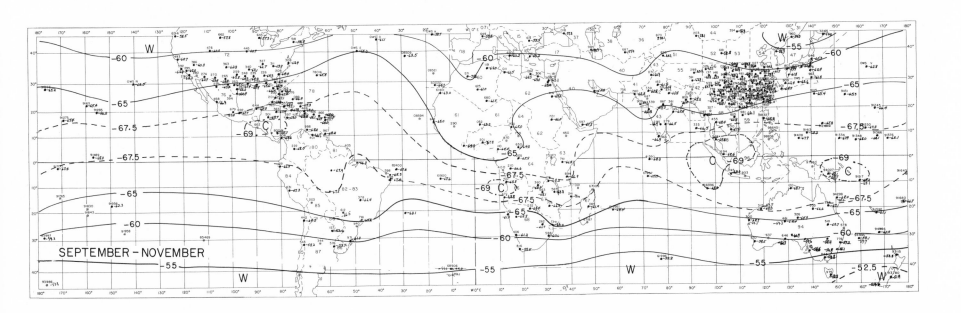

SEPTEMBER – NOVEMBER

79   MEAN TEMPERATURE AND WIND FIELDS

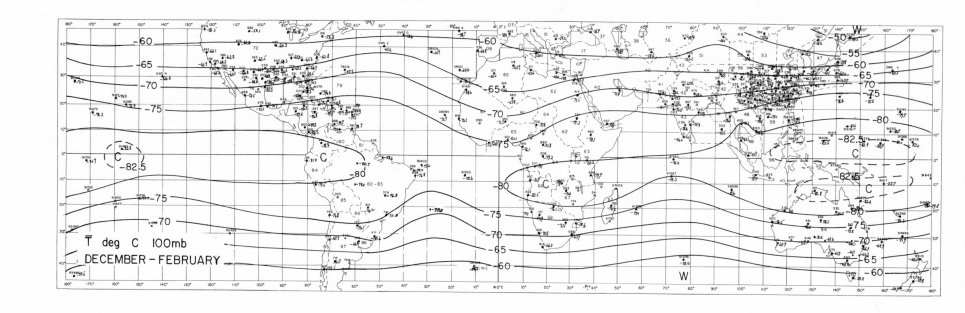

T̄ deg C 100mb
DECEMBER - FEBRUARY

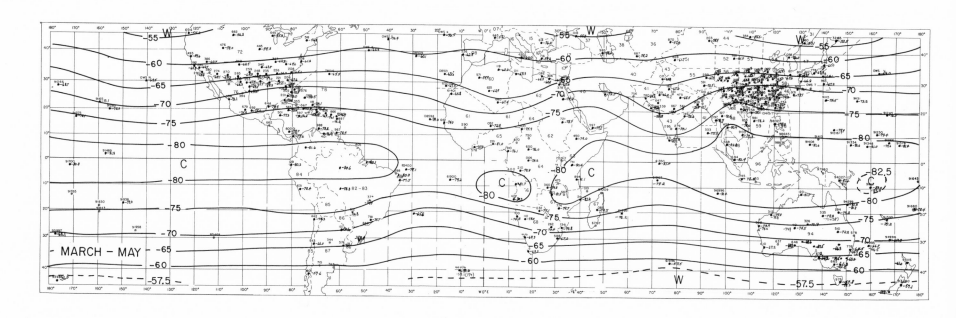

MARCH - MAY

Plate 3.10

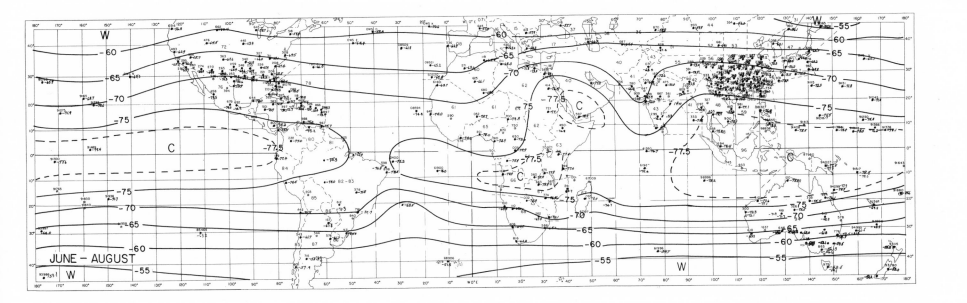

JUNE — AUGUST

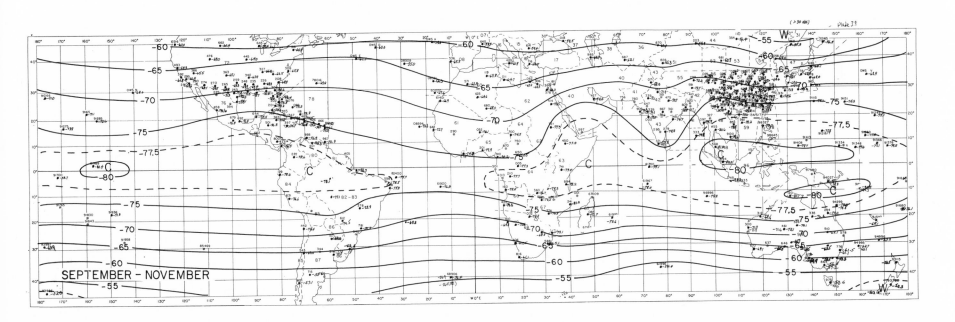

SEPTEMBER — NOVEMBER

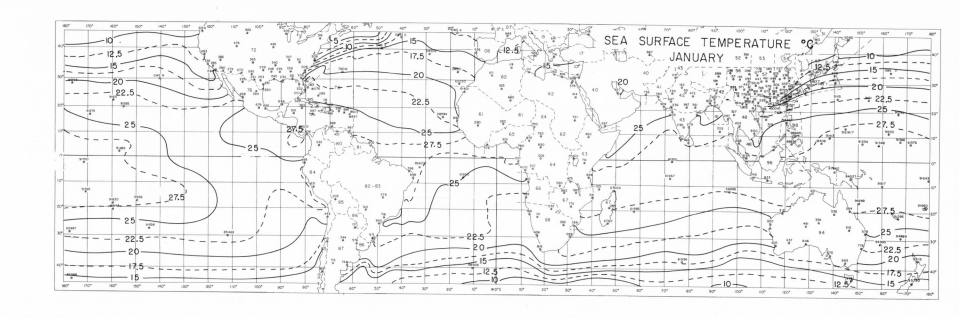

SEA SURFACE TEMPERATURE °C
JANUARY

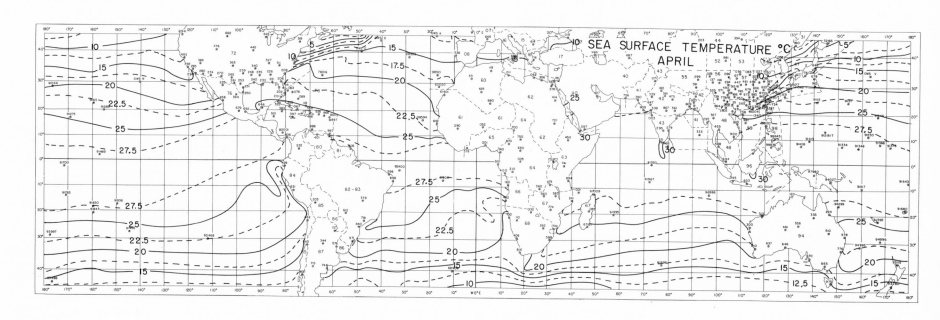

SEA SURFACE TEMPERATURE °C
APRIL

Plate 3.11

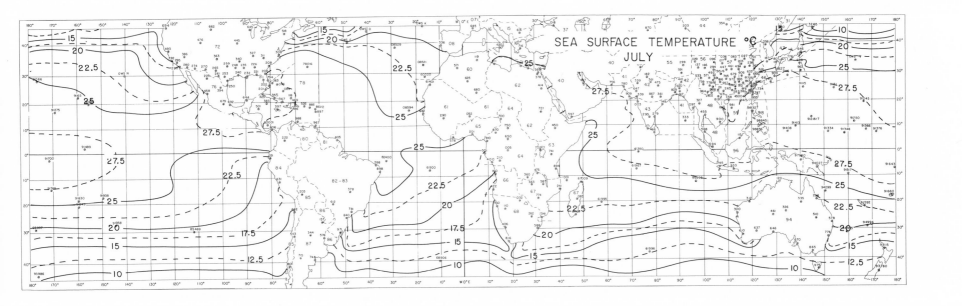

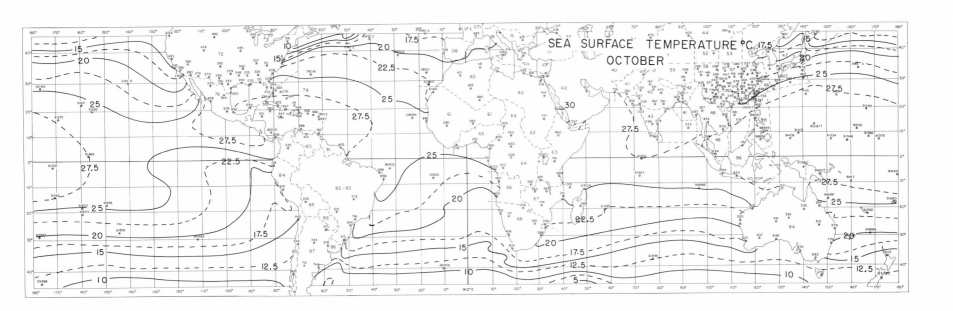

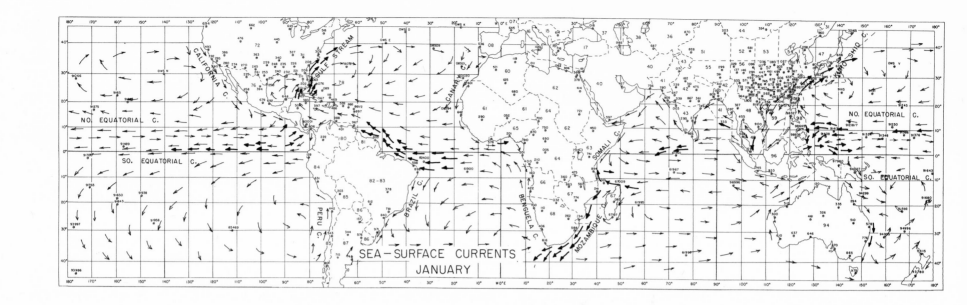

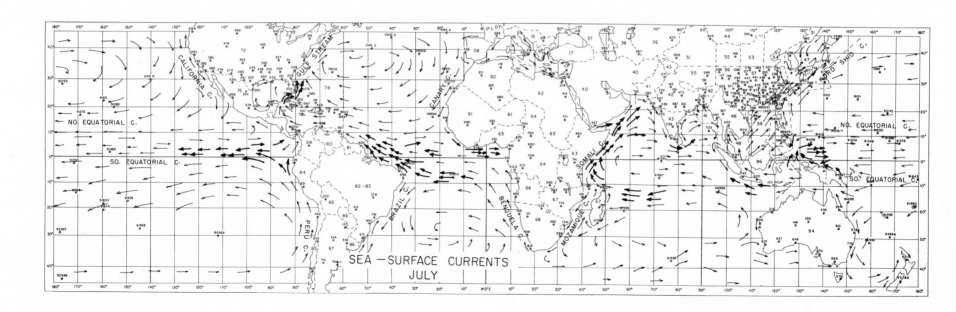

Plate 3.12

84   MEAN TEMPERATURE AND WIND FIELDS

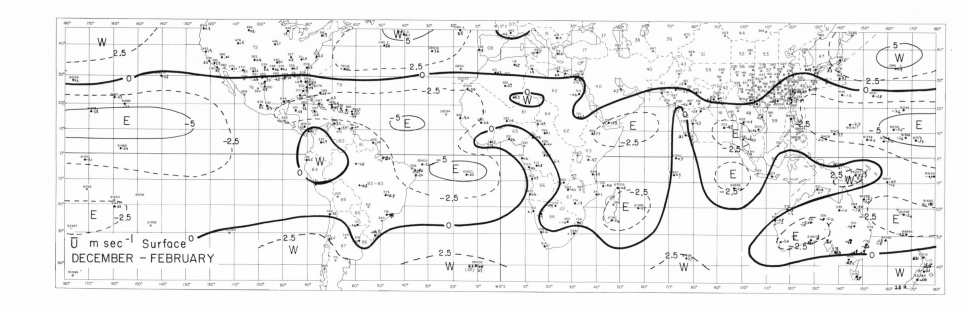

Ū m sec⁻¹ Surface
DECEMBER – FEBRUARY

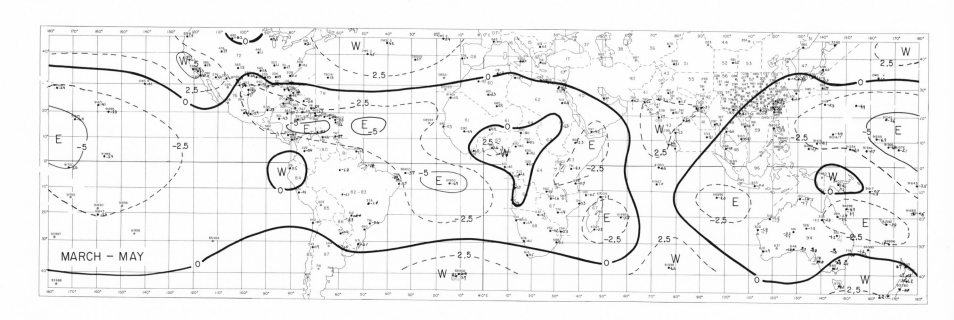

MARCH – MAY

Plate 3.13

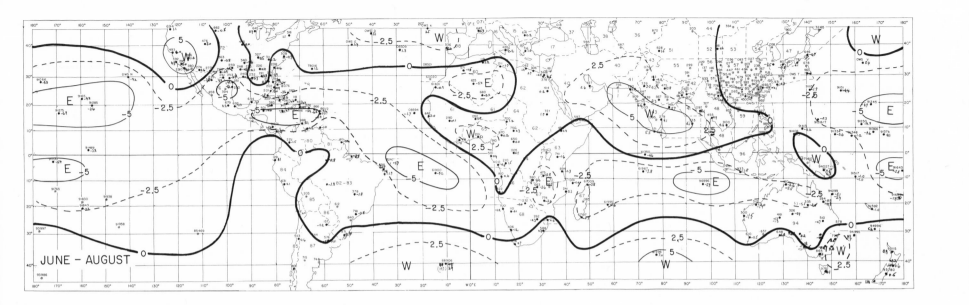

JUNE – AUGUST

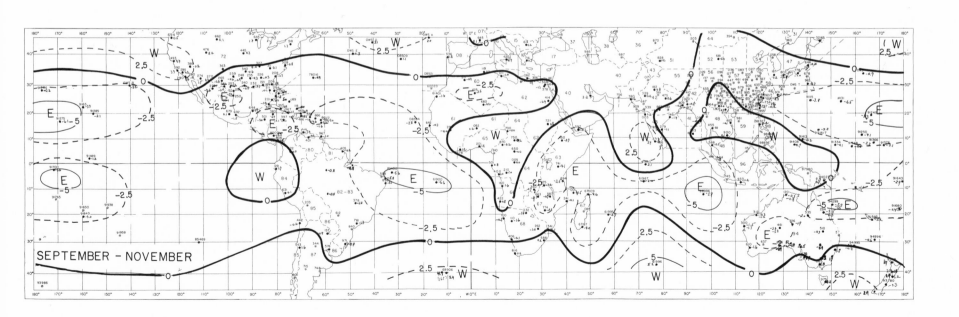

SEPTEMBER – NOVEMBER

87   MEAN TEMPERATURE AND WIND FIELDS

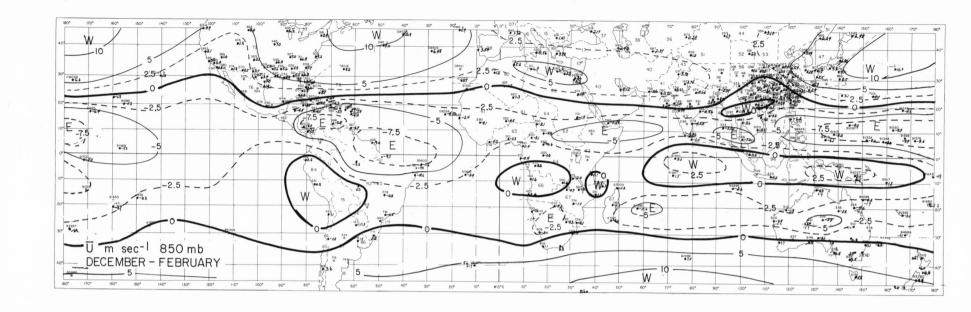

$\bar{U}$ m sec⁻¹ 850 mb
DECEMBER – FEBRUARY

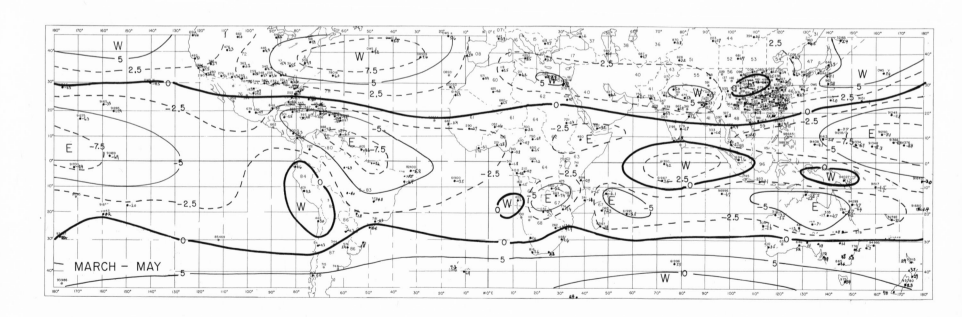

MARCH – MAY

Plate 3.14

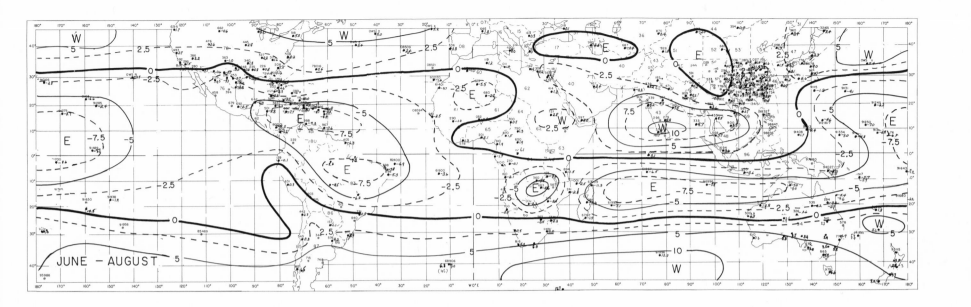

JUNE – AUGUST

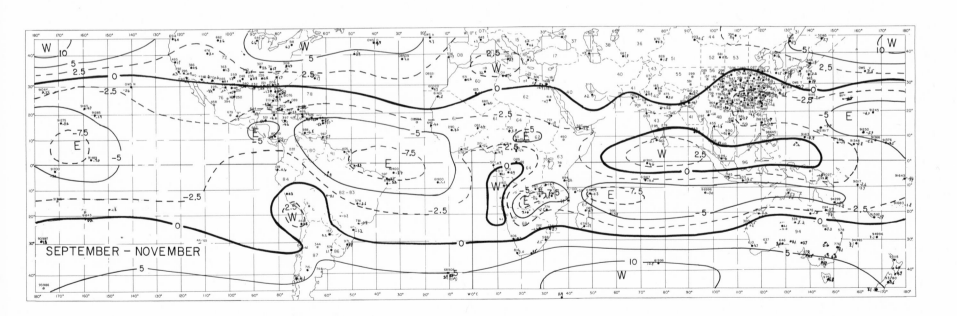

SEPTEMBER – NOVEMBER

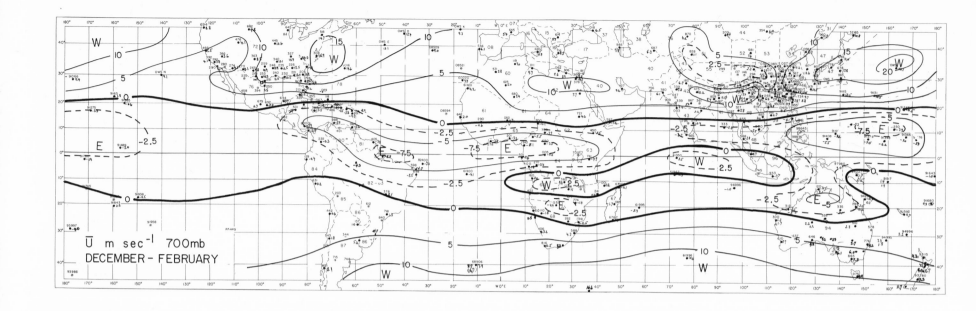

$\bar{U}$ m sec^{-1} 700mb
DECEMBER - FEBRUARY

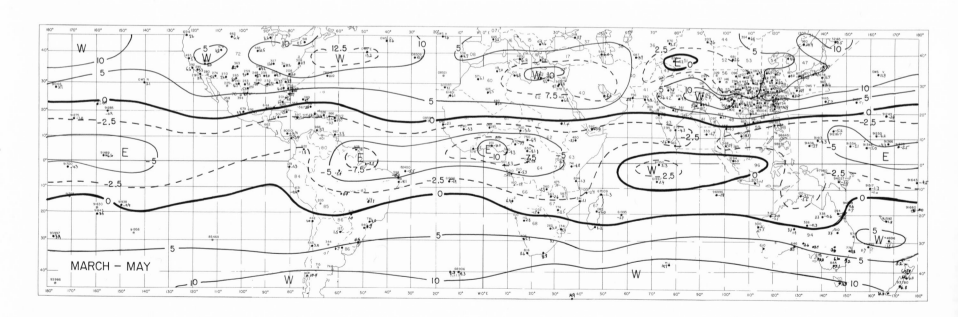

MARCH - MAY

Plate 3.15

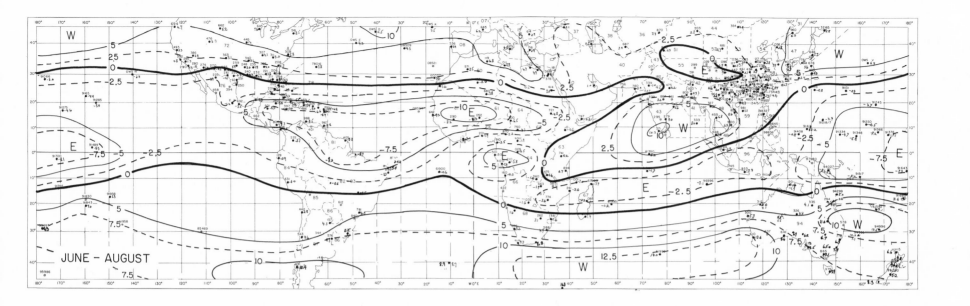

JUNE – AUGUST

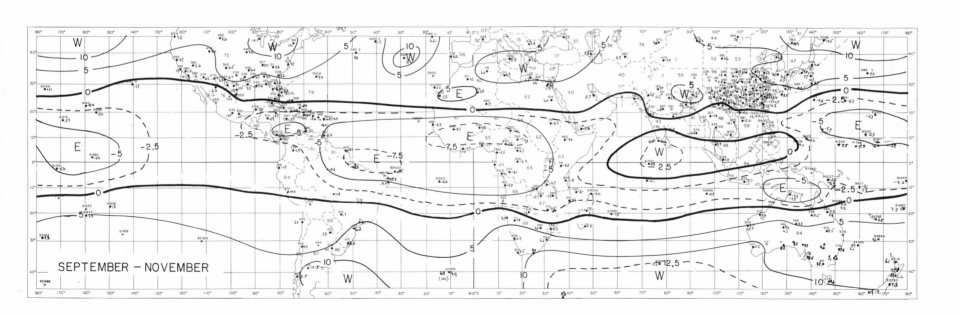

SEPTEMBER – NOVEMBER

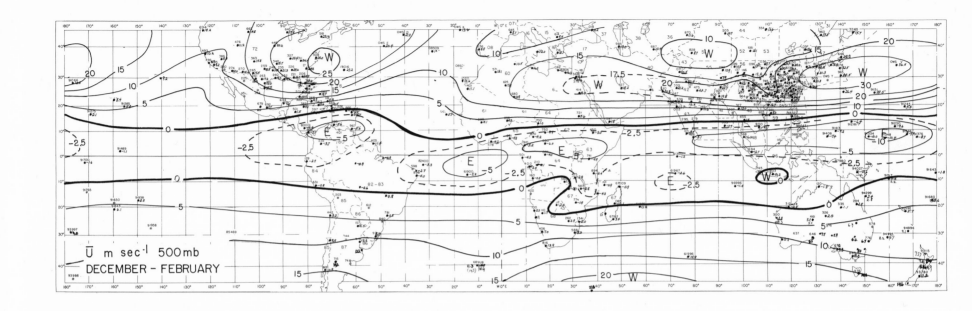

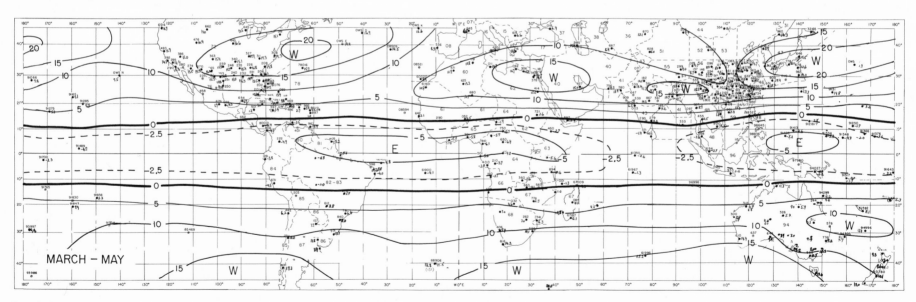

Plate 3.16

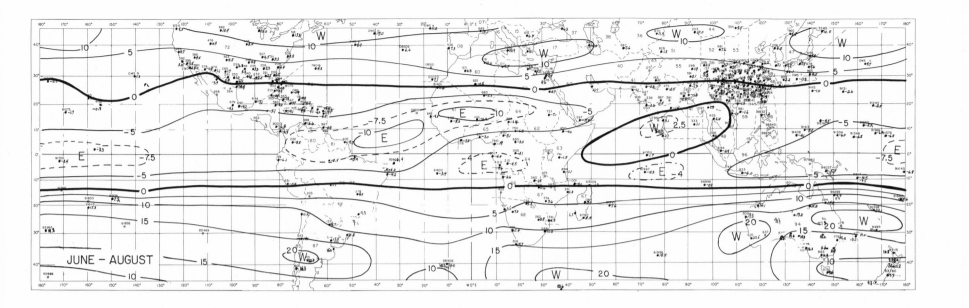

JUNE - AUGUST

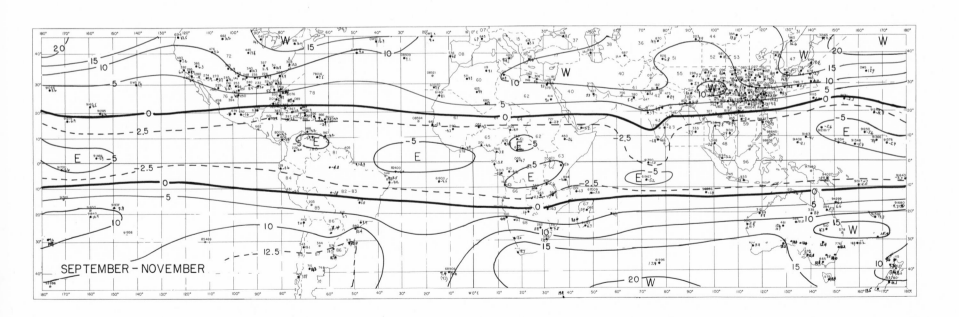

SEPTEMBER - NOVEMBER

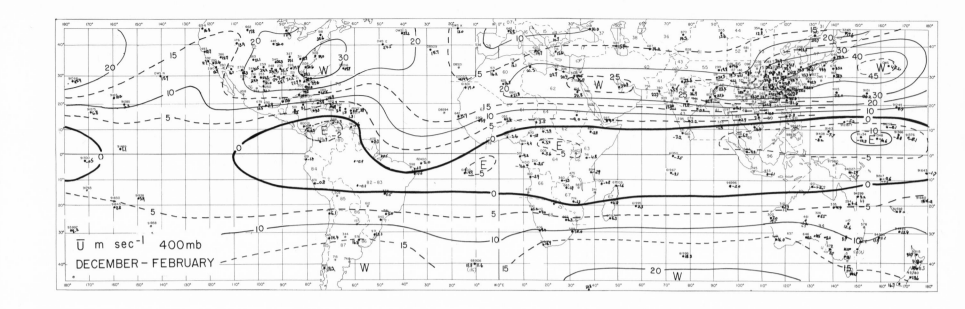

$\overline{U}$ m sec^{-1} 400mb
DECEMBER – FEBRUARY

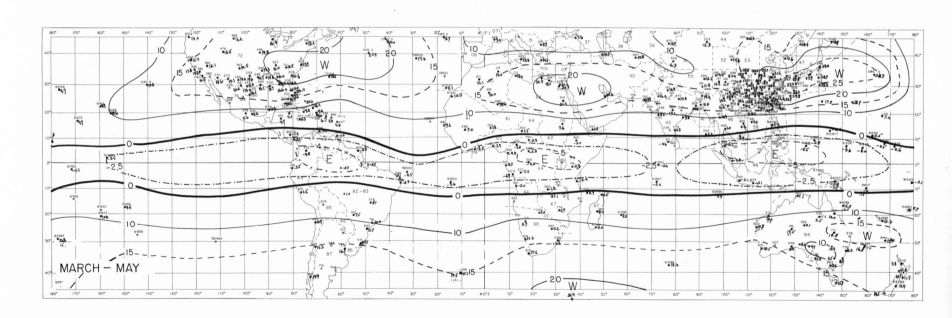

MARCH – MAY

Plate 3.17

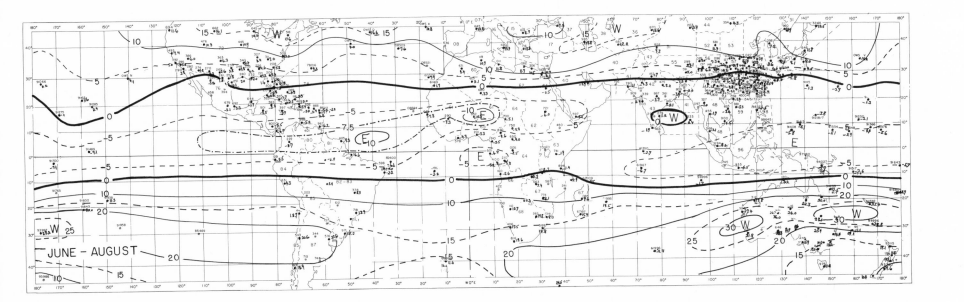

JUNE – AUGUST

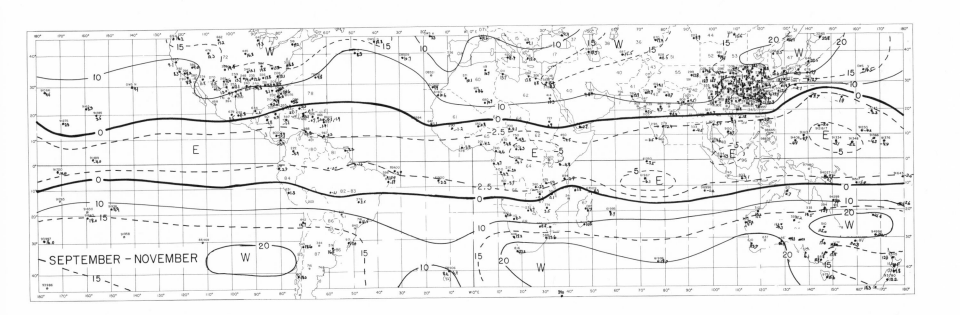

SEPTEMBER – NOVEMBER

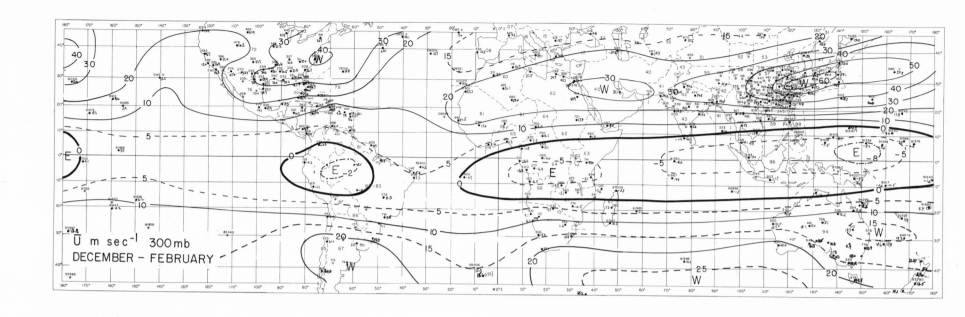

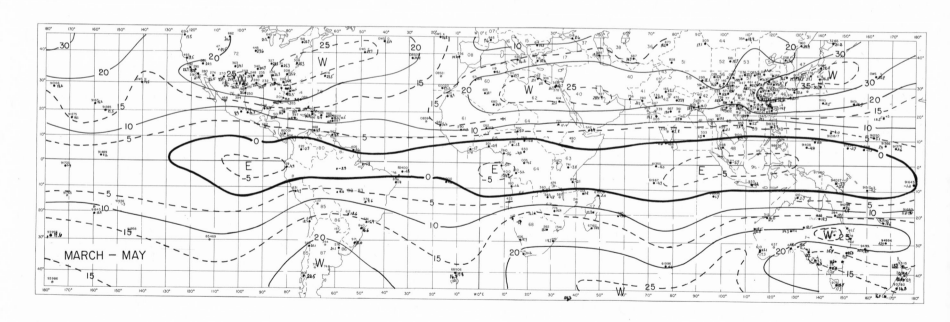

Plate 3.18

96   MEAN TEMPERATURE AND WIND FIELDS

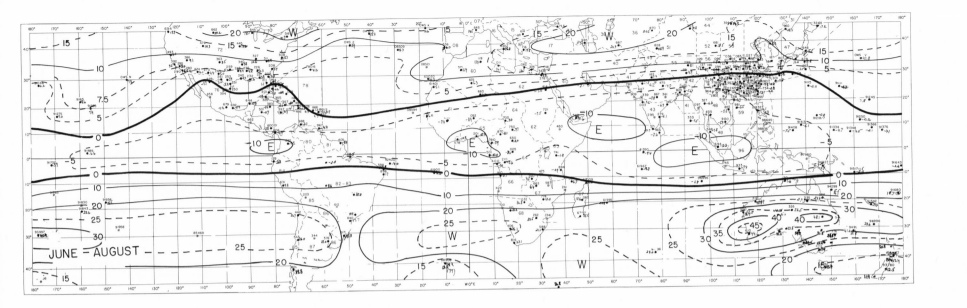

JUNE - AUGUST

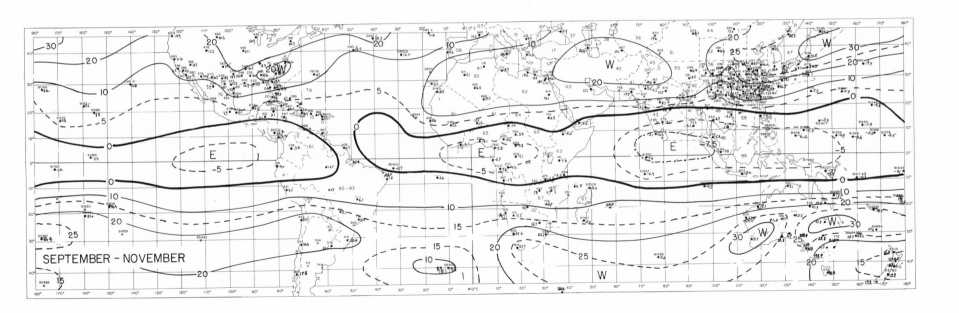

SEPTEMBER - NOVEMBER

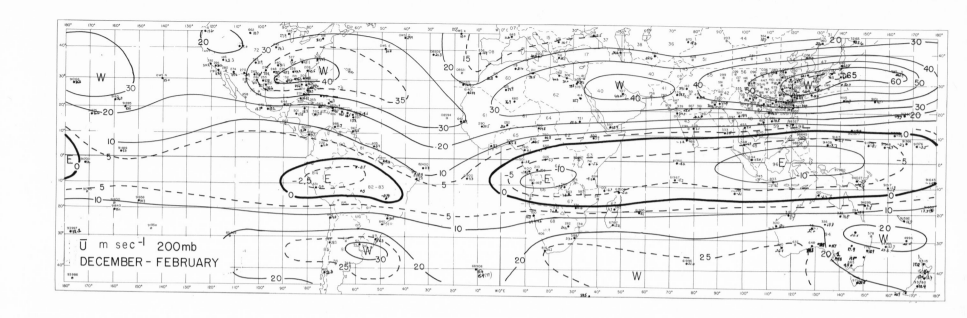

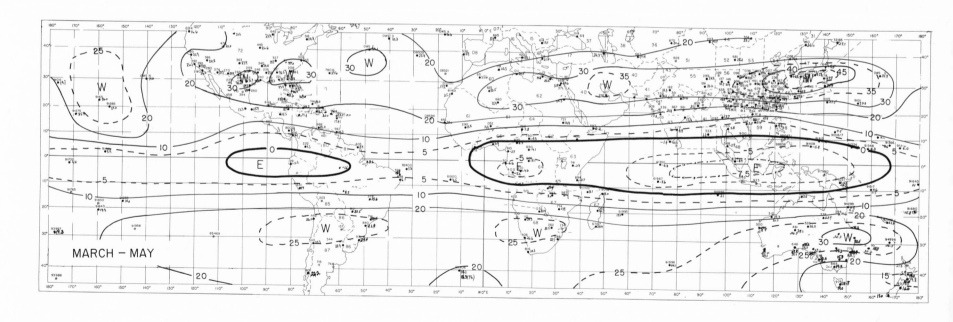

Plate 3.19

98   MEAN TEMPERATURE AND WIND FIELDS

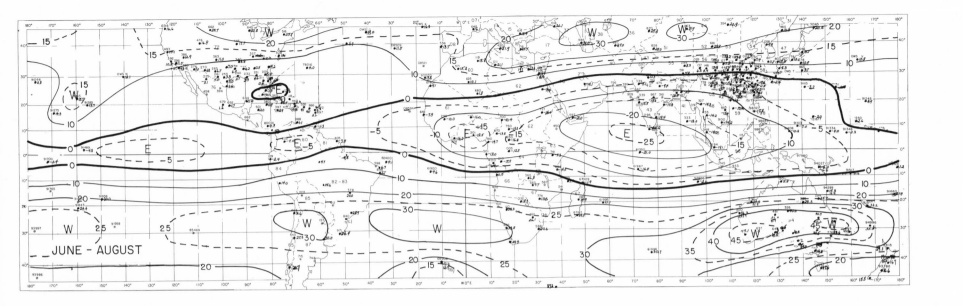

JUNE – AUGUST

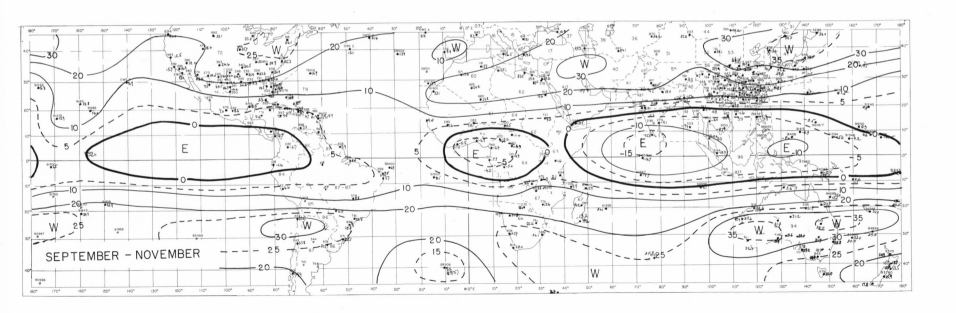

SEPTEMBER – NOVEMBER

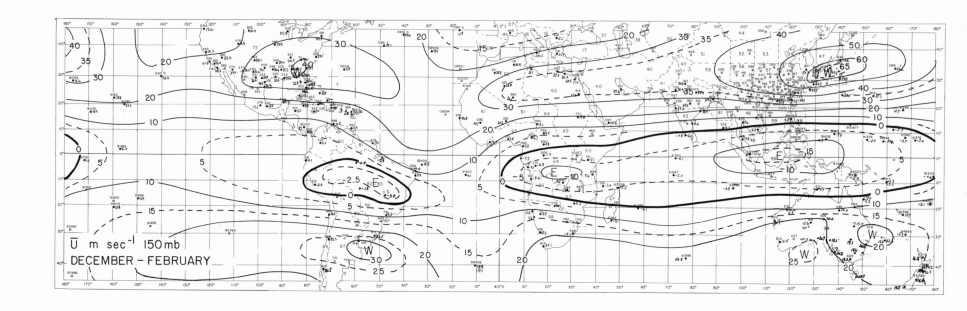

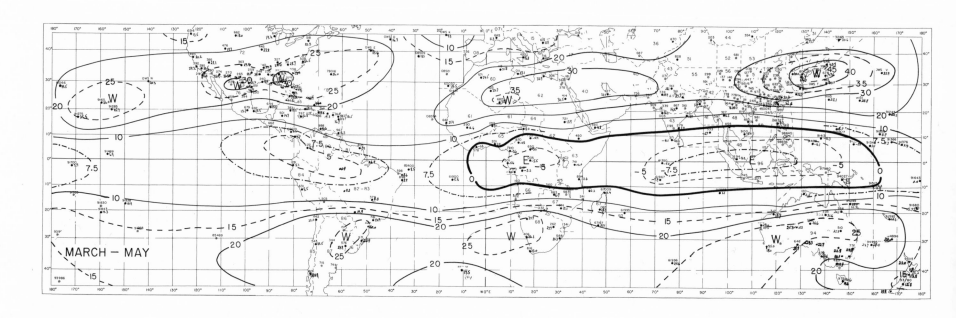

Plate 3.20

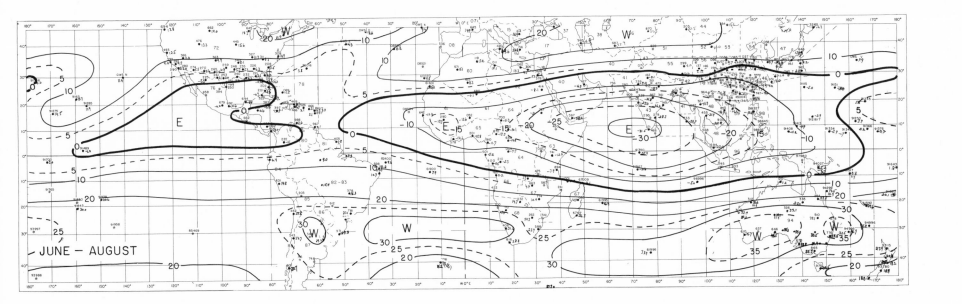

JUNE – AUGUST

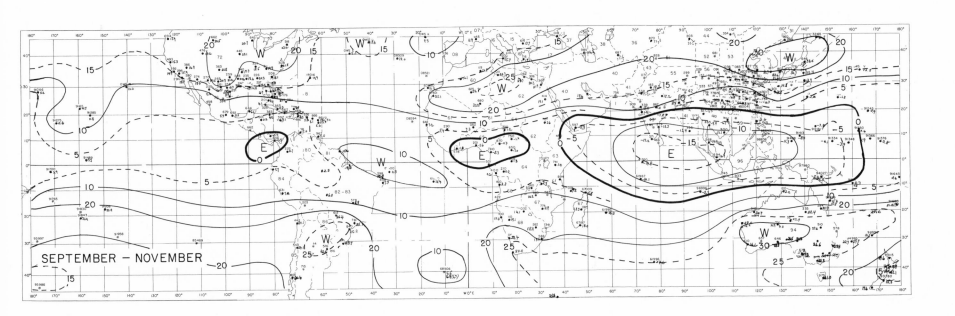

SEPTEMBER – NOVEMBER

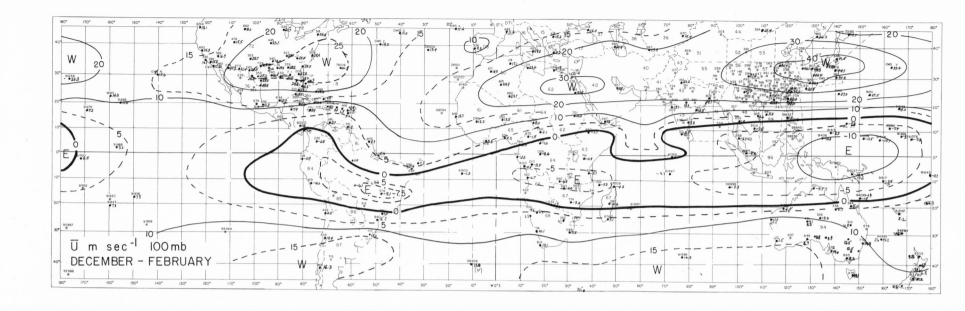

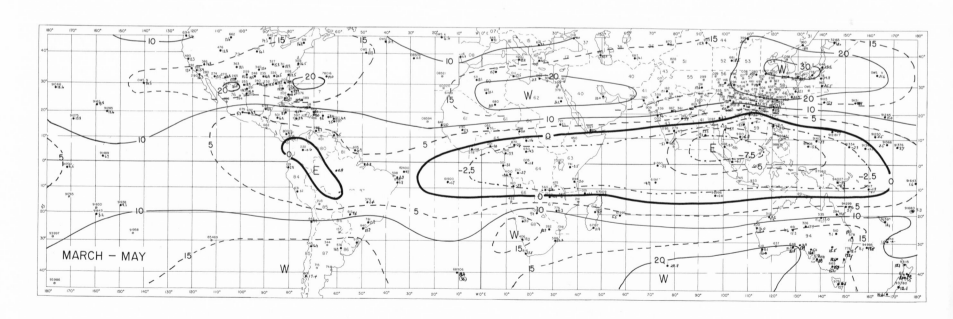

Plate 3.21

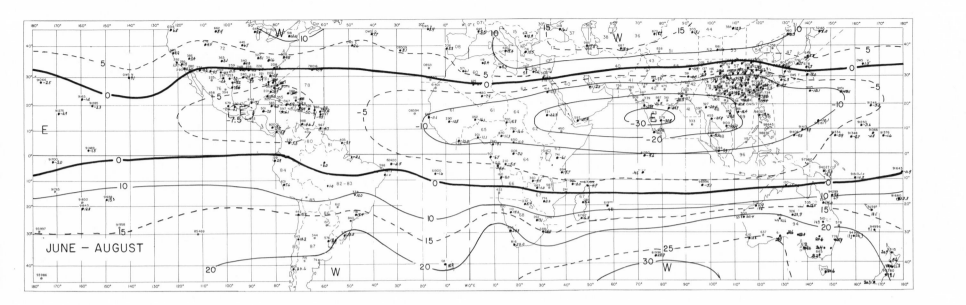

JUNE – AUGUST

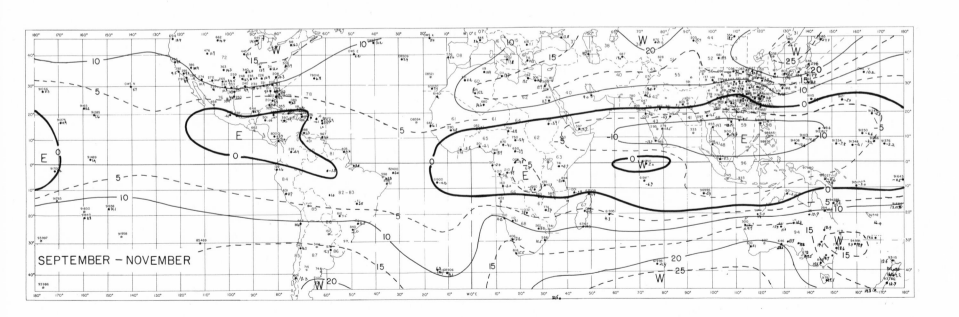

SEPTEMBER – NOVEMBER

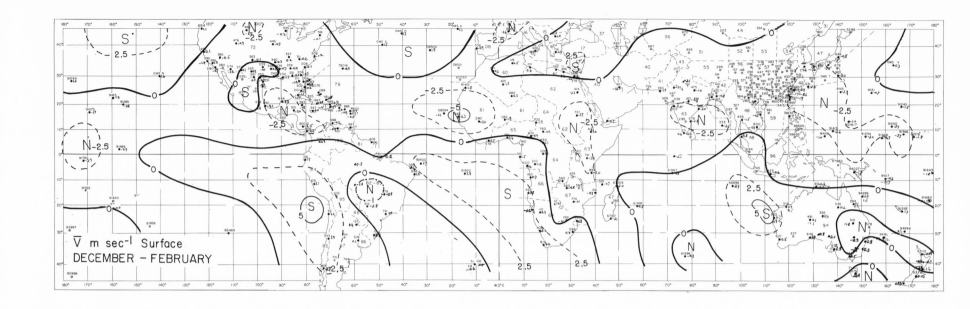

$\overline{V}$ m sec^{-1} Surface
DECEMBER – FEBRUARY

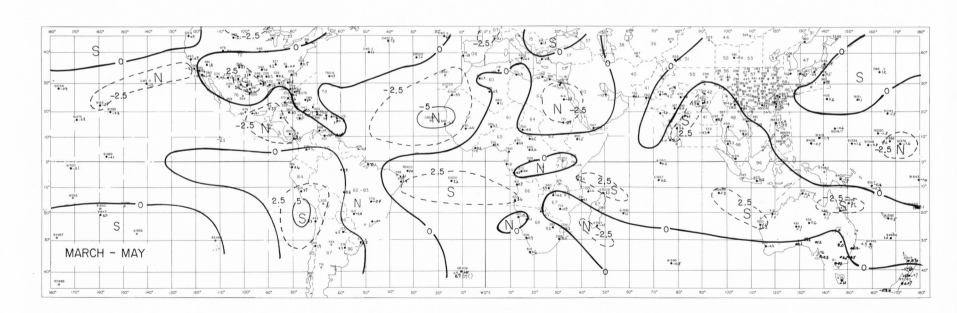

MARCH – MAY

Plate 3.22

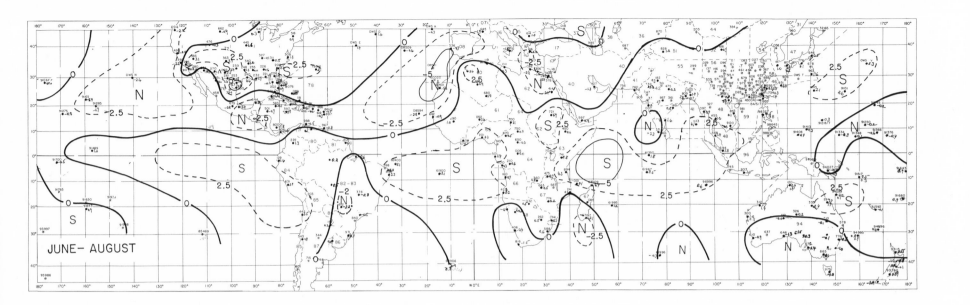

JUNE – AUGUST

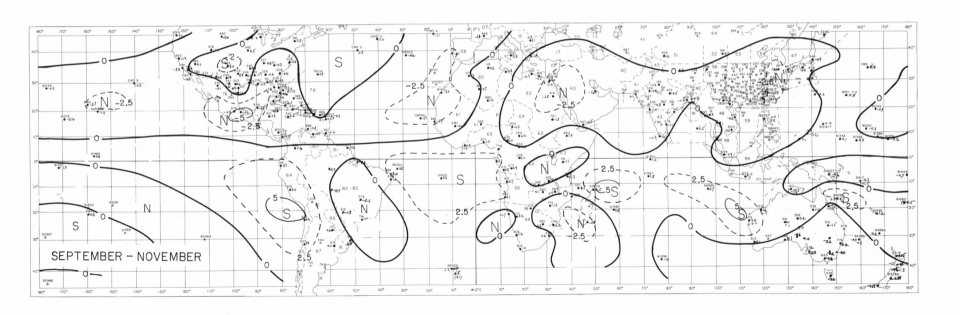

SEPTEMBER – NOVEMBER

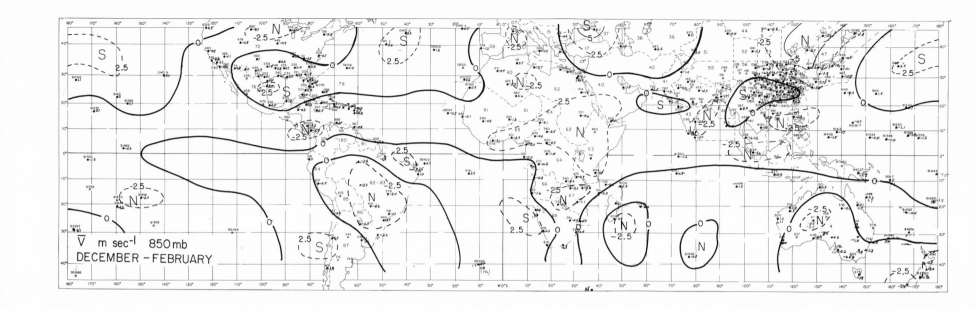

$\overline{V}$ m sec⁻¹ 850mb
DECEMBER – FEBRUARY

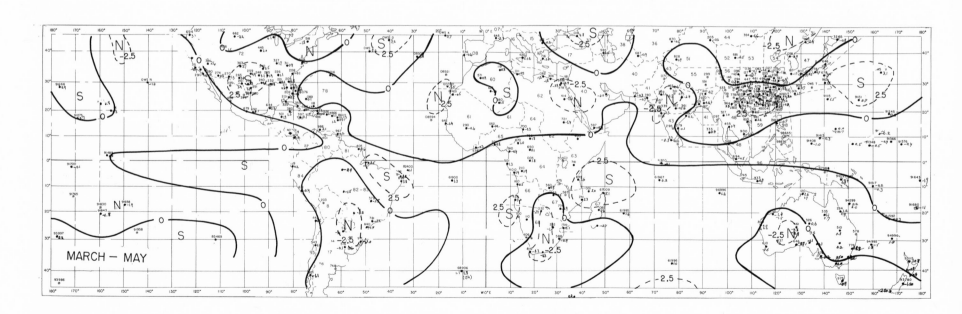

MARCH – MAY

Plate 3.23

106   MEAN TEMPERATURE AND WIND FIELDS

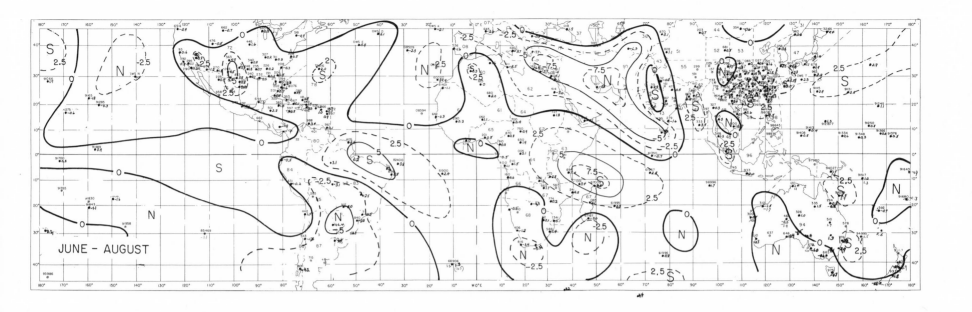

JUNE - AUGUST

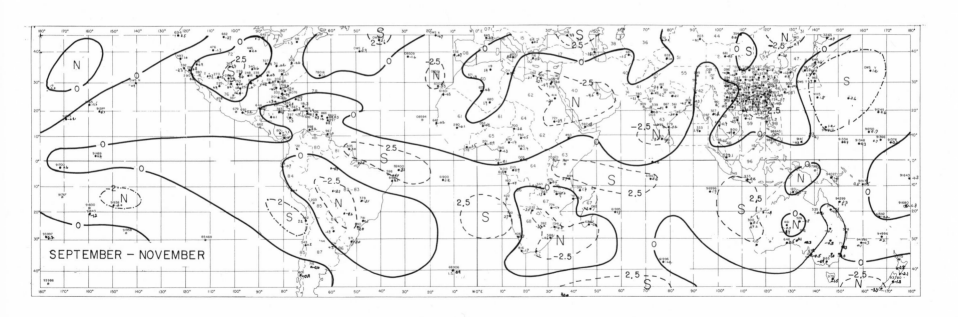

SEPTEMBER - NOVEMBER

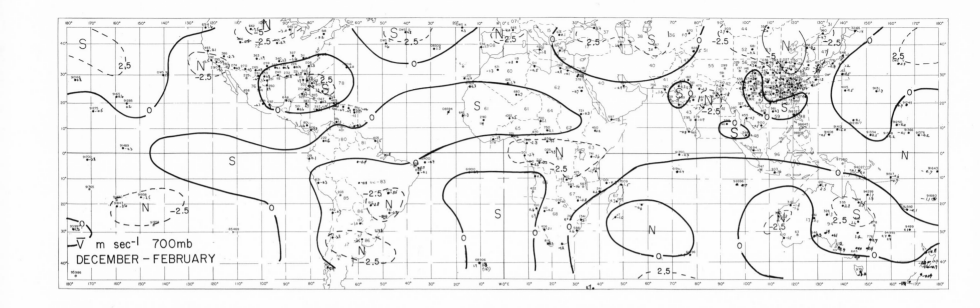

$\overline{V}$ m sec⁻¹ 700mb
DECEMBER – FEBRUARY

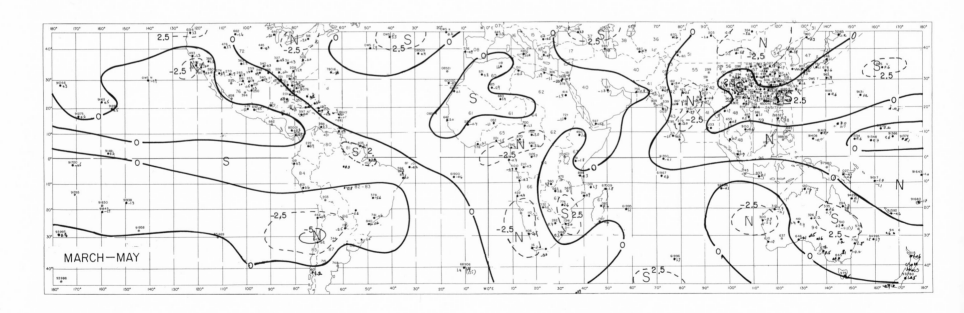

MARCH – MAY

Plate 3.24

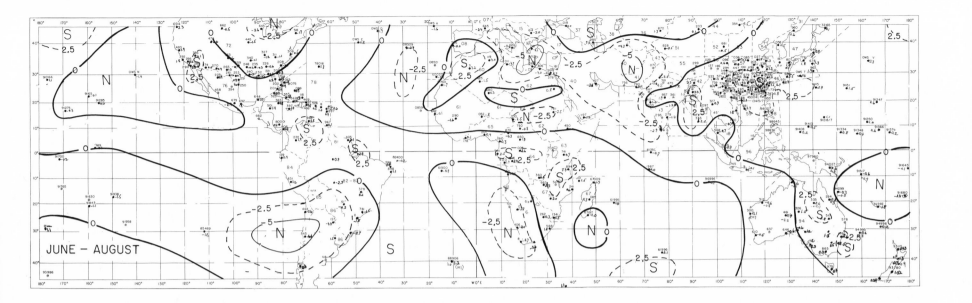

JUNE – AUGUST

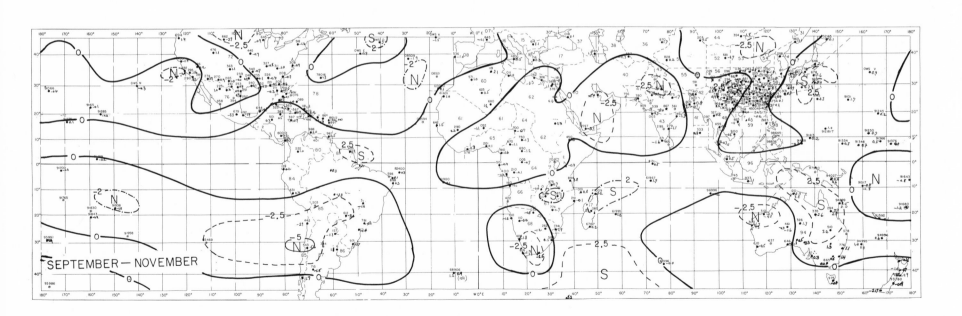

SEPTEMBER – NOVEMBER

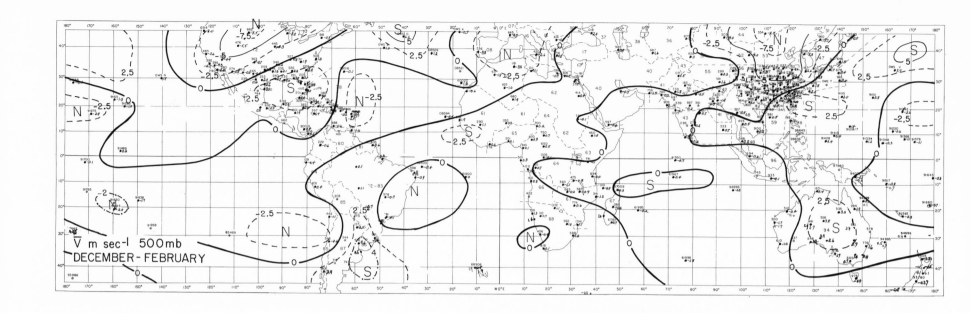

$\overline{V}$ m sec⁻¹ 500mb
DECEMBER-FEBRUARY

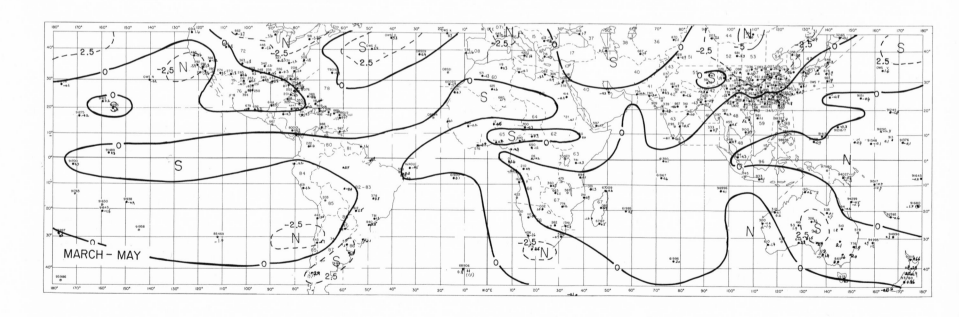

MARCH - MAY

Plate 3.25

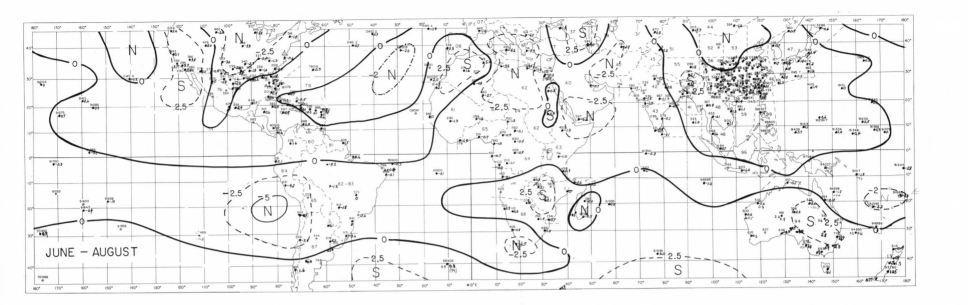

JUNE – AUGUST

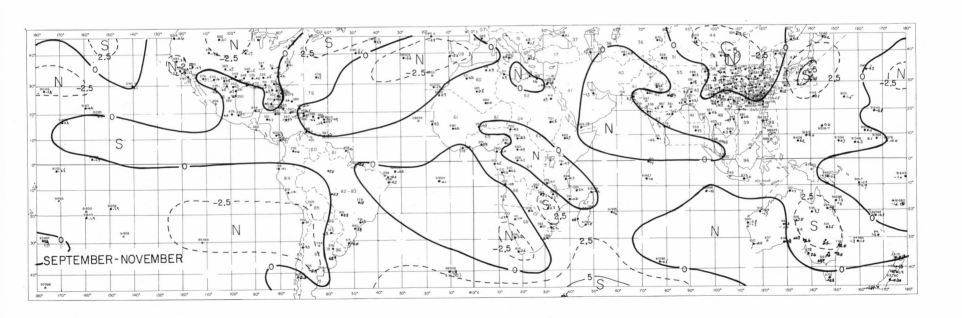

SEPTEMBER–NOVEMBER

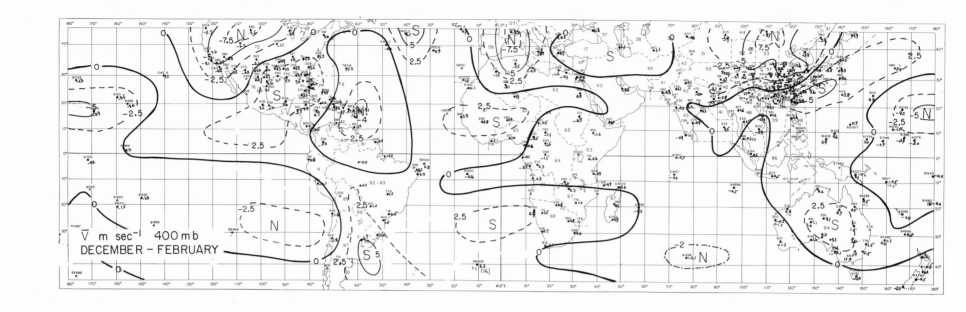

$\overline{V}$ m sec⁻¹ 400 mb
DECEMBER – FEBRUARY

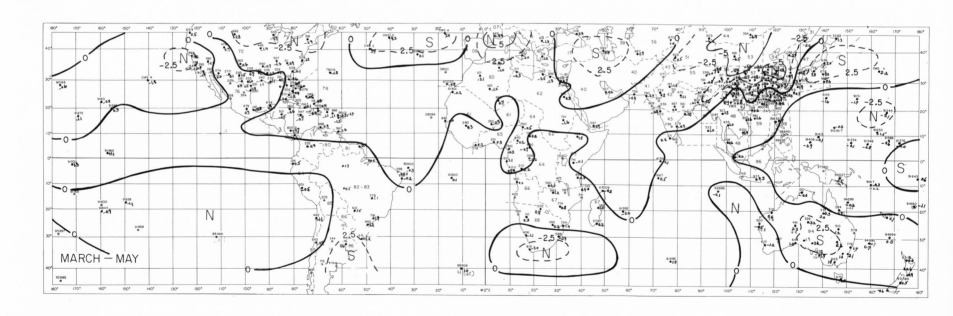

MARCH – MAY

Plate 3.26

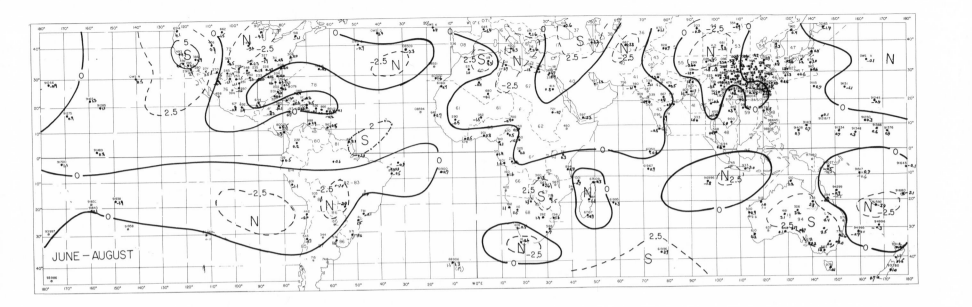

JUNE — AUGUST

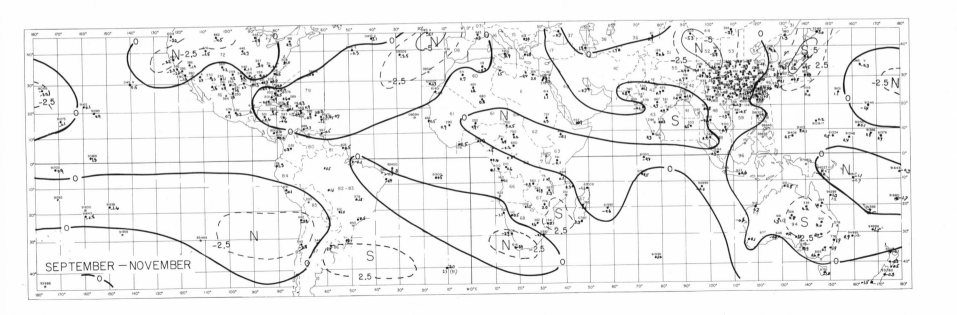

SEPTEMBER — NOVEMBER

113   MEAN TEMPERATURE AND WIND FIELDS

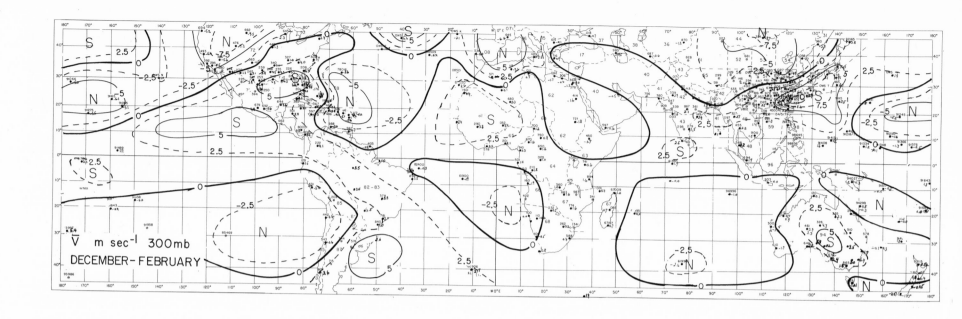

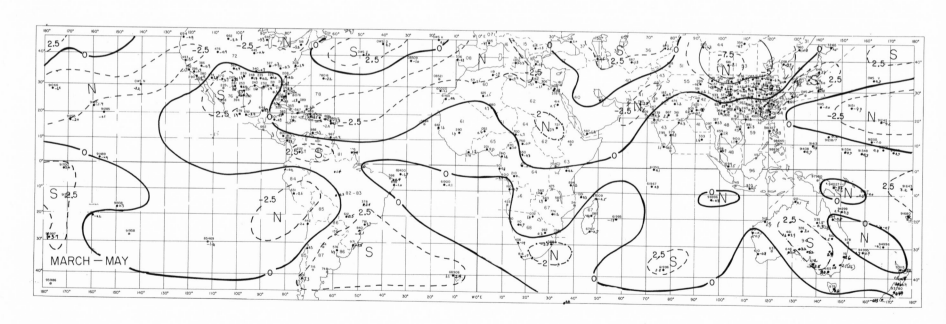

Plate 3.27

114   MEAN TEMPERATURE AND WIND FIELDS

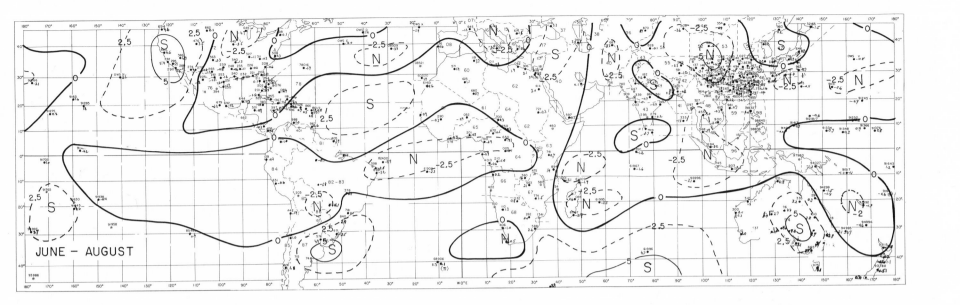

JUNE — AUGUST

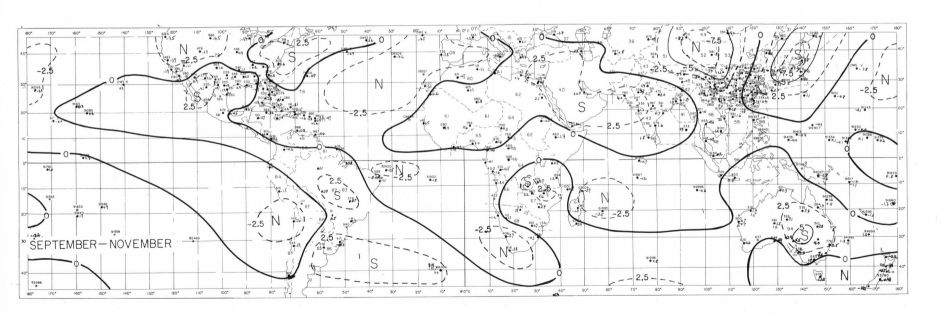

SEPTEMBER—NOVEMBER

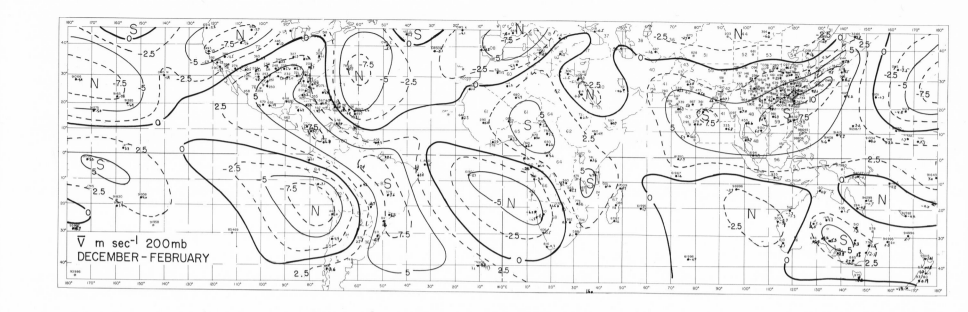

$\overline{V}$ m sec⁻¹ 200mb
DECEMBER – FEBRUARY

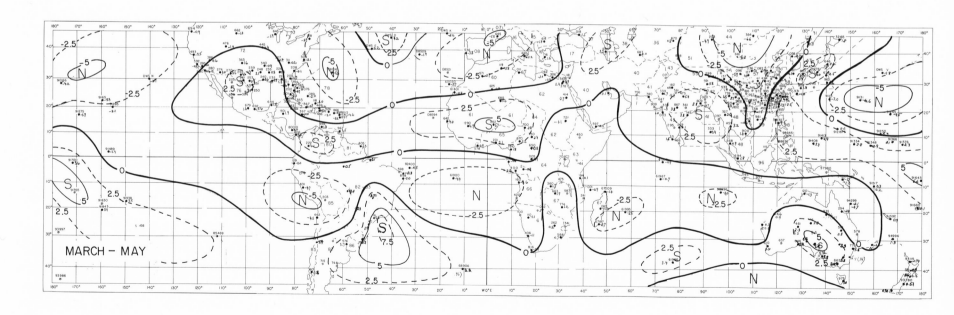

MARCH – MAY

Plate 3.28

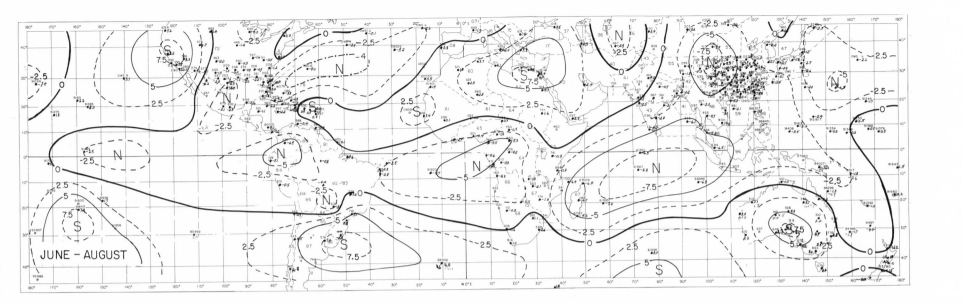

JUNE – AUGUST

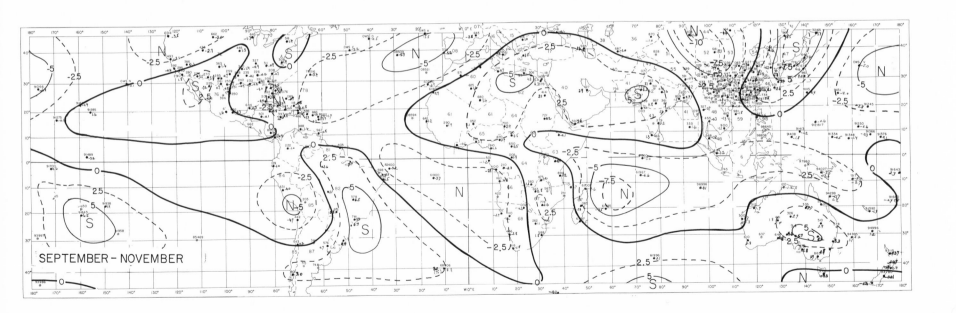

SEPTEMBER – NOVEMBER

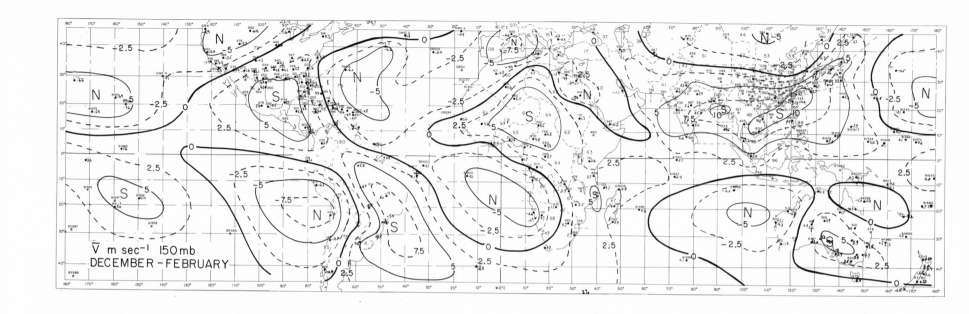

$\overline{V}$ m sec⁻¹ 150 mb
DECEMBER – FEBRUARY

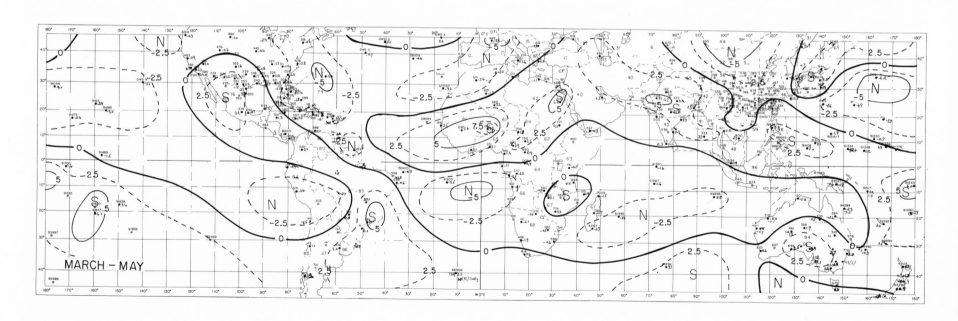

MARCH – MAY

Plate 3.29

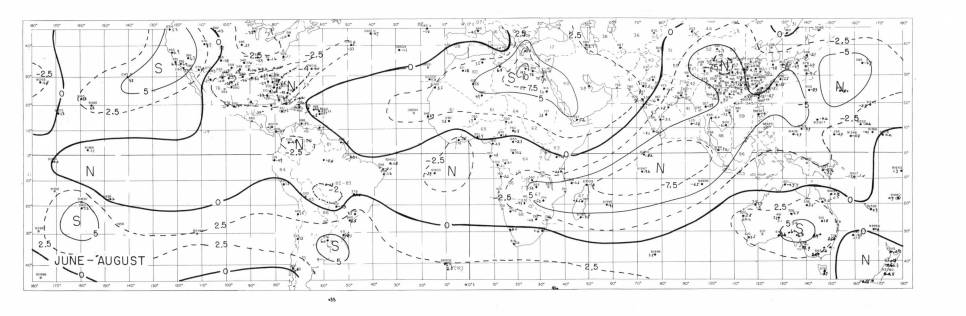

JUNE - AUGUST

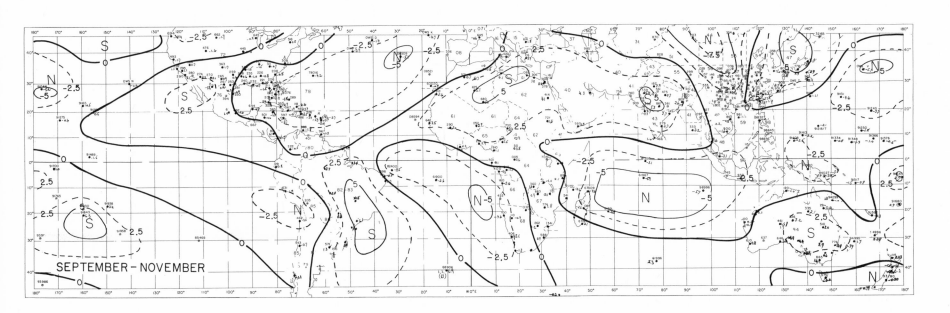

SEPTEMBER - NOVEMBER

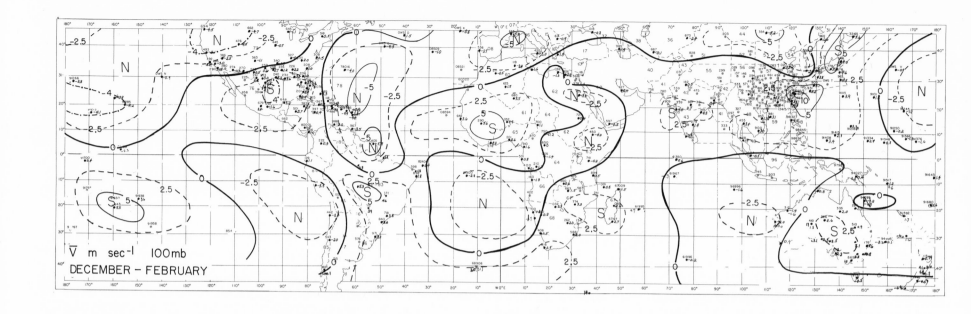

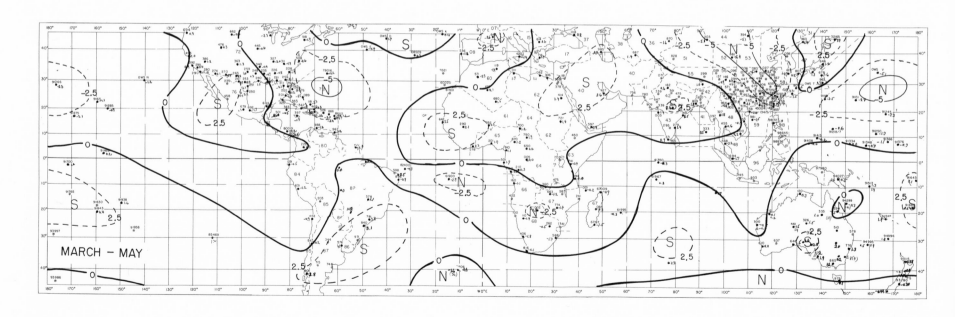

Plate 3.30

120   MEAN TEMPERATURE AND WIND FIELDS

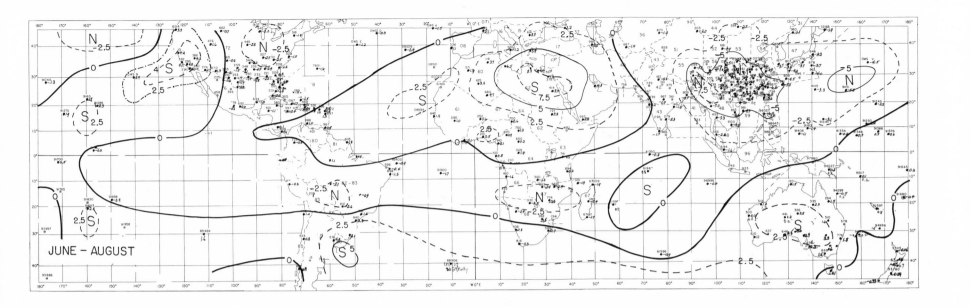

JUNE – AUGUST

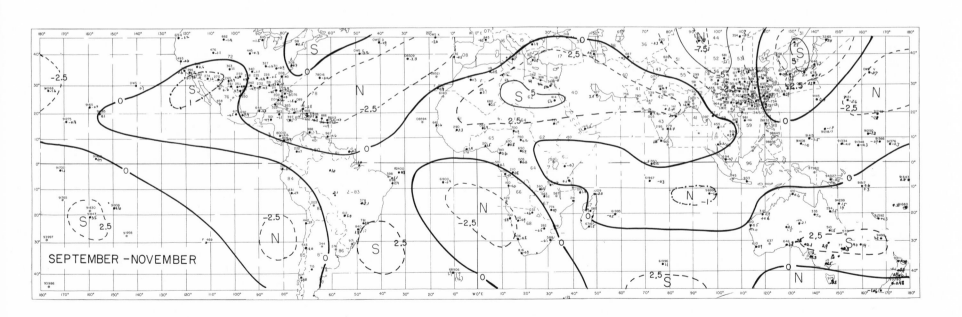

SEPTEMBER – NOVEMBER

STREAMLINES 850mb DEC–FEB

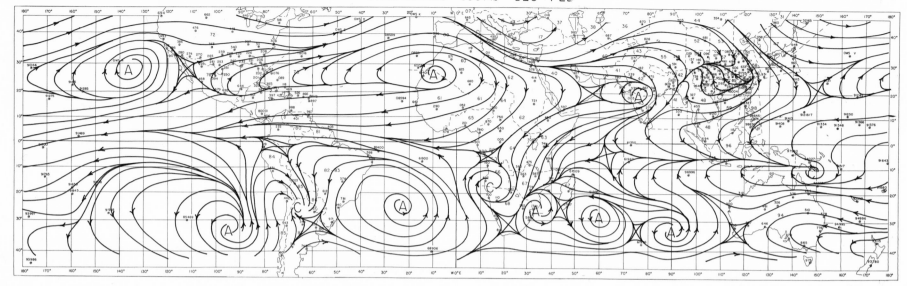

STREAMLINES 200mb — DECEMBER – FEBRUARY

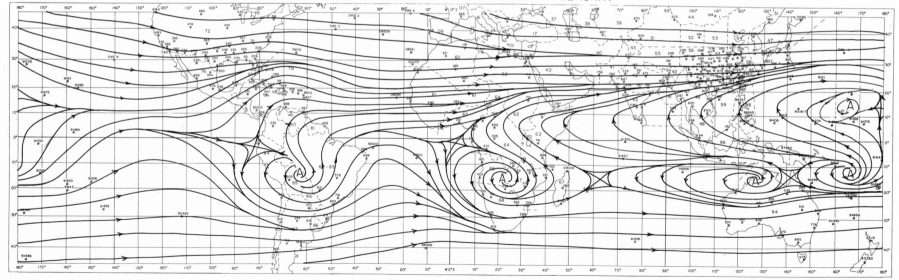

Plate 3.31

STREAMLINES 850mb JUN–AUG

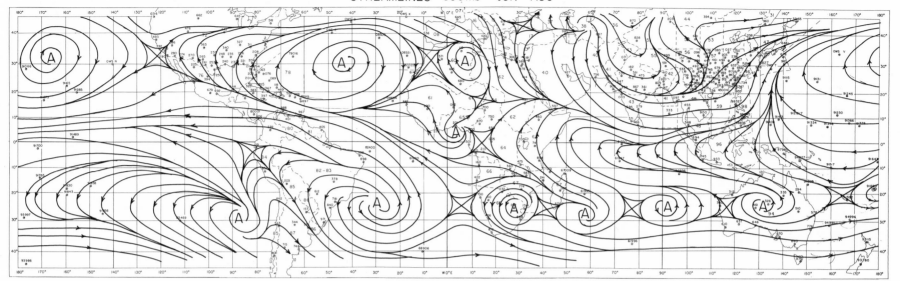

STREAMLINES 200mb JUN–AUG

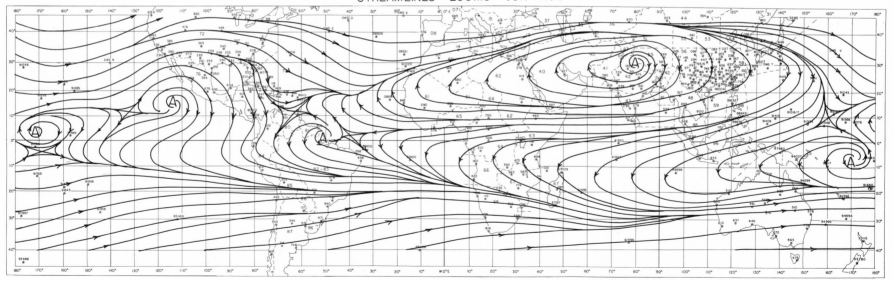

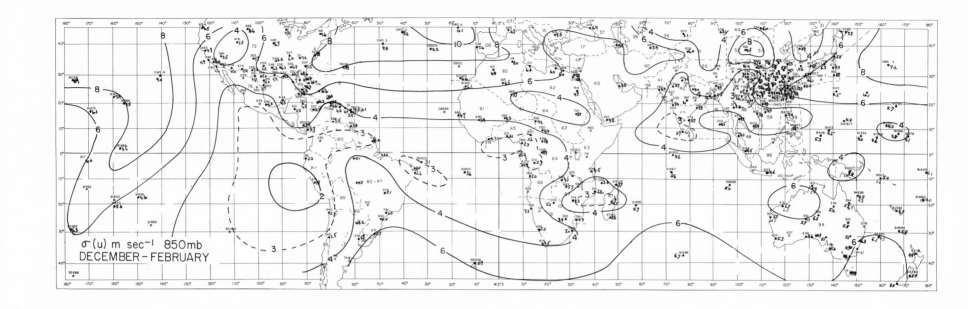

σ(u) m sec⁻¹ 850mb
DECEMBER-FEBRUARY

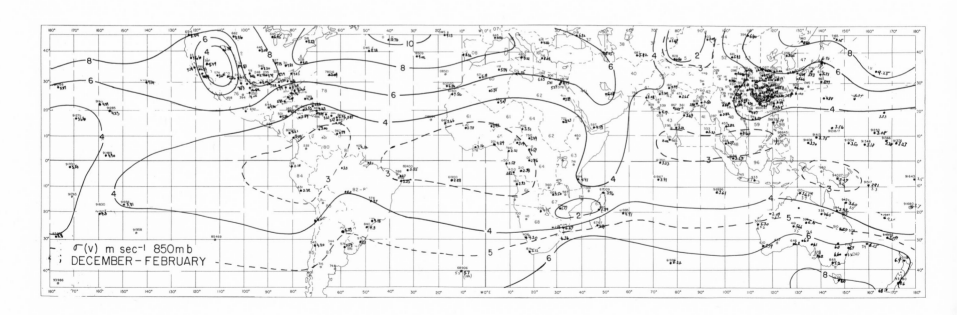

σ(v) m sec⁻¹ 850mb
DECEMBER-FEBRUARY

Plate 3.32

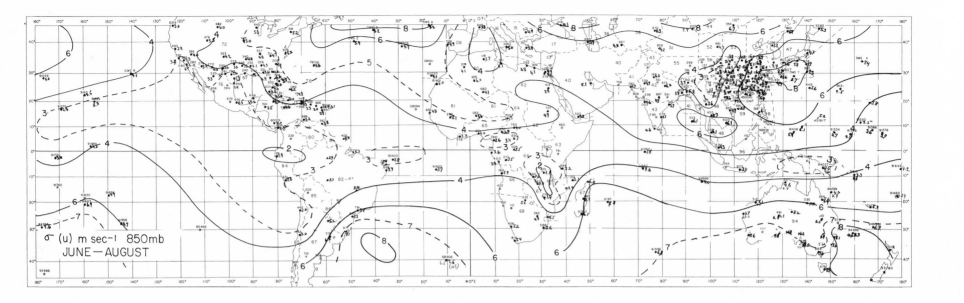

σ (u) m sec⁻¹ 850mb
JUNE—AUGUST

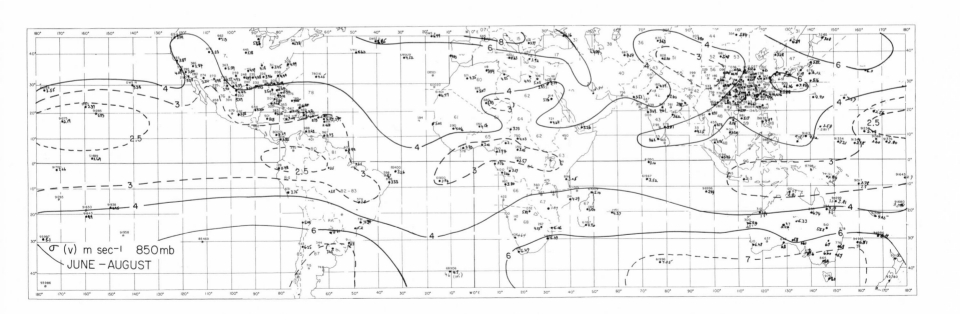

σ (v) m sec⁻¹ 850mb
JUNE—AUGUST

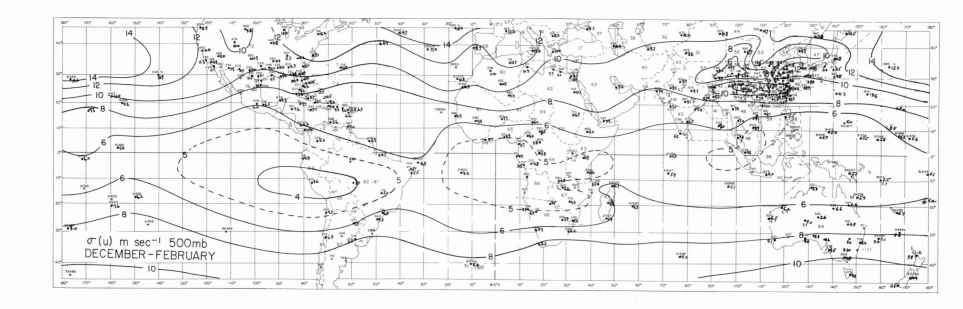

σ(u) m sec⁻¹ 500mb
DECEMBER–FEBRUARY

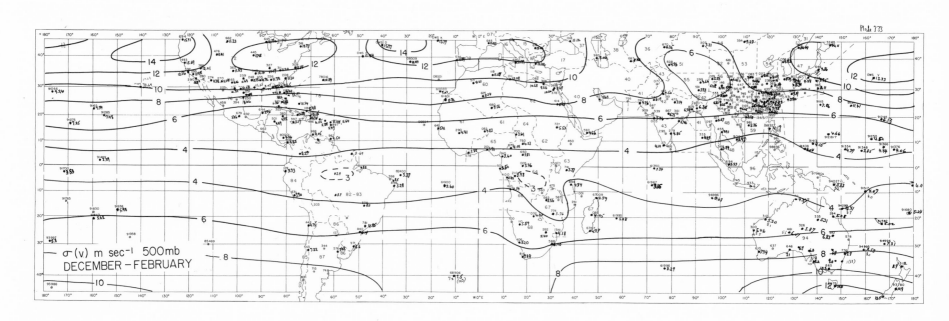

σ(v) m sec⁻¹ 500mb
DECEMBER–FEBRUARY

Plate 3.33

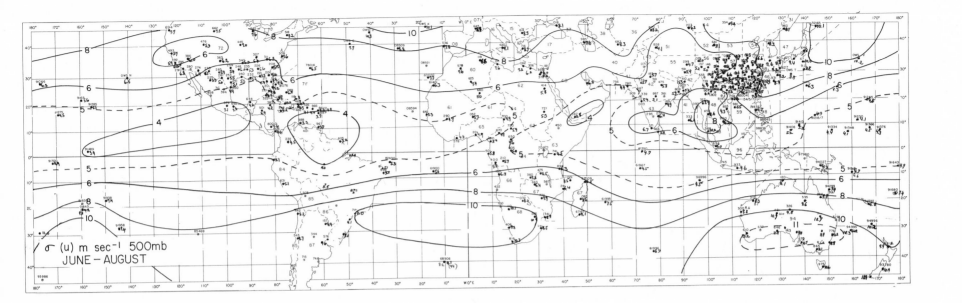

σ (u) m sec⁻¹ 500mb
JUNE−AUGUST

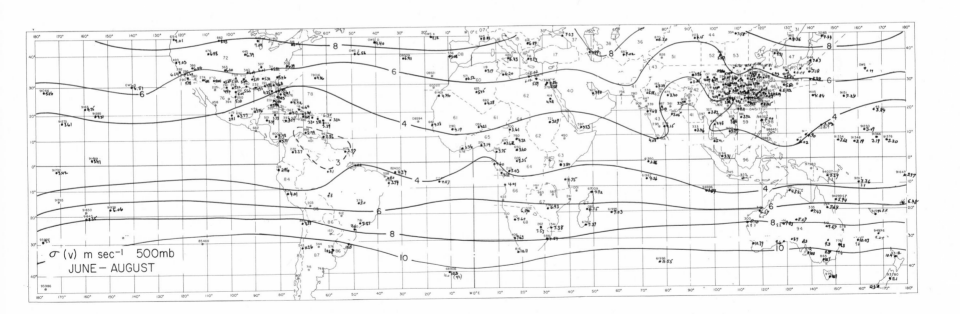

σ (v) m sec⁻¹ 500mb
JUNE−AUGUST

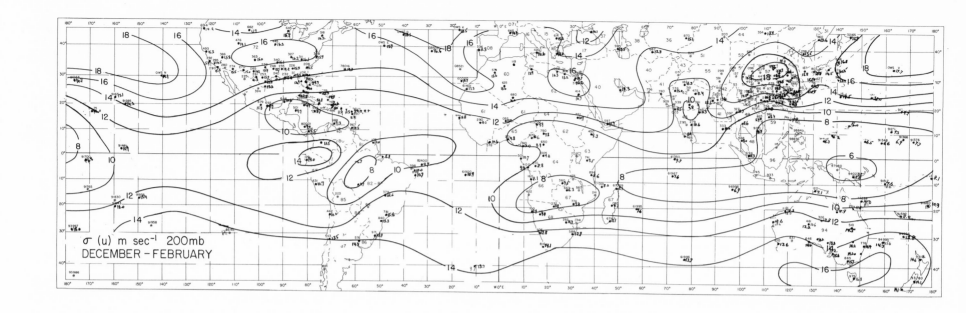

σ (u) m sec⁻¹ 200mb
DECEMBER – FEBRUARY

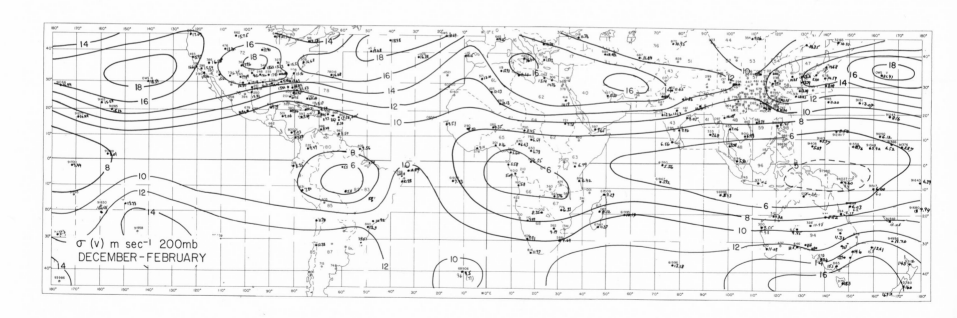

σ (v) m sec⁻¹ 200mb
DECEMBER – FEBRUARY

Plate 3.34

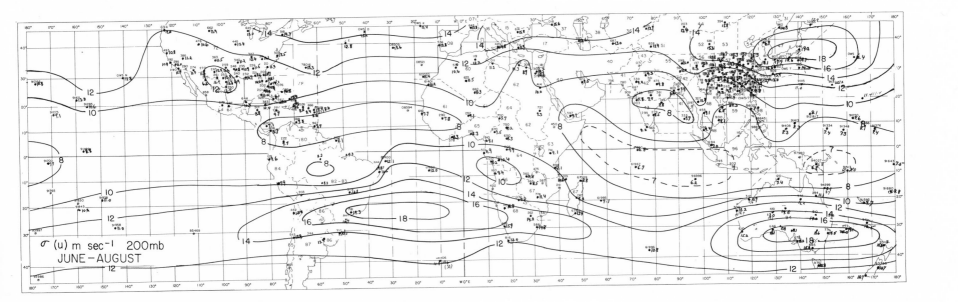

σ (u) m sec⁻¹   200mb
JUNE–AUGUST

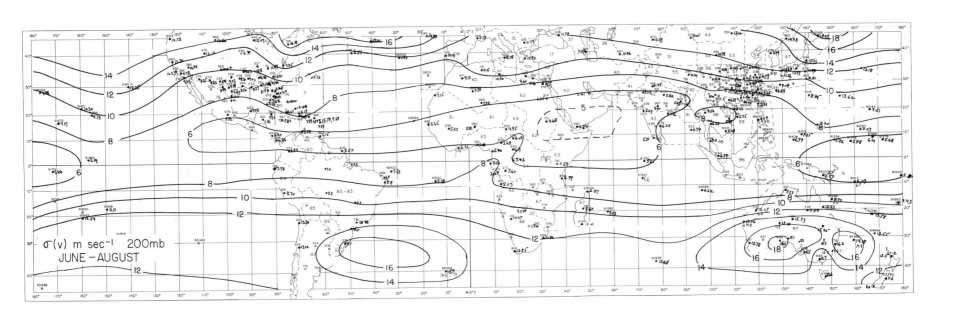

σ (v) m sec⁻¹   200mb
JUNE–AUGUST

References

Buch, H. 1954. *Hemispheric Wind Conditions during the Year 1950*. Final Report, part 2. Cambridge: Massachusetts Institute of Technology, Department of Meteorology, Planetary Circulations Project, Contract no. AF19(122)–153. 126 pp.

Charney, J. G. 1969. A further note on large-scale motion in the tropics. *J. Atmos. Sci.* 26:182–185.

Crutcher, H. L. 1959. *Upper Wind Statistics Charts of the Northern Hemisphere (850, 700 and 500 mb levels)*. Office of the Chief of Naval Operations, Navair 50–1C–535, vols. I and II. Washington, D.C.: Government Printing Office.

————. 1961. Meridional cross-sections of upper winds over the Northern Hemisphere. Technical Paper no. 41. Washington, D.C.: U.S. Department of Commerce, National Weather Records Center, U.S. Weather Bureau. 307 pp.

————. 1966. *Components of the 1000 mb Winds (or Surface Winds) of the Northern Hemisphere*. Navair 50–1C–51, published by direction of the Chief of Naval Operations, Washington, D.C.: U.S. Government Printing Office,

Ferrel, W. 1859. The motions of fluids and solids relative to the Earth's surface, *Math. Mon.* 1:140, 210, 300, 366, 397.

Findlater, J. 1969. A major low level air current near the Indian Ocean during the northern summer. *Quart. J. Roy. Meteorol. Soc.* 95:362–380.

Goldie, N., J. G. Moore, and E. E. Austin. 1958. Upper air temperature over the world. *Geophys. Mem.* vol. 13, no. 101. London: HMSO. 101 pp.

Gordon, A. H. 1953. Seasonal changes in the mean pressure distribution over the world and some inferences about the general circulation. *Bull. Amer. Meteorol. Soc.* 34:357–367.

Guterman, E. G. 1967, 1970. *Atlas of Wind Characteristics in the Southern Hemisphere*, vols. 1 and 2. Moscow: Institute of Aeroclimatology.

Haurwitz, B., and J. M. Austin. 1944. *Climatology*. New York: McGraw-Hill. 400 pp.

Heastie, H., and P. M. Stephenson. 1960. Upper winds over the world. *Geophys. Mem.* vol. 13, no. 103. London: HMSO. 217 pp.

Hydrographic Office, U. S. Navy. 1944. *World Atlas of Sea Surface Temperatures*, 2nd edition. Washington, D.C.: U.S. Hydrographic Office, H.O. pub. no. 225, 1954 (reprint).

Kendrew, W. G. 1961. *The Climates of the Continents*. Oxford: Oxford University Press. 580 pp.

Manabe, S., and J. Smagorinsky. 1967. Simulated climatology of a general circulation model with a hydrological cycle. Part II: Analysis of the tropical atmosphere. *Mon. Weath. Rev.* 95:155–169.

Manabe, S., J. L. Holloway, Jr., and H. M. Stone. 1970. Tropical circulation in a time-integration of a global model of the atmosphere. *J. Atmos. Sci.* 4:580–613.

Obasi, G. O. P. 1963. Atmospheric momentum and energy calculations for the Southern Hemisphere during the IGY. Report no. 6. Cambridge: Massachusetts Institute of Technology, Department of Meteorology, Planetary Circulations Project. 353 pp.

Palmén, E., and L. A. Vuorela. 1963. On the mean meridional circulations in the Northern Hemisphere during the winter season. *Quart. J. Roy. Meteorol. Soc.* 89:131–138.

Petterssen, S. 1956. *Weather Analysis and Forecasting*, Vol. I. New York: McGraw-Hill. 422 pp.

Riehl, H. 1954. *Tropical Meteorology*. New York: McGraw-Hill. 392 pp.

Thompson, B. W. 1965. *The Climate of Africa*. Nairobi: Oxford University Press. 132 pp.

Tucker, G. B. 1959. Mean meridional circulations in the atmosphere. *Quart. J. Roy. Meteorol. Soc.* 85:209–224.

U.S. Department of Commerce, 1967. *World Weather Records, 1951–1960*. Washington, D.C.: Government Printing Office. Vols. 1–6.

Vuorela, L. A., and I. Tuominen. 1964. On the mean zonal and meridional circulations and the flux of moisture in the Northern Hemisphere during the summer season. *Pure and Appl. Geophys.* 57:167–180.

Wellington, J. H. 1955. *Southern Africa: A Geographical Study*. Vol. I, *Physical Geography*. Cambridge: Cambridge University Press. 508 pp.

4

The Angular Momentum Budget and the Maintenance of the Zonal Wind

4.1 Introduction

The absolute angular momentum, taken about the earth's axis, of a unit mass of air is given by

$$M = (\Omega r \cos \varphi + u) r \cos \varphi. \tag{4.1}$$

The symbols have their usual meteorological meanings (Table 1.1). The term $\Omega r^2 \cos^2 \varphi$, usually called the Ω-*momentum*, is the angular momentum which would be present if the atmosphere were in solid rotation with the angular velocity of the earth. The term $ur \cos \varphi$ is called the *relative angular momentum*, since it depends on the zonal wind speed u relative to the surface of the earth. The variation of r with height in the atmosphere has little effect on the value of M. It is usually replaced by a constant equal to the radius of the earth. Such a substitution is necessary, in the pressure system of coordinates, to preserve the principles of the conservation of angular momentum and of energy (Lorenz 1967, pp. 17–18).

The principle of the conservation of angular momentum arises from Newton's second law of motion, and states that the angular momentum of a system will change only if external torques are applied. For the earth-atmosphere system, taken as a whole, the only appreciable external torques are the gravitational tidal torques. These torques are of no significance on the time scales which are of interest here. If the atmosphere alone is considered, it is subject to large torques exerted through its interaction with the surface of the earth.

4.2 The Angular Momentum Equation

The angular momentum equation for the atmosphere, in spherical and pressure coordinates, is

$$\frac{\partial M}{\partial t} + \frac{1}{a \cos \varphi} \left(\frac{\partial Mu}{\partial \lambda} + \frac{\partial}{\partial \varphi} Mv \cos \varphi \right)$$

$$+ \frac{\partial M\omega}{\partial p} = -\frac{\partial \phi}{\partial \lambda} + a \cos \varphi \, F_\lambda. \tag{4.2}$$

Here r has been replaced by the radius of the earth in the expression for M, i.e.,

$$M = (\Omega a \cos \varphi + u) a \cos \varphi.$$

131

If we make the reasonable assumption that molecular and turbulent effects act primarily to transfer horizontal momentum in the vertical direction, the frictional term in equation (4.2) can be rewritten to give

$$\frac{\partial M}{\partial t} + \frac{1}{a \cos \varphi} \left(\frac{\partial Mu}{\partial \lambda} + \frac{\partial}{\partial \varphi} Mv \cos \varphi \right)$$
$$+ \frac{\partial}{\partial p} (M\omega + ga \cos \varphi \, \tau_\lambda) = -\frac{\partial \phi}{\partial \lambda}. \tag{4.3}$$

The equation serves to define the components of the horizontal and quasivertical angular momentum flux densities as

$$J_\lambda = Mu$$
$$J_\varphi = Mv$$
$$J_p = M\omega + ga \cos \varphi \, \tau_\lambda. \tag{4.4}$$

A general formulation of the angular momentum balance of the general circulation of the atmosphere would be afforded by the time-averaged form of equation (4.3),

$$\frac{\overline{\partial M}}{\partial t} + \frac{1}{a \cos \varphi} \left(\frac{\partial \overline{J_\lambda}}{\partial \lambda} + \frac{\partial}{\partial \varphi} \overline{J_\varphi} \cos \varphi \right) + \frac{\partial \overline{J_p}}{\partial p} = -\frac{\overline{\partial \phi}}{\partial \lambda}, \tag{4.5}$$

where

$$\overline{J_\lambda} = a^2 \cos^2 \varphi \, \Omega \overline{u} + a \cos \varphi (\overline{u}\,\overline{u} + \overline{u'u'})$$
$$\overline{J_\varphi} = a^2 \cos^2 \varphi \, \Omega \overline{v} + a \cos \varphi (\overline{u}\,\overline{v} + \overline{u'v'})$$
$$\overline{J_p} = a^2 \cos^2 \varphi \, \Omega \overline{\omega} + a \cos \varphi (\overline{u}\,\overline{\omega} + \overline{u'\omega'}) + ga \cos \varphi \, \overline{\tau}_\lambda. \tag{4.6}$$

The flux terms have been decomposed into expressions representing the flux of Ω-momentum by the time-averaged motion, the flux of relative angular momentum by the mean motions, and the flux of relative angular momentum due to eddy motion.

An evaluation of these equations would require a knowledge of the terms at grid points over the globe and at various pressure levels. The difficulties in obtaining reasonable values for these terms, together with the zonal nature of the atmospheric circulation, has led most investigators to the zonally averaged form of equation (4.5) as

$$\frac{\overline{\partial [M]}}{\partial t} + \frac{1}{a \cos \varphi} \frac{\partial}{\partial \varphi} [\overline{J_\varphi}] \cos \varphi + \frac{\partial}{\partial p} [\overline{J_p}] = -\left[\frac{\overline{\partial \phi}}{\partial \lambda}\right]. \tag{4.7}$$

Here the flux terms are expanded in the form

$$[\overline{J_\varphi}] = a^2 \cos^2 \varphi \, \Omega[\overline{v}] + a \cos \varphi ([\overline{u}][\overline{v}] + [\overline{u}^*\overline{v}^*] + [\overline{u'v'}])$$
$$[\overline{J_p}] = a^2 \cos^2 \varphi \, \Omega[\overline{\omega}] + a \cos \varphi ([\overline{u}][\overline{\omega}] + [\overline{u}^*\overline{\omega}^*] + [\overline{u'\omega'}] + g[\overline{\tau}_\lambda]), \tag{4.8}$$

which represent the flux of Ω-momentum by the temporally and longitudinally averaged motions, and the flux of relative angular momentum by the mean meridional motion, the standing and transient eddies, and the "frictional" stresses.

We have mentioned that the only appreciable external torques acting on the atmosphere occur at the earth's surface. Integrating equations (4.5) or (4.7) over the mass of the atmosphere enclosed by the latitudes φ_1, φ_2, the earth's surface, and the "top" of the atmosphere gives $(dm = a^2 \cos \varphi \, d\lambda \, d\varphi \, dp/g)$

$$\frac{\overline{\partial}}{\partial t} \int M \, dm + 2\pi a \int_0^{p_0} \{[\overline{J_\varphi}(\varphi_2)] \cos \varphi_2 - [\overline{J_\varphi}(\varphi_1)] \cos \varphi_1\} \frac{dp}{g}$$
$$+ 2\pi a^3 \int_{\varphi_1}^{\varphi_2} \cos^2 \varphi [\overline{\tau}_\lambda(p_0)] \, d\varphi$$
$$= -a^2 \int_0^{2\pi} \int_{\varphi_1}^{\varphi_2} \int_0^{p_0} \frac{\overline{\partial \phi}}{\partial \lambda} \cos \varphi \, d\lambda \, d\varphi \, \frac{dp}{g}. \tag{4.9}$$

This equation asserts that the change of angular momentum within the region is a consequence of the flux of angular momentum across the λ, p surfaces at latitudes φ_1 and φ_2, the torque exerted on the atmosphere by frictional effects at the surface ($p = p_0$), and the so-called mountain torque.

In tropical regions, the surface winds blow from east to west. The frictional interaction with the earth's surface therefore exerts a torque on the earth in the negative direction (i.e., in the direction opposite the earth's rotation which we have taken as the positive direction in our definition of angular momentum). Conversely, the earth must be exerting a positive torque on the atmosphere, and thus the tropical region is a source of angular momentum for the atmosphere. It is contrary to observation that the angular momentum of the tropical region increases continually; there must be a flux of angular momentum horizontally out of the region. Similarly in middle latitudes, where westerly winds prevail, the earth exerts a negative torque on the atmosphere and acts as a sink for angular momentum. By the same reasoning, a horizontal angular momentum flux into this region must exist. On a broad scale, therefore, the atmosphere must act to transfer angular momentum from tropical to middle latitudes.

If the integration of equation (4.9) is taken over the entire mass of the atmosphere,

$$\overline{\frac{\partial}{\partial t} \int M \, dm} + 2\pi a^3 \int_{-\pi/2}^{\pi/2} \cos^2 \varphi [\overline{\tau}_\lambda(p_0)] \, d\varphi$$

$$= -a^2 \int_0^{2\pi} \int_{-\pi/2}^{\pi/2} \int_0^{p_0} \frac{\partial \overline{\phi}}{\partial \lambda} \cos \varphi \, d\lambda \, d\varphi \, \frac{dp}{g}.$$

On the long term the angular momentum of the atmosphere does not change, and the equation reduces to the condition that the external torques, through surface friction and the mountain torque terms, must be zero. For a nonmountainous earth, a condition on the momentum balance of the atmosphere requires both surface easterly and westerly winds to exist.

One of our goals in this chapter will be to investigate the nature of the vertical and horizontal fluxes of angular momentum and of the external torques acting on the atmosphere through the surface stress and the mountain torque terms.

4.3 Horizontal Fluxes of Angular Momentum

As has been mentioned, the tropical regions, where surface easterlies predominate, are the source regions for atmospheric angular momentum while middle latitudes, where surface westerlies prevail, are sink regions. The atmosphere acts so as to balance these external torques. This requires a vertical flux of angular momentum from the surface layers in the tropical source region, a horizontal poleward flux of angular momentum to middle latitudes, and a subsequent downward flux of angular momentum to the surface layers where angular momentum is removed from the atmosphere.

The poleward transport of angular momentum by the atmosphere has been a topic of interest to meteorologists for many years. Lorenz (1967) gives a very interesting account of the early theories of the general circulation. Early workers ascribed this transport of momentum

to mean meridional motions, and it was not until 1926 that Jeffreys suggested that the transport could be dependent on large-scale atmospheric eddies. These have their representation on surface maps as closed areas of high and low pressure, and on upper air maps as "waves" in the flow. Previously these eddies had been considered as perturbations on the mean flow which had little importance in the large-scale general circulation of the atmosphere.

After World War II it was realized that observations were available to test Jeffreys' hypothesis directly. Several groups began to study the problem, notably Priestley in Australia (Priestley 1948, 1949, 1950, 1951a, b); Starr's group at M.I.T. (Starr 1948; Widger 1949; Starr and White 1952, 1954; Buch 1954; this work has recently been summarized by Lorenz 1967 and Starr 1968); and Bjerknes and Mintz at UCLA (Bjerknes 1948, 1955; Mintz 1951).

The results of such studies proved Jeffreys to be correct in his assertion that the eddies were an integral part of the general circulation. This is particularly true at higher latitudes where mean meridional motions are small. In equatorial regions where the Hadley circulation is relatively strong, mean motions play an important part in the transport of angular momentum.

Our main interest is in the meridional flux $\overline{J}_\varphi$ of the angular momentum. As given in equation (4.6), it may be written as

$$\overline{J}_\varphi = a^2 \cos^2 \varphi \, \Omega \overline{v} + a \cos \varphi (\overline{u}\overline{v} + \overline{u'v'}).$$

The data necessary to evaluate this term in the tropical regions have been given in map form in Chapter 3 with the exception of the term $\overline{u'v'}$.

This term has been evaluated from the raw data, and maps of $\overline{u'v'}$ are given in Plates 4.1–4.9. It should be noted that this eddy flux term is obtained from daily values of u and v at rather widely spaced stations. It does not, therefore, resolve short-period and small-scale motions. These submeasurement scales of motions can be assigned to a "friction" term which would then represent molecular and "turbulent" momentum fluxes. The contribution to the horizontal flux by such a friction term is thought to be much smaller than that due to the other terms so that it can be neglected for our purposes.

Maps of $\overline{u'v'}$ at the surface and the 850, 700, 500, 400, 300, 200, 150, and 100-mb levels are given for four seasons in Plates 4.1–4.9. Considering the fact that $\overline{u'v'}$ is a quadratic function involving the meridional wind component, the maps show reasonably coherent patterns. The clearest patterns occur at the highest levels, where magnitudes are largest.

Maps of $\overline{u'v'}$ at the surface are difficult to analyze due to the patchy nature of the patterns. This no doubt reflects the effect of many local influences. The amplitudes of the fluxes are generally one-half or less of those appearing at 850 mb.

At 850 mb, there is still considerable patchiness in the patterns, although several notable features are apparent such as the persistent southward flux over Australia, South Africa, and South America. In June–August, there is a tendency for this negative flux to occur on the western side of these continents and a less negative or positive flux to occur on the eastern sides of the continents. It appears to be associated with the wave pattern in the mean winds which was noted in the map discussion of Chapter 3. This suggests that this eddy flux contribution may be associated, to some extent, with temporal oscillations of the large-scale planetary wave motion.

The less detailed structure of the eddy flux patterns over the oceanic areas is more than likely a consequence of the relatively sparse data in these regions as compared to continental areas. Broadly speaking, the fluxes are poleward in each hemisphere, although the direction of the flux is by no means a simple function of latitude. The reasons for the rather large longitudinal variations are not clear.

At 700 mb there is a region of northward flux over North Africa and a region of southward flux over western China in all seasons. The southward flux over South Africa and Australia has increased with altitude, particularly in June–August. Considerable seasonal variation is apparent, with larger fluxes occurring in the middle-latitude regions of the winter hemisphere.

At 500 mb the magnitudes of the fluxes are larger, and the patterns are considerably smoother and more zonal in character. The seasonal variation in the magnitudes of the fluxes is notable. In general the fluxes are poleward in each hemisphere, with the exception of the region over Asia, which exhibits persistent negative fluxes.

The patterns at 400 and 300 mb are similar to those at 500 mb, although peak flux values are often almost double those at the lower level. In June–August there is evidence of a net flux from the winter to the summer hemisphere.

At 200 mb, the pattern of alternating fluxes noted at lower levels in the Southern Hemisphere is again apparent, together with a similar pattern in the Northern Hemisphere. Maxima of the fluxes occur over continental regions, and the pattern is clearest in the winter hemisphere. The interhemispheric flux is a prominent feature of the maps at this level. The net flux of momentum is from the winter and spring to the summer and autumn hemispheres.

At 150 mb, cross-equatorial fluxes are large and show the same seasonal variation as those at 200 mb. There is a large region of southward flux over the Red Sea in September–November and in March–May.

The pattern at 100 mb is essentially a scaled-down version of that found at 150 mb. There are some regions where the analysis is uncertain (e.g., over China) because of the limited numbers of observations.

Cross sections at 10-degree latitude intervals from 40°N to 40°S are given in Figures 4.1 and 4.2 for the December–February and June–August seasons, and at the equator for the equinoxes in Figures 4.1 and 4.2. The following points are of note:

(a) The transient eddy momentum flux shows marked longitudinal variations at each latitude. Poleward of 10°, the fluxes are predominately poleward in direction; they attain their maximum values in the 300–200-mb layer and are strongest in the winter season. The pattern of fluxes in the Southern Hemisphere is smoother and more regular than that of the Northern Hemisphere, and at 30°S, for instance, shows three maxima apparently associated with the continents.

(b) The region within 10 degrees of the equator, unlike more poleward regions, shows momentum fluxes which change direction with season while retaining remarkably similar patterns and magnitudes. The regions of maximum flux are located at higher elevations, about 200–150 mb, than those of higher latitudes. The flux is directed from the winter hemisphere to the summer hemisphere predominantly.

Cross sections of $\overline{u'v'}$ along the 150°E, 150°W, 80°W, 20°E, and

80°E meridians of longitude are given in Figure 4.3 for December–February and in Figure 4.4 for June–August. These cross sections clearly show the poleward flux of momentum at more poleward latitudes, as well as the equatorial maxima which occur at somewhat greater elevations. The equatorial maximum (or maxima) of the flux appears as a distinct feature of the cross sections, and suggests the operation of a markedly different process of momentum transport in this region. The equatorial flux of momentum is generally from the winter to the summer hemisphere, but as at 80°W in December–February, it may be in the opposite direction locally or, as at 20°E, show two maxima of opposite sign.

A comparison of these cross sections with those of $\bar{u}$ clearly shows that the equatorial momentum flux is primarily down the angular momentum gradient—as opposed to the countergradient flux at middle latitudes. A comparison of the maps of $\overline{u'v'}$ and of $\bar{u}$ at 150 mb (Plates 3.20 and 4.8) shows that this down-gradient flux is a prominent feature in the equatorial region at most longitudes. As we commented earlier (Newell et al. 1970), this flux may be a phenomenon predicted by Eady (1950) which he termed "quasibarotropic driven turbulence."

Cross sections of the meridional flux $\overline{J_\varphi}$ around latitude circles may also be obtained. The longitudinal distribution of the Ω-momentum flux is proportional to $\bar{v}$ as given in the cross sections of Chapter 3. This term will be larger, in general, than the remaining terms. The net flux across a latitude circle due to the transport of Ω-momentum will be zero as $g^{-1} \int_0^{2\pi} \int_0^{p_0} va \cos \varphi \, d\lambda \, dp = 0$, that is there is no net mass flux across the latitude circle in a season. The remaining terms in $\overline{J_\varphi}$, which represent the flux of relative angular momentum, do not have such integral constraints.

The flux of relative angular momentum due to the mean motions is proportional to $\bar{u}\bar{v}$. Cross sections of this term at 20°N, 0°, and 20°S are given in Figure 4.5 for the December–February and June–August seasons. The cross sections show that

(a) The relative angular momentum flux by the mean motions, like that by the transient eddies, is concentrated primarily in the upper troposphere and shows marked latitudinal and seasonal variations.

(b) The values are generally considerably greater than the corresponding values for the transient eddies, although they show a greater variation in sign around a latitude circle.

(c) At the equator, the flux behaves in a manner very similar to the transient eddy flux. The fluxes change sign in the two seasons in the same sense as the transient eddy flux, although here the largest values are clearly seen in the June–August season. Maximum values are found in the Eastern Hemisphere.

(d) The momentum flux by the mean motion in the monsoon region is very important. The enhanced northerly and easterly motions characteristic of the monsoon circulation result in a very large momentum flux into the Northern Hemisphere.

Another aspect of the atmospheric momentum flux can be considered in terms of the data on a pressure surface. We consider the 200-mb surface for the December–February season. The flux of Ω-momentum on the pressure surface will, in general, be much larger than the flux of relative angular momentum. The Ω-momentum flux across a latitude circle on a pressure surface will have a net contribution which will be balanced by opposing fluxes at other levels so that the zero mass flux condition is satisfied in the vertical. The Ω-momentum fluxes, in the zonal and meridional directions, are simply $a^2 \cos^2 \varphi \, \Omega$ times the values of $\bar{u}$ and $\bar{v}$ given on the maps of Chapter 3. The horizontal streamlines of Ω-momentum will closely resemble the streamline maps of the horizontal flow given in Chapter 3. The relative angular momentum transport by the mean motions and the transient eddies are given in Plate 4.10.

The terms in the meridional flux of relative angular momentum show both positive and negative values. The flux of relative angular momentum by the mean motions is generally poleward in both hemispheres, with the largest values occurring in the Northern Hemisphere. The transequatorial flux is generally negative in the Eastern Hemisphere and positive in the Western Hemisphere at this level.

The meridional flux due to the transient eddies is generally much smaller than that due to the mean motions, except in the vicinity of the equator and in parts of the Southern Hemisphere where values are comparable. The transequatorial flux by the transient eddies is predominantly negative.

Figure 4.1

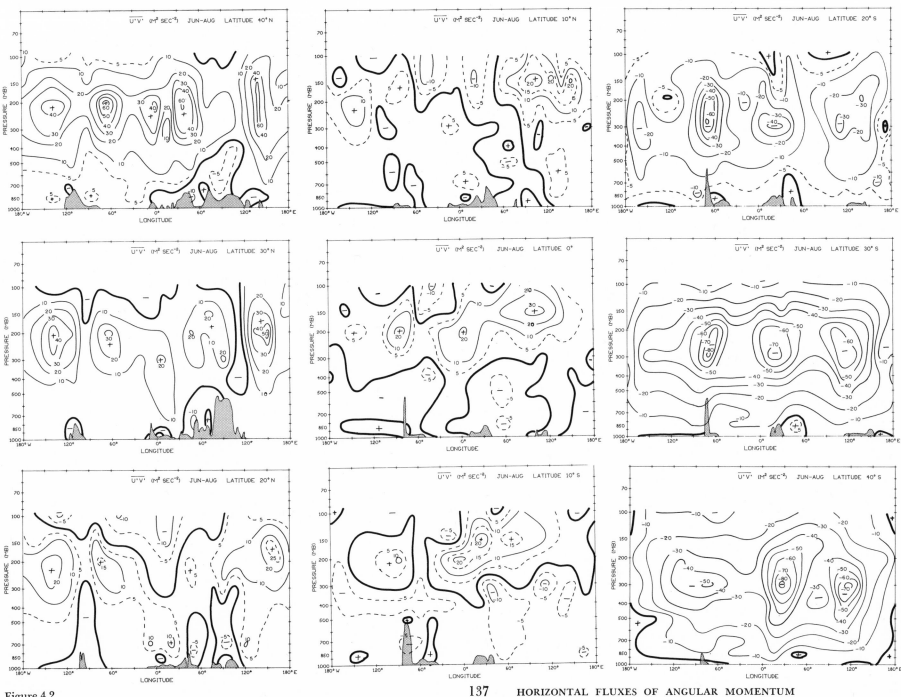

Figure 4.2

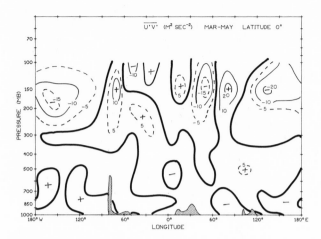

Figure 4.1 (continued)

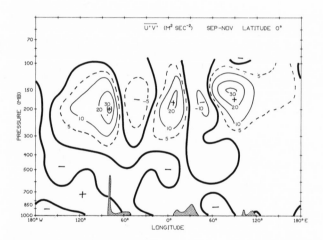

Figure 4.2 (continued)

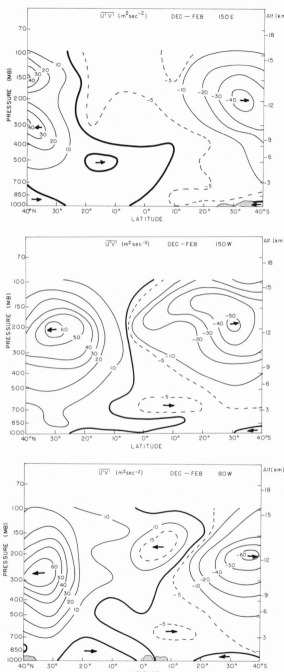

Figure 4.3

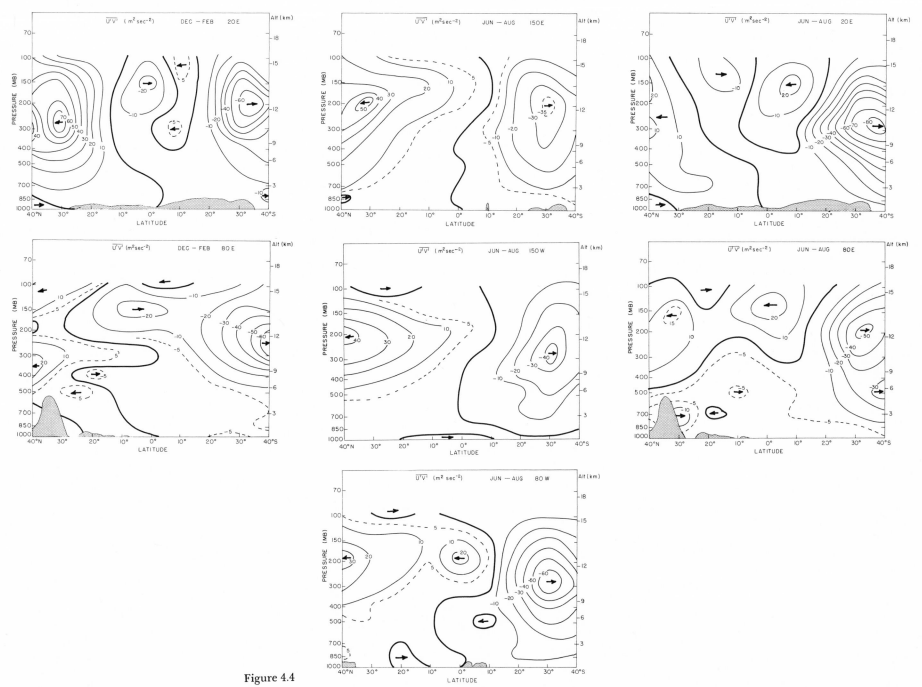

Figure 4.4

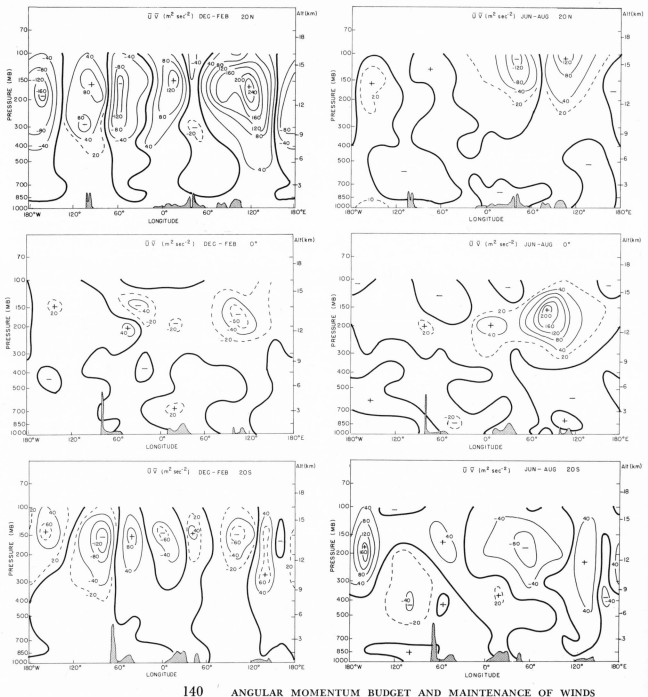

Figure 4.5

The value of

$$\frac{\partial \bar{J}_p}{\partial p} = -\frac{1}{a \cos \varphi} \left(\frac{\partial}{\partial \lambda} \left(\bar{J}_\lambda + a \cos \varphi \, \bar{\phi} \right) + \frac{\partial}{\partial \varphi} \bar{J}_\varphi \cos \varphi \right)$$

at 200 mb for December–February is presented in Plate 4.11. Regions of positive values imply divergence of the vertical flux and hence, in effect, a region of a convergence of the horizontal flux terms. The local time change of total angular momentum has been neglected in this equation. The resulting pattern is controlled basically by the fluxes of Ω-momentum and shows predominantly horizontal divergence in the Southern Hemisphere tropical region and convergence in the Northern Hemisphere. This is in agreement with the expected angular momentum transport associated with the Hadley cell motions.

4.4 The Zonally Averaged Angular Momentum Balance

In the previous section we discussed the horizontal flux of angular momentum by the transient eddies and gave some consideration to the longitudinal variation of the angular momentum flux and to the angular momentum budget of the 200-mb level. A considerably simpler approach to the angular momentum budget of the atmosphere is in terms of the zonally averaged equation. This averages out the longitudinal variations of the terms but permits a simpler two-dimensional treatment of the equation. Such a treatment preserves the "zonal" nature of the atmospheric circulation, but renders interpretation of the processes somewhat less direct. The terms in the equation can be obtained with reasonable accuracy from the usual meteorological data. Terms involving ω are the exception, however, and must be obtained indirectly.

Before considering the terms in equation (4.7) we consider the integrated value of the zonally averaged angular momentum in various regions of the atmosphere.

4.4.1 The Angular Momentum Content of the Atmosphere

The time-averaged angular momentum content of the "latitude belt" bounded by the latitudes φ_1, φ_2 is given by

$$\int \overline{M} \, dm = \int a^2 \cos^2 \varphi \, \Omega \, dm + \int a \cos \varphi \, \bar{u} \, dm.$$

As the mass contained in a region does not vary appreciably (as measured by the mean surface pressure), it is the relative angular momentum which is expected to govern any changes in the angular momentum content of the belt (Priestley 1951b). The relative angular momentum content of the region may be obtained as

$$\frac{2\pi a^3}{g} \int_0^{p_0} \int_{\varphi_1}^{\varphi_2} [\bar{u}] \cos^2 \varphi \, d\varphi \, dp,$$

and this term may be evaluated from the data given in Table 3.1. The results are given in Table 4.1 for the four seasons of this study. The table shows that

(a) The values of the integrated relative angular momentum have the characteristic pattern of positive angular momentum in middle latitudes, associated with westerly winds, and negative angular momentum in tropical regions, associated with easterly winds. There is a clear seasonal trend similar to that of the winds, that is, a maximum of westerly momentum in the winter hemisphere which is less intense and situated farther poleward in the summer hemisphere.

(b) The total angular momentum content of the atmosphere and the values for each hemisphere vary markedly. The maximum of westerly momentum for the globe occurs in the December–February season, in conjunction with strong Northern Hemisphere westerlies. The minimum value occurs in June–August in conjunction with relatively strong westerlies in the Southern Hemisphere which are offset by the strong region of easterly momentum in the Northern Hemisphere. This region of strong easterly momentum is sufficient to give a negative value for the hemispheric angular momentum content in this season. This is a reflection of the strong "monsoon" easterlies which occur at this time. The relative angular momentum content of the two hemispheres is comparable in most seasons, but it shows a strong imbalance, with a value of about 35 to 1, in the June–August season. Similar results have been obtained by Priestley (1948, 1951b) for January and July; by Lamb (1959), who found a ratio of 25; and by Kidson and Newell (1969), who obtained a ratio of 16 between the hemispheres in this season.

(c) It is interesting to note that if the atmosphere gains or loses

Table 4.1. Relative angular momentum content of 10° latitude belts. (Units 10^{30} gm cm^2 sec^{-1})

Lat.	Dec–Feb	Mar–May	June–Aug	Sept–Nov
80°N	2	2	2	2
70	17	13	8	15
60	58	38	30	53
50	141	94	90	139
40	252	198	121	181
30	354	264	29	140
20	209	150	−98	4
10	−23	−31	−145	−89
0	−63	−77	−126	−85
10	−43	−33	−32	−33
20	34	107	200	143
30	145	193	292	225
40	184	208	243	213
50	153	178	190	186
60	53	78	95	89
70	7	19	22	21
80°S	1	3	3	2
Total N.H.	978	690	−26	402
S.H.	502	715	950	803
Globe	1480	1405	924	1205
Ratio $\frac{\text{S.H.}}{\text{N.H.}}$	0.5	1.0	36.5	2.0

angular momentum, the earth must lose or gain the corresponding amount of angular momentum. This may supposedly be accomplished by a gain or loss of angular momentum by the oceanic circulation, by a change in the rate of rotation or the shape of the earth, or by a combination of similar factors (see Munk and MacDonald 1960, for example).

4.4.2 Horizontal Transport of Angular Momentum

As given in equation (4.8), the zonally averaged horizontal transport of angular momentum may be written in the form

$$[\overline{J}_\varphi] = a^2 \cos^2 \varphi\ \Omega[\bar{v}] + a \cos \varphi([\bar{u}][\bar{v}] + [\bar{u}^*\bar{v}^*] + [\overline{u'v'}])$$

The transport of Ω-momentum by the mean motions can be obtained from the values of $[\bar{v}]$ given in Table 3.3.

The replacement of r by a in the expression for $[\overline{J}_\varphi]$, together with the condition of zero mass flux, implies that $\int a^2 \cos^2 \varphi\ \Omega\ [\bar{v}]\ dp = 0$. In actuality there should be a small net horizontal flux of Ω-momentum due to the variation of r, and to the net horizontal transfer of mass. Both of these effects are small compared to the uncertainty of the observations.

Values of $[\bar{u}^*\bar{v}^*]$ and $[\overline{u'v'}]$ are given in Tables 4.2 and 4.3, at the end of the chapter, for the four 3-month seasons of our study. Above 100 mb, only the transient eddy flux term is available in the tropical region and was obtained from latitude band means. The effect of the biennial oscillation was minimized by giving equal weight to data from even and odd years for the period 7/1957–6/1963. In the extratropical Southern Hemisphere, the data are not sufficient to evaluate any of the terms above 100 mb.

Graphs of the transient eddy momentum flux term $[\overline{u'v'}]$ are given in Figure 4.6 for the four seasons. The patterns are dominated by a primary maximum of poleward flux at middle latitudes in each hemisphere. These maxima change position and intensity with season, but the Southern Hemisphere maximum always exceeds that of the Northern Hemisphere in magnitude. Poleward of these middle-latitude features there is a region of oppositely directed flux in each hemisphere. Values in the Southern Hemisphere are much greater than those in the corresponding Northern Hemisphere region. The equatorial maximum

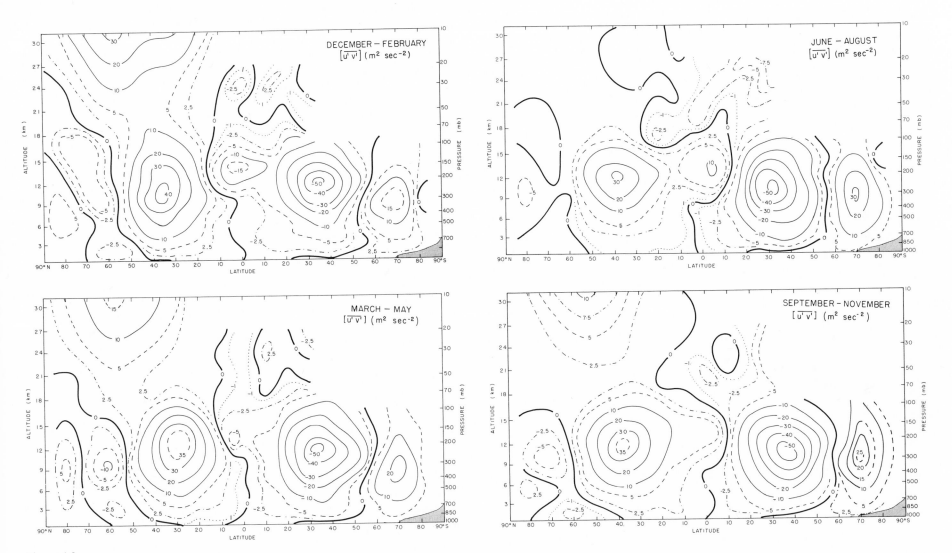

Figure 4.6

of eddy momentum flux, which we noted in the discussion of the maps and cross sections, is well represented in the zonally averaged values. The direction of this flux is from the winter and spring to the summer and fall hemispheres. Stratospheric eddy fluxes are available only for the Northern Hemisphere. Maximum values are associated with the polar night jet.

A comparison of the graphs of $[\overline{u'v'}]$ and of those of $[\bar{u}]$ (Figure 3.13), shows that the equatorial flux of momentum is generally down the angular momentum gradient. The fluxes are strongly countergradient in the lower temperate latitudes.

Graphs of the standing eddy flux of momentum $[\bar{u}^*\bar{v}^*]$, for the four seasons, are given in Figure 4.7. Except in the equatorial upper troposphere and in the Northern Hemisphere stratosphere, these fluxes are much smaller than the corresponding transient eddy fluxes. In the troposphere, the standing eddy term is generally more important in the Northern Hemisphere middle latitudes than in the corresponding Southern Hemisphere region, as noted by Obasi (1963a,b; 1965). The maximum standing eddy flux in the equatorial region is of the same sign as that of the transient eddy maximum. At northern middle latitudes, the standing and transient eddy fluxes reinforce one another, while at southern middle latitudes they generally oppose one another. The most marked seasonal changes in the standing eddy fluxes occur in the equatorial upper troposphere and in the Northern Hemisphere middle stratosphere.

Figure 4.8 gives the "mean meridional" momentum flux term $[\bar{u}]\,[\bar{v}]$. As would be expected, this term is small in extratropical regions and has its maximum values in regions associated with the upper tropospheric branches of the Hadley and Ferrel cells. The seasonal variations are clearly associated with the seasonal behavior of the Hadley and Ferrel cell circulations (Figure 3.19). There is a significant transequatorial flux in most seasons. At middle latitudes there is a region of flux convergence in each hemisphere. The amplitude of this flux is comparable to, or greater than, that of the standing eddy fluxes in most regions. It exceeds that of the transient eddy flux in equatorial regions. The only appreciable stratospheric contribution from this term occurs in conjunction with the December–February polar night jet.

Figure 4.9 gives values of $[\overline{uv}]$, the sum of the mean meridional and eddy flux terms. The overall pattern is dominated by a large poleward momentum flux maximum in each hemisphere. These maxima are associated mainly with the transient eddy flux. In the upper troposphere, relative angular momentum flows away from a region near 10°N in December–February and away from a region near 10°S in June–August. Relative angular momentum flows into a region near 50–60° in each hemisphere. The importance of the standing eddy and mean meridional fluxes is evident in the extension of the middle-latitude flux region into the equatorial region where a second flux maximum exists. As we have seen in Figures 4.1, 4.2, and 4.5, this trans-equatorial flux is not uniform about the latitude circle but shows a definite longitudinal maximum in the Eastern Hemisphere.

The poleward flux associated with the middle-latitude maximum has been the subject of many investigations, for example those of Starr and White (1954), Buch (1954), Obasi (1963a, b; 1965) and Yoshida (1967). The transequatorial flux has received less attention, although it has been noted in the work of Obasi. It was stressed by Tucker (1965) in his detailed study of the momentum flux near the equator, which was based on data from 7 stations, and was noted by Henning (1968) in an analysis of data from 14 stations.

The transequatorial flux is from the winter and spring to the summer and fall hemispheres. A comparison of these graphs with those of the mean zonal wind $[\bar{u}]$ (Figure 3.13) shows that, at the height of the maximum, the flux is into the region of easterlies. This transequatorial flux serves to export momentum from the hemisphere in which the largest low-level easterlies are found, that is, from the hemisphere in which the largest transfer of angular momentum from the earth to the atmosphere is occurring.

The vertical integrals of the three components of the horizontal momentum transport are shown in Figure 4.10 for the four seasons. The vertical integral of the total horizontal momentum transport is shown in Figure 4.11. The importance of the transient eddy term in middle latitudes and of the mean motion term in the equatorial region is clearly apparent, as is the greater magnitude of the standing eddy term in the Northern Hemisphere as compared to that in the Southern

Figure 4.7

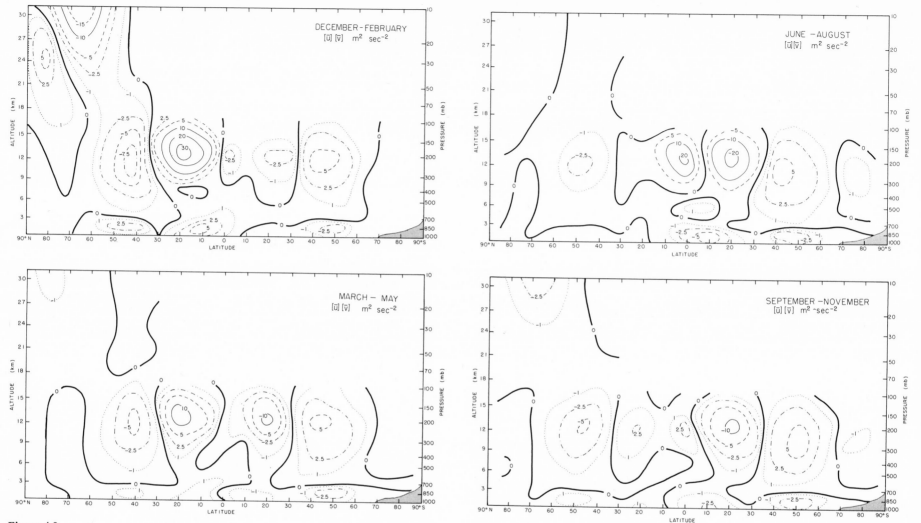

Figure 4.8

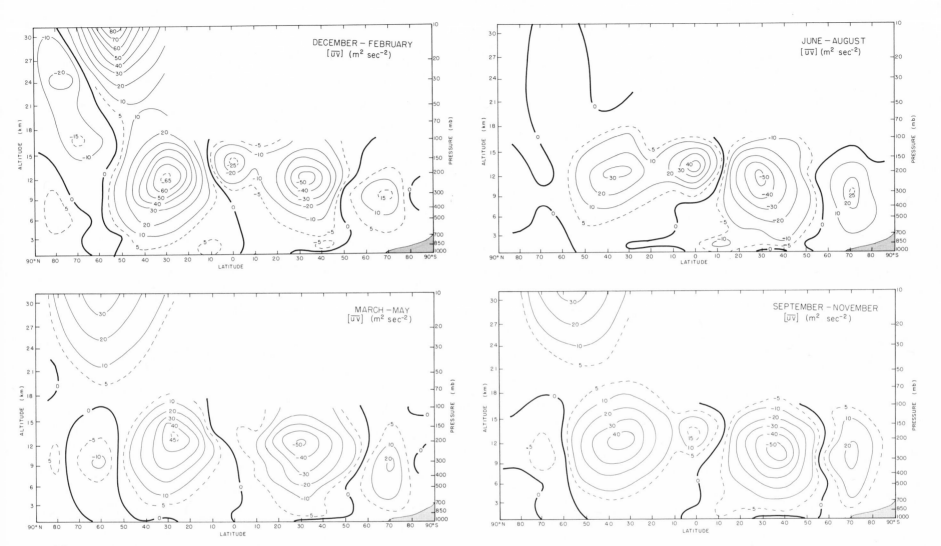

Figure 4.9

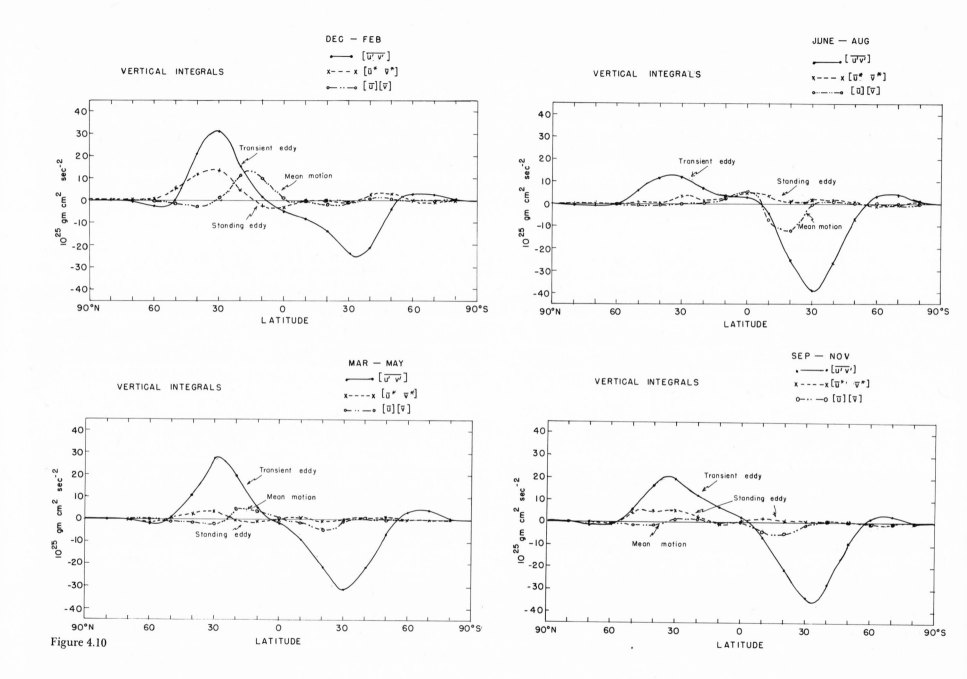

Figure 4.10

Hemisphere. Figure 4.11 shows the seasonal variation of the total momentum transport. Largest values occur in the winter and spring in each hemisphere. They are accompanied by an appreciable transequatorial flux from the winter and spring to the summer and fall hemispheres. The seasons of maximum net poleward flux also correspond to the seasons of maximum hemispheric relative angular momentum content, as shown in Table 4.1.

4.4.3 Vertical Transport of Angular Momentum

We have seen in the previous section that the maximum horizontal angular momentum transport occurs in the upper troposphere. As the atmosphere gains or loses angular momentum only through contact with the ground, it is evident that there must be an upward transport of angular momentum in the tropical region of surface easterlies, and a downward transport in the middle latitude region of surface westerlies.

As given in equation (4.8), the vertical flux of angular momentum may be written in the form

$$[\bar{J}_p] = a^2 \cos^2 \varphi \, \Omega[\bar{\omega}] + a \cos \varphi([\bar{u}][\bar{\omega}] + [\bar{u}^*\bar{\omega}^*] + [\overline{u'\omega'}] + g[\bar{\tau}_\lambda]).$$

Unfortunately none of these terms can be measured directly. Terms involving $[\bar{\omega}]$ are obtained along with the values of $[\bar{v}]$, which are themselves obtained indirectly in extratropical regions. The term involving $\bar{\omega}^*$ may be calculated from a knowledge of $\bar{u}$ and $\bar{v}$ at grid points over the globe. The term involving ω' requires daily values of the parameters and cannot be obtained with either any facility or much accuracy. The term $g[\bar{\tau}_\lambda]$ can be only crudely estimated.

The relative importance of several of these terms may be assessed from the data at hand. The importance of the vertical transport of the earth's Ω-momentum far outweighs that of the vertical transport of relative angular momentum in most of the atmosphere. An estimate of this term can be obtained from the data of Table 3.4.

A Hadley or Ferrel cell is capable of important vertical fluxes of Ω-momentum. The flux across a pressure surface between latitudes φ_1 and φ_2 is given by

$$2\pi a^2 \int_{\varphi_1}^{\varphi_2} a^2 \cos^3 \varphi \, \Omega[\bar{\omega}] \, d\varphi.$$

In a Hadley circulation a net upward transport of angular momentum across a pressure level will occur as upward motion at low latitudes, where $[\bar{\omega}]$ and $\cos^3 \varphi$ are relatively large, will not be offset by downward motion at higher latitudes, where $\cos^3 \varphi$ will be smaller. By the same reasoning, a Ferrel cell will accomplish a net downward flux of Ω-momentum. By way of illustration, the distribution of the vertical flux of Ω-momentum with latitude across the 400-mb level is shown in Figure 4.12. The boundary between the Hadley and Ferrel cells is exhibited, on the diagram, by the minima of the curves at 20–30° of latitude on either side of the equator. The net effect of the Hadley circulation is seen to be a vertically upward flux of momentum across the 400-mb level. Similarly, the effect of the Ferrel circulations is a net downward flux of momentum. These vertical fluxes transport momentum from the surface source to the 200-mb level, where the maximum horizontal transport occurs and thence downward to the sink at middle latitudes.

The vertical flux of relative angular momentum by the zonally averaged motion, in terms of $[\bar{u}] [\bar{\omega}]$, can be evaluated from the data given in Chapter 3. It is much smaller than the flux of Ω-momentum.

Grid point values of $\bar{u}$ and $\bar{v}$ were used to calculate $\bar{\omega}$ through the continuity equation, and thence $[\bar{u}^*\bar{\omega}^*]$ for the tropical region. The value of $\bar{\omega}$ was set to zero at 1000 mb, and the values at the remaining levels were obtained by integration. These calculated values were adjusted in a linear fashion so that $\bar{\omega}$ was also zero at 100 mb, i.e.,

$$\bar{\omega}_a(p) = \bar{\omega}(p) - \left(\frac{1000 - p}{900}\right) \bar{\omega}(100).$$

The largest values of $[\bar{u}^*\bar{\omega}^*]$ so obtained were of the order of 5×10^{-2} mb cm sec^{-2}. This is similar in magnitude to values quoted by Starr and Dickinson (1963) in a discussion of this term. The contribution to the vertical transport of angular momentum is minor, and is estimated only with fair accuracy.

The remaining term in the vertical angular momentum flux involves the transient eddy and turbulent fluxes. The transient eddy flux requires daily values of u and ω for its evaluation, and these were not available for our study. Starr and Dickinson (loc. cit.) have discussed

149   THE ZONALLY AVERAGED ANGULAR MOMENTUM BALANCE

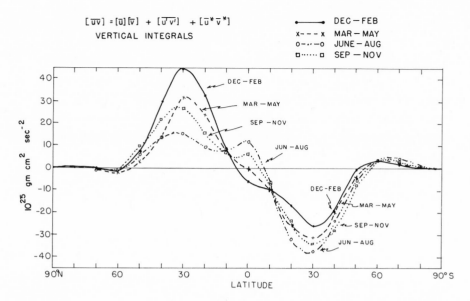

$$[\overline{uv}] = [\overline{u}][\overline{v}] + [\overline{u'v'}] + [\overline{u^*\overline{v}^*}]$$
VERTICAL INTEGRALS

DEC—FEB
MAR—MAY
JUNE—AUG
SEP—NOV

Figure 4.11

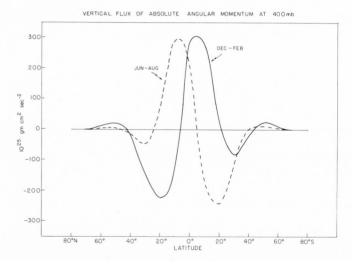

VERTICAL FLUX OF ABSOLUTE ANGULAR MOMENTUM AT 400 mb

Figure 4.12

this term and quote computations by Molla and Loisel (1962), who used the adiabatic method to calculate daily values of ω, and who found that between 20°N and 80°N both the transient and standing eddy components were small compared to the remaining terms in the vertical transport. Sheppard (1953) has stressed that in middle latitudes it is common experience that $\overline{u'v'}$ and $\overline{v'w'}$ are both positive. The implication is that for the synoptic scales of motion, $\overline{u'w'}$ is also positive. He suggested that smaller scale motions (which we have represented by the term $g[\bar{\tau}_\lambda]$ might be responsible for middle-latitude downward flux and examined cumulus convection as a possibility (Sheppard 1958). Radar observations were used to further probe this hypothesis by Newell (1959), and this work has been continued by others. Priestley (1967) examined the role of frontal systems in this downward flux of angular momentum with encouraging results.

Most recently, Starr et al. (1970) have estimated the vertical eddy flux of angular momentum for the Northern Hemisphere through equation (4.7). Taking an average over time such that $\overline{\partial M/\partial t} \sim 0$, and assuming that the mountain torque term gives a contribution essentially at the surface, the zonally averaged angular-momentum equation becomes

$$\frac{1}{a\cos\varphi}\frac{\partial}{\partial\varphi}[\overline{J}_\varphi]\cos\varphi + \frac{\partial}{\partial p}[\overline{J}_p] = 0.$$

This equation may be solved in terms of a stream function χ such that

$$\chi = \int_0^p 2\pi a\cos\varphi[\overline{J}_\varphi]\frac{dp}{g}$$

$$[\overline{J}_\varphi] = \frac{g}{2\pi a\cos\varphi}\frac{\partial\chi}{\partial p}$$

$$[\overline{J}_p] = \frac{g}{2\pi a^2\cos\varphi}\frac{\partial\chi}{\partial\varphi}.$$

The vertical momentum flux can thus be evaluated when $[\overline{J}_\varphi]$ is known. The terms in $[\overline{J}_p]$ involving $[\bar{\omega}]$ are also known if $[\bar{u}]$ and $[\bar{v}]$ are known, and the eddy flux of momentum $[\bar{u}^*\bar{\omega}^*] + [\overline{u'\omega'}] + g[\bar{\tau}_\lambda]$ can be obtained by subtraction.

The result depends critically on a correct and independent evaluation of $[\bar{v}]$. As values of $[\bar{v}]$ in extratropical regions are not obtained independently in this study, and in fact are estimated from the zonal-momentum equation by assuming the vertical eddy flux is small, no attempt can be made to evaluate this term here.

Starr et al. (1970) find the expected upward flux of angular momentum in tropical regions and the downward flux in middle latitudes; but they find also the result implied by Sheppard's argument, that a region of upward momentum flux exists in the middle-latitude troposphere. This flux acts to transfer angular momentum into the middle-latitude jet.

For purposes of illustration, graphs of χ evaluated from $[\bar{J}_\varphi]$ are given in Figure 4.13. Because of the indirect method which was used to evaluate $[\bar{v}]$, the vertical flux of momentum by the eddies is confined to the boundary layer. The graphs clearly show the transport of momentum from the source region in the tropics to the sink region in middle latitudes.

4.4.4 The Transfer of Angular Momentum between the Earth and the Atmosphere

In the previous two sections, information concerning the horizontal and vertical fluxes of angular momentum has been presented. As has been mentioned repeatedly, the fluxes of angular momentum in the atmosphere are such as to balance the external torques, and these torques arise where the atmosphere is in contact with the earth.

Integrating the momentum equation (4.5) gives

$$\overline{\frac{\partial}{\partial t} \int M \, dm} + \int \frac{1}{a \cos \varphi} \frac{\partial}{\partial \varphi} \bar{J}_\varphi \cos \varphi \, dm + \int a \cos \varphi \, \bar{\tau}_\lambda (p_0) \, dS$$

$$+ \int \frac{\partial \bar{\phi}}{\partial \lambda} \, dm = 0 \qquad (4.10)$$

Terms involving ω are taken to be zero at the surface, that is, at $p = p_0$.

The term $\int a \cos \varphi \, \bar{\tau}_\lambda (p_0) \, dS$ represents the average rate of transfer of angular momentum from the earth to the atmosphere due to frictional torques (dS is an element of surface area). This term may be estimated from surface wind data.

The term $\int \partial \phi / \partial \lambda \, dm$ represents the so-called mountain torque. If the integration is carried out around latitude circles, this term will be

zero unless the geopotential surface exhibits discontinuities. Obviously this can be the case only where mountains intersect the ϕ surface.

These two terms serve to provide the only appreciable external torques on the atmosphere and hence are the only mechanisms which act as sources and sinks of atmospheric angular momentum.

1 THE MOUNTAIN TORQUE
(By Zen-Kay Jao, Department of Meteorology, Massachusetts Institute of Technology)

The term

$$\iiint \frac{\partial \bar{\phi}}{\partial \lambda} a^2 \cos \varphi \, d\lambda \, d\varphi \frac{dp}{g} = \iiint \frac{\partial \bar{p}}{\partial \lambda} a^2 \cos \varphi \, d\lambda \, d\varphi \, dz$$

represents the torque on the atmosphere due to differences of pressure across mountain ranges. The expression can be written in terms of pressures on the eastern and western sides of the mountain ranges:

$$\iint (p_\mathrm{w} - p_\mathrm{e}) a^2 \cos \varphi \, d\varphi \, dz. \qquad (4.11)$$

An evaluation of this equation in detail would require a knowledge of the pressure at each point of the mountain surface as well as a knowledge of the detailed topography. Obviously these details are not known, and approximate methods must be used to estimate the expression.

The contribution of the mountain torque term to the momentum balance of the atmosphere has received relatively little attention since the pioneering study by White (1949), in which the term was estimated for the Northern Hemisphere between 25 and 65°N. White used mean monthly surface and 10,000-ft. pressures to estimate the pressure differences across mountain ranges which were more than 5 degrees of latitude and 5 degrees of longitude in extent. Lorenz (1954) pointed out that the height field (for the mountains) need not be specified at any more points than the pressure field, and he presented a simple method for the mountain torque computation. More recently, Yeh and Chu (1958) extended the mountain torque calculations to the equator. The results of these studies have shown that the mountain torque is by no means insignificant in the Northern Hemisphere. Poleward of about 20°N it is generally of the same sign as the surface stress and is roughly one-half as large.

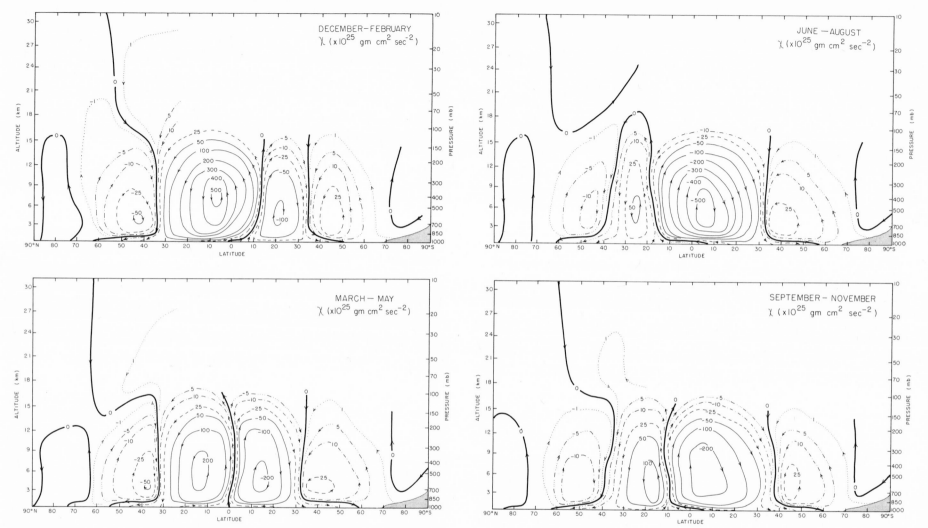

Figure 4.13

No comparable studies have so far been reported for the Southern Hemisphere, although Hutchings and Thompson (1962) have found that the torque due to the New Zealand Alps was approximately 10 percent of the surface torque within their range of latitudes. In this section, estimates of the mean monthly mountain torque term are obtained as a function of latitude in the Southern Hemisphere.

The expression (4.11) is evaluated by using the simplifying assumptions that the differences in pressure may be obtained from the appropriate sea-level pressures, and that the pressure is constant up to the average height of the mountain range and is zero above this height. These assumptions are only rough approximations to reality, but they should be sufficient at least to show the importance of this term in the angular momentum balance. It was assumed that only mountain ranges of broad latitudinal or longitudinal extent formed barriers across which significant differences in pressure could occur. Following the procedure adopted by White (1949), all mountains whose extent is less than 5° of latitude or 5° of longitude were neglected. Antarctica has been omitted because pressure observations are lacking except in a few regions near the coast.

Monthly mean pressure data for individual stations are available in the *World Weather Records* published by ESSA, U.S. Department of Commerce. The 10-year period 1951 to 1960 was chosen for the calculation. If the mean sea-level pressure of a station was not given, a reduction to sea level using the station pressure was used. A constant lapse rate of $0.6°C/100$ m was assumed, together with a standard sea-level temperature of $15°C$ to obtain the approximate equation

$$p_0 = p(1 - 0.00002184\, z)^{-5.255};$$

p_0 is mean sea-level pressure, p the station pressure, and z the station height. For station heights below 200 m, the value is within 1 mb of that obtained for the WMO recommended reduction formula given in *World Weather Records* (vol. 5, p. 237). The monthly mean sea-level pressures thus obtained were used to determine the pressure difference across the continents at 2-degree latitude intervals. Linear interpolation along straight lines joining the nearest two stations was used where necessary. Topographical profiles across South Africa, South America,

Australia, Madagascar, and Indonesia were constructed from 1:5,000,000 topographical maps. If $\Delta p = p_w - p_e$ denotes the mean sea-level pressure difference across the mountain range at a given latitude, the integrated torque for a one-degree latitude strip ($a\Delta\varphi = 111$ km) is approximately $\Delta p \times H \times a\cos\varphi \times 111$ km, where H is the mean height of the mountain range at this latitude and a is the radius of the earth.

Monthly values of the mountain torque term at 2-degree intervals of latitude, for the region 1°S to 51°S, have been plotted in Figure 4.14. The graph has been extended through a second cycle so as to clearly show the period of positive torque in the summer season. It may be seen that the maximum mountain torques occur at the beginning of the summer and winter seasons, but that significant differences exist in the nature of these maxima. During the summer season (December, January and February) the mountain torque is positive in the region from 3°S to 51°S so that the earth abstracts angular momentum from the atmosphere over almost the entire Southern Hemisphere. During winter the mountain torque is predominantly negative, and the earth supplies angular momentum to the atmosphere at a rate which reaches a maximum in July. In the region from 41°S to 49°S, the mountain torque is positive, and the mountains remove angular momentum from the atmosphere in every month of the year. Hutchings and Thompson (1962) obtained a like result for this latitude region in their analysis of New Zealand data. The average monthly torque exerted on the atmosphere by the mountain ranges of the South Island (41°S–46°S) was mainly positive.

The annual mean mountain torque of the Southern Hemisphere is such that the average boundary between eastward and westward torques occurs at 25°S. The largest negative torque of 0.3×10^{25} gm cm² sec⁻² is found at 20°S, while largest positive torque of 0.2×10^{25} gm cm² sec⁻² is observed at 41°S. The sign of the average torque is usually the same as that for the surface stress computed by Priestley (1951b) but, as will be seen later, its magnitude is only about a tenth as large.

2 THE SURFACE STRESS   The surface stress may be estimated from the surface wind **V** using an equation of the form

$$\boldsymbol{\tau} = \rho_0 C_D\, |\mathbf{V}|\, \mathbf{V};$$

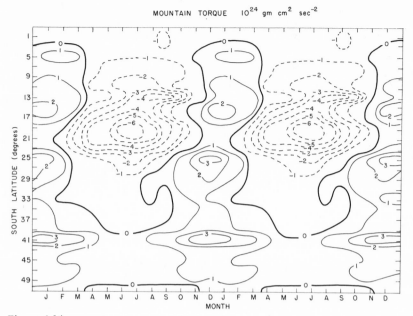

MOUNTAIN TORQUE $\quad 10^{24}$ gm cm^2 sec^{-2}

Figure 4.14

C_D is a drag coefficient, ρ_0 is the air density, and $|\mathbf{V}|$ is the surface wind speed. The drag coefficient C_D depends on the wind speed, but the form of the relation is not known with great accuracy. For conditions of near neutral stability over the open sea, Sheppard (1958) gives the relation

$$C_D = (0.8 + 0.00114\,|\mathbf{V}|) \times 10^{-3}, \tag{4.12}$$

where the wind speed $|\mathbf{V}|$ at the anemometer level is measured in cm sec^{-1}.

We are concerned with the component of the surface stress in the zonal direction only (equation 4.10), as this is the component which affects the angular momentum. Because of the nonlinear nature of the surface stress, this parameter cannot be obtained simply from the mean wind components, and the wind distribution at individual grid points must be considered. The direct calculations of the surface stress in this study are performed using the mean wind components $\bar{u}$ and $\bar{v}$ and the daily variability of the wind components as measured by $\sigma(u)$ and $\sigma(v)$. It was assumed that the daily values of u and v were normally distributed and were uncorrelated. Crutcher (1959) shows that this is usually a satisfactory approximation to the wind distribution. In this case

$$\bar{\tau}_\lambda = \overline{\rho_0 C_D\,|\mathbf{V}|\,u} \sim \bar{\rho}_0 \iint P(u, v) C_D(|\mathbf{V}|)\,u\,|\mathbf{V}|\,du\,dv,$$

where

$$P(u, v) = \frac{1}{\sqrt{2\pi}\,\sigma(u)\,\sigma(v)}\,\mathrm{e}^{-\frac{(u - \bar{u})^2}{2\sigma^2(u)}}\,\mathrm{e}^{-\frac{(v - \bar{v})^2}{2\sigma^2(v)}}$$

and

$$|\mathbf{V}| = (u^2 + v^2)^{1/2}.$$

The integration was performed numerically in steps of $0.25\,\sigma$ for u and v. Sheppard's relation for the drag coefficient was used. The resulting values were averaged around latitude circles and are shown in Figure 4.14.

The latitudinal distribution of the surface stress is similar to that found by Priestley (1951b) (the curve labeled τ_0 in Figure 4.16) who obtained values from surface winds over the oceans using a value of

0.0013 for the drag coefficient. This value would correspond to a wind speed of about 5 m sec^{-1} in Sheppard's relation (equation 4.12). The magnitude of the stresses calculated by Priestley is about 50 percent greater than that calculated by the method described above, and as will be seen in the next section, our values are too small to provide the source of angular momentum required to balance the horizontal fluxes given in section 4.4.2. The calculation was repeated using the value of 0.0013 for the drag coefficient that was used by Priestley, but this led to a decrease in the calculated values of the stress (Figure 4.15).

The calculated surface stress is apparently too small, and this may be due to errors either in the drag coefficient or in the wind distribution. The drag coefficient given by Sheppard is appropriate to neutral conditions over the oceans, and a larger value may be more suitable over land. In our case, however, it appears that the principal source of error lies in our underestimating the strength of the mean zonal surface wind over the oceans. Our winds were determined almost exclusively from land stations and are consistently lower than Riehl's (1954) values obtained from ship observations for the mid-season months. The two sets of profiles are plotted in Figure 4.15, where it can be seen that the Northern Hemisphere easterlies are weaker than those of Riehl by about 1 m sec^{-1} in both seasons, while the difference is closer to 1.5 m sec^{-1} in the Southern Hemisphere. Our surface wind speeds may therefore be underestimated by a factor of about one-third. To obtain a rough estimate of the effect of this increase on the surface torque, the stress calculations were repeated with all wind components and their standard deviations increased by this amount (the drag coefficient given by Sheppard was again used; see Figure 4.15). The increase in the wind speeds and standard deviations led to a nearly threefold increase in the surface torque. This assumption obviously leads to an overestimate of the torque in middle latitudes. In view of the inadequacy of these surface stress calculations, free use will be made of Priestley's work in the discussion of the momentum budget.

4.4.5 The Momentum Balance

In previous sections we have discussed the fluxes of angular momentum and the surface stress and mountain torque terms. The simplest way of

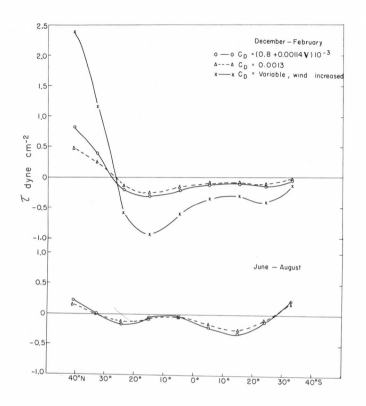

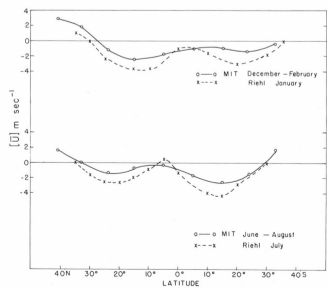

Figure 4.15

155   THE ZONALLY AVERAGED ANGULAR MOMENTUM BALANCE

comparing the deduced torques with the flux of momentum in the atmosphere is in terms of the vertically integrated form of equation (4.7). This may be written as

$$\frac{\partial}{\partial t} \int_0^{p_0} 2\pi a \cos\varphi \, [\overline{M}] \frac{dp}{g}$$

$$+ \frac{\partial}{\partial \varphi} \int_0^{p_0} 2\pi (a^2 \cos^2\varphi \, \Omega[\bar{v}] + a \cos\varphi \, [\overline{uv}]) \cos\varphi \frac{dp}{g}$$

$$= -2\pi a^2 \cos^2\varphi ([\bar{\tau}_\lambda(p_0)] + [\bar{\tau}_M]),$$

where the mountain torque term has been written as $[\bar{\tau}_M]$. As a first approximation, the time change of the angular momentum may be taken to be zero, and this, together with the condition for zero mass flux across a latitude circle, gives

$$\frac{1}{a} \frac{\partial}{\partial \varphi} \int_0^{p_0} 2\pi a^2 \cos^2\varphi \, [\overline{uv}] \frac{dp}{g} = -2\pi a^2 \cos^2\varphi ([\bar{\tau}_\lambda(p_0)] + [\bar{\tau}_M]). \quad (4.13)$$

This formulation of the momentum balance equation does not require a knowledge of the vertical fluxes of momentum. It states simply that the gradient of the vertically integrated momentum transport is balanced by the torques due to the surface stresses and the presence of mountains.

The vertical integral of the momentum transport has been given in Figure 4.11. The gradient is shown in Figure 4.16 as the curve labeled τ_F. The other curves are the surface stress τ_0 derived from Priestley's (1951b) results and the mountain torque contribution, τ_M. We note that

(a) There is a good qualitative agreement between the results calculated from momentum-flux data and the independent stress results of Priestley.

(b) Priestley's results and ours show the transfer of angular momentum from earth to the atmosphere occurring in the tropical regions. Maximum values occur on either side of the equator, and change amplitude and position with season. Both results also show a relative minimum near the equator in all seasons. It is particularly noteworthy that both sets of results show a negative stress, i.e., a transfer of momentum from atmosphere to earth, near the equator in June–August.

Patches of westerly winds which would be expected to transfer momentum in this sense are clearly evident on the $\bar{u}$ maps of Chapter 3. Nevertheless, the zonally averaged wind is easterly, and this is usually taken to imply a transfer of angular momentum from earth to atmosphere.

A number of explanations of this discrepancy are possible. First, the zonally averaged wind at this latitude may be biased by the distribution of observing stations so that an easterly zonal wind is inferred whereas a small westerly wind actually exists. This explanation is supported by Riehl's wind profile shown in Figure 4.15. Alternately, the zonally averaged wind may be actually easterly while the stress is in the negative direction. This would require an appropriate correlation between the wind speed, the u component, and/or the drag coefficient.

(c) In middle latitudes, both results show the transfer of angular momentum from the atmosphere to the earth. The rate of transfer varies somewhat with season and is clearly greatest in the Southern Hemisphere in all seasons except possibly in December–February.

(d) The vertically integrated momentum transports of Figure 4.11 show marked transequatorial transport in all seasons except March–May. The implication of equation (4.13) is that a transport across the equator implies an imbalance in the transfer of momentum between the earth and the atmosphere in the two hemispheres. Thus the transport of momentum from the Southern Hemisphere to the Northern Hemisphere in June–August and September–November implies a greater transport of momentum from the earth to the atmosphere in the Southern Hemisphere in these seasons, with the excess being transported into the Northern Hemisphere. The opposite situation is apparent in December–February. Such a situation is also implied by the independent surface stress data given by Priestley. According to the graphs of integrated relative momentum fluxes (Figure 4.10) the processes which are responsible for the transequatorial transport differ with season. The mean motion and standing eddy transports are of most importance in June–August; transient eddy transport in September–November; and transient and standing eddy transport in December–February. The transport is from the hemisphere in which the maximum surface easter-

156   ANGULAR MOMENTUM BUDGET AND MAINTENANCE OF WINDS

lies are found (Figure 3.13) and in which the most intense Hadley cell occurs (Figure 3.19). This transequatorial flux is down the gradient of the integrated total angular momentum.

In general we may say that the broad-scale features of the momentum balance are reasonably well measured. We have presented maps and figures of the various components of the momentum transports which clearly show the relative importance of the different mechanisms in the maintenance of the atmospheric momentum balance in different regions of the atmosphere. Our integrated results agree well with the surface stress computation of Priestley.

4.5 The Maintenance of the Zonal Wind
In the previous section we considered the angular momentum balance from the standpoint of the components of the horizontal and vertical fluxes of momentum, and we compared the integrated values of the fluxes with the observed surface stresses which must be balanced by them. In this section we follow a different approach and investigate the maintenance of the zonal flow.

4.5.1 The Zonal-Wind Equation
The time-averaged equation for the zonal-wind component is directly related to the angular momentum equation (4.3), from which it may be obtained, in the form

$$\overline{\frac{\partial u}{\partial t}} + \frac{1}{a \cos \varphi} \frac{\partial \overline{uu}}{\partial \lambda} + \frac{1}{a \cos^2 \varphi} \frac{\partial}{\partial \varphi} \overline{uv} \cos^2 \varphi$$

$$+ \frac{\partial}{\partial p} (\overline{u\omega} + g\overline{\tau}_\lambda) - f\overline{v} + \frac{1}{a \cos \varphi} \frac{\partial \overline{\phi}}{\partial \lambda} = 0. \tag{4.14}$$

The turbulent stress term in the vertical has been retained. This equation may be thought of as the equation of the relative angular momentum, if multiplied by $a \cos \varphi$. The Coriolis term represents the transformation of momentum between Ω-momentum and relative angular momentum in this case, while the pressure gradient represents the redistribution of relative angular momentum by pressure forces.

The equation may be further decomposed into the form

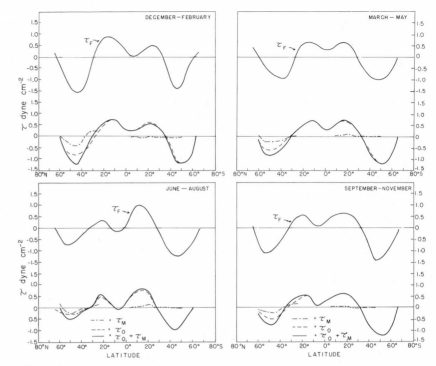

Figure 4.16

$$\overline{\frac{\partial u}{\partial t}} + \frac{\bar{u}}{a \cos \varphi} \frac{\partial \bar{u}}{\partial \lambda} + \frac{\bar{v}}{a \cos \varphi} \frac{\partial}{\partial \varphi} \bar{u} \cos \varphi + \bar{\omega} \frac{\partial \bar{u}}{\partial p} - f\bar{v}$$

$$+ \frac{1}{a \cos \varphi} \frac{\partial \bar{\phi}}{\partial \lambda} + \frac{1}{a \cos \varphi} \frac{\partial \overline{u'^2}}{\partial \lambda} + \frac{1}{a \cos^2 \varphi} \frac{\partial}{\partial \varphi} \overline{u'v'} \cos^2 \varphi$$

$$+ \frac{\partial}{\partial p} (\overline{u'\omega'} + g\bar{\tau}_\lambda) = 0. \qquad (4.15)$$

To a first approximation the time rate of change of the zonal wind may be taken to be zero for the seasonal average. The equation thus states that there must be a balance between the advection of momentum by the mean winds, the Coriolis term, the pressure gradient force, and the divergence of the eddy momentum fluxes.

Some of these terms can be estimated on a pressure surface from the data presented in Chapter 3. We present here only the term

$$G = -\left(\frac{1}{a \cos \varphi} \frac{\partial \overline{u'^2}}{\partial \lambda} + \frac{1}{a \cos^2 \varphi} \frac{\partial}{\partial \varphi} \overline{u'v'} \cos^2 \varphi \right).$$

This term represents the "acceleration" of the zonal mean wind due to the convergence of eddy momentum in the horizontal. As can be seen from Plate 4.12, the value of this quantity varies markedly with latitude in December–February at 200 mb. In regions where G is negative, implying a divergence of momentum by the eddy fluxes, the mean westerlies in these regions would be decelerated if this effect was not offset by the remaining terms in equation (4.15). In particular we expect the Coriolis term to oppose the eddy divergence in this broad region, as $\bar{v}$ is primarily positive there (see Plate 3.19). Similar remarks apply to other regions. The value of G has its largest values in the winter hemisphere, as would be expected, and also shows the largest longitudinal variation in this hemisphere. The zonally averaged representation of the terms in the zonal-momentum equation, which are presented next, must be recognized to be averages of terms which vary appreciably in longitude, as exemplified by G.

4.5.2 The Maintenance of the Zonally Averaged Wind

Equation (4.14) may be zonally averaged, decomposed into mean motion and eddy terms, and written as

$$\frac{\partial [\bar{u}]}{\partial t} = -\frac{[\bar{v}]}{a \cos \varphi} \frac{\partial}{\partial \varphi} [\bar{u}] \cos \varphi - [\bar{\omega}] \frac{\partial [\bar{u}]}{\partial p} + f[\bar{v}]$$
$$\qquad\qquad (1) \qquad\qquad\qquad (2) \qquad\qquad (3)$$

$$-\frac{1}{a \cos \varphi} \left[\frac{\partial \bar{\phi}}{\partial \lambda} \right] - \frac{1}{a \cos^2 \varphi} \frac{\partial}{\partial \varphi} ([\bar{u}^* \bar{v}^*] + [\overline{u'v'}]) \cos^2 \varphi$$
$$\qquad (4) \qquad\qquad\qquad\qquad (5)$$

$$-\frac{\partial}{\partial p} ([\bar{u}^* \bar{\omega}^*] + [\overline{u'\omega'}]) + K_z \frac{\partial^2 [\bar{u}]}{\partial z^2}. \qquad (4.16)$$
$$\qquad (6) \qquad\qquad\qquad (7)$$

The turbulent stress divergence has been written in terms of an eddy diffusion coefficient K_z and the gradient of the zonal wind, i e.,

$$\frac{\partial}{\partial p} g[\bar{\tau}_\lambda] = -\frac{1}{\rho} \frac{\partial [\bar{\tau}_\lambda]}{\partial z} = -K_z \frac{\partial^2 [\bar{u}]}{\partial z^2}.$$

As before, terms (1) and (2) represent advection of the averaged momentum by the mean flow, and terms (5), (6), and (7) represent the divergence of the eddy and turbulent momentum fluxes. Term (3) in this case represents the effect of the Coriolis force applied to the ageostrophic meridional motion. Term (4) will be zero everywhere except at low levels where mountain ranges intersect the ϕ surface.

Most of these terms can be evaluated from the data presented earlier in this chapter or in Chapter 3. Graphs of terms (1), (2), and (7) are presented in Figure 4.17 for the tropical regions. Elsewhere they are small. Graphs of terms (3) and (5) for the globe are presented in Figure 4.18. Term (6) cannot be evaluated with sufficient accuracy and is not presented here. Term (4) will be zero everywhere except at the surface and is similarly omitted. The only directly measured term in the equation is term (5), labeled G in the diagram. Terms involving $[\bar{v}]$ are obtained indirectly except in the tropical region.

We may obtain a somewhat different viewpoint of the momentum balance of the atmosphere from an investigation of the terms in equation (4.16) than was obtained in the previous section. Here we assume that the terms on the right-hand side of the equation balance, to a first approximation, as a typical value for the time derivative is 10^{-4} cm

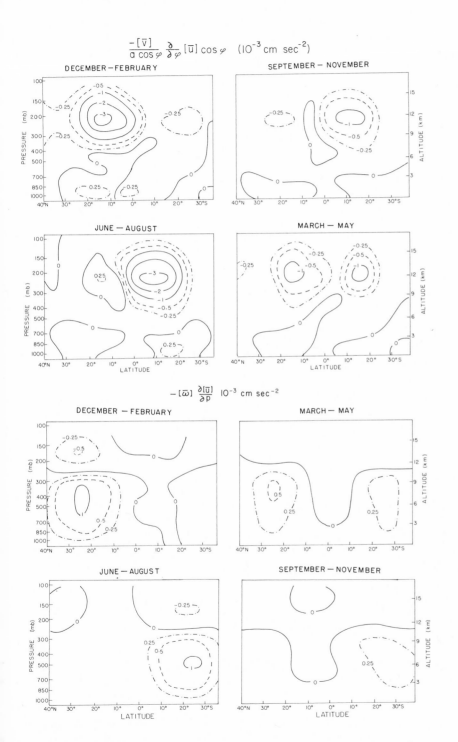

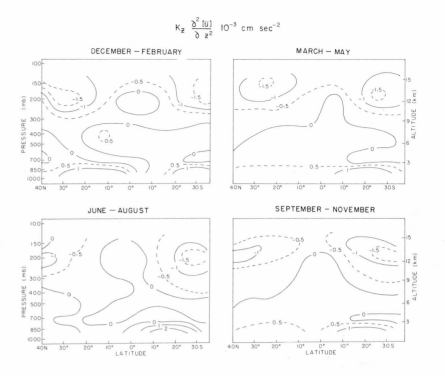

Figure 4.17

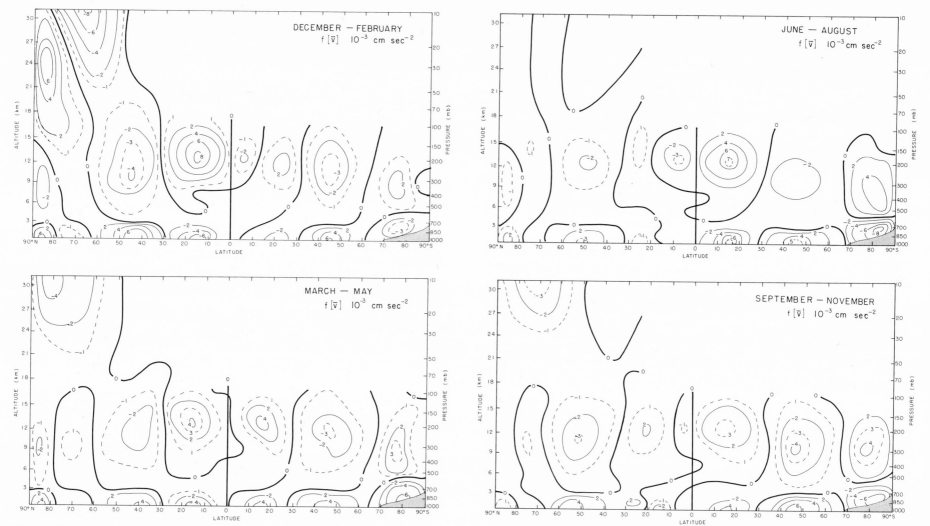

Figure 4.18

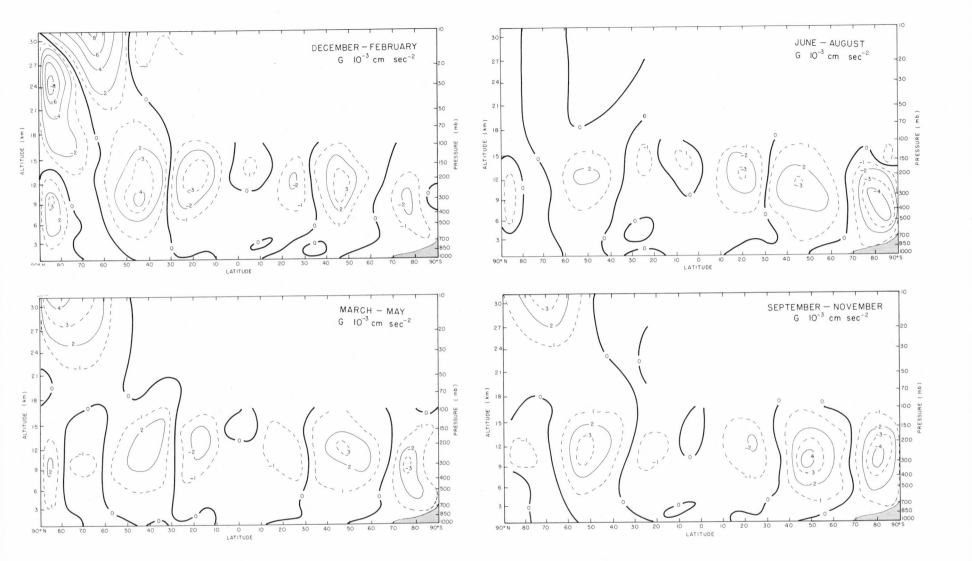

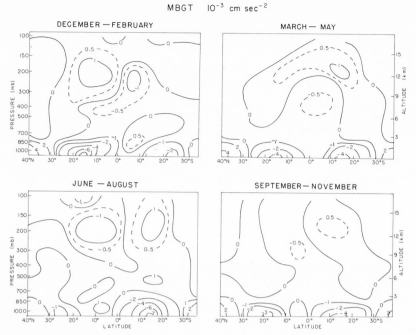

MBGT 10^{-3} cm sec^{-2}

Figure 4.19

sec^{-2} while typical values of the remaining terms are 10^{-3} cm sec^{-2}. The terms may be thought of as acting to "accelerate" or "decelerate" the mean zonal flow depending on sign. There must be a cooperative relationship among the terms to maintain a balance.

Figure 4.19 gives values of MBGT, which is the sum of all terms on the right-hand side of equation (4.16), with the exception of terms (4), (6), and (7), which are either poorly known or unavailable. These terms should be small except in the surface boundary layer. Despite the fact that the balance among the terms is achieved with only moderate accuracy away from the boundary layer, we may gain an insight into the relative importance of the terms in the maintenance of the zonal wind.

It is immediately apparent from the figures that the Coriolis term (term (3)) and the eddy flux divergence term (term (5)) are of great importance in all seasons. The advection of momentum by the meridional motions (term (1)) is of importance in the solsticial seasons while the vertical advection of momentum is generally of lesser importance.

In December–February the Coriolis term has its largest values in the tropical upper troposphere, where the Hadley circulation is strong. The positive value in the Northern Hemisphere and the negative value in the Southern Hemisphere are associated with the upper branch of the large cell which exists in the region at this time. The change in sign of $2\Omega \sin \varphi$ across the equator, and the subsequent increase of this value with latitude, is very important for the relative sign and strength of term (3) at different latitudes. At lower levels, the flow in the lower branch of the Hadley cell also produces appreciable values. Poleward of these regions, the contribution of the Ferrel circulations to this term is obvious. Again there is the appropriate difference in sign of the contribution from the upper and lower branches of the cells. There is also an appreciable contribution to this term in the region of the polar night jet.

In general there is an acceleration of the zonal flow in the upper branches of direct cells and in the lower branch of indirect cells, together with a deceleration of the zonal flow in the lower branches of the direct and the upper branches of the indirect cells in either hemisphere. If a large cell straddles the equator, as in December–February, there will be a change of sign of the term from one side of the equator to the other.

In the tropical upper troposphere in December–February, the effect of the Hadley cell, through the Coriolis term, is offset by advection of momentum by the meridional circulation, together with the strong horizontal flux divergence associated with the poleward transport of momentum into the subtropical jet by the eddy motions. In the middle-latitude region a similar situation exists in terms of the opposition of the Coriolis and eddy divergence terms in the upper branch of the Ferrel circulation. This opposition of the two processes is also seen in the stratosphere in the region of the polar night jet and in the direct circulation in the polar regions.

The same general situation prevails in June–August. The Coriolis contribution is even more marked in the Southern Hemisphere Hadley circulation than it was in the Northern Hemisphere in December–February. Again the advection of momentum by the Hadley circulation aids the eddy divergence in offsetting the large contribution from the Coriolis term in the upper troposphere. In March–May and September–November similar comments apply. Here, however, the Hadley and Ferrel circulations are more symmetric with respect to the equator, as are the patterns of eddy flux divergence.

There is a notable imbalance between the terms in the lower levels, where the Coriolis contributions, which arise in the lower branches of the meridional cells, are not compensated by the remaining terms. This situation is understandable in terms of the absence of the vertical eddy flux and surface stress terms in the cross sections. These terms must act to accelerate the mean flow in the tropical region of surface easterlies and to decelerate the mean flow in the middle-latitude regions of surface westerlies. The necessity for this contribution to the momentum budget is clearly apparent in MBGT. A typical value of the net easterly acceleration of the lowest layer is about 5×10^{-3} cm sec^{-2}. If the layer is taken to include a mass of 100 gm cm^{-2} (100 mb), the corresponding stress is 0.5 dyne cm^{-2}, which is comparable to values found by other techniques.

The parameterized representation of the turbulent stress divergence given by term (7) at least shows the proper sign of the acceleration required near the surface. The term is only poorly approximated, as K_z is not well known. Estimates range from 1 to 20 m^2 sec^{-1} in the troposphere, and a value of 10 m^2 sec^{-1} was used to calculate the term here. The values are only gross estimates, and the magnitudes are not to be considered accurate.

4.6 Concluding Remarks

Our investigation of the angular momentum balance of the atmosphere has included the estimation of the terms in the angular-momentum equation, as well as an investigation of the terms appearing in the time-averaged zonal-wind equation. As has been noted, the atmosphere gains angular momentum in regions of tropical surface easterlies and loses momentum in regions of middle-latitude surface westerlies, while the horizontal transport of momentum occurs mainly in the upper troposphere. The vertical flux of momentum from the lower to the upper atmosphere in tropical regions must be accomplished by turbulent and eddy transports at the lowest levels, but a net vertical flux of Ω-momentum apparently is responsible for the bulk of the vertical transport away from the surface layers. A similar downward flux occurs at middle latitudes. The basic pattern is then one of vertical transport of Ω-momentum by the meridional motions in the tropical regions and horizontal transport of relative angular momentum by the meridional motions in the Hadley circulation, and by the eddy fluxes. There is a subsequent downward transport of momentum by the Ferrel circulation into the friction layer in middle latitudes, where angular momentum flows into the earth.

In terms of the zonal-wind equation, or what is equivalent, the equation for the relative angular momentum, the momentum balance is represented by the balance between the advection of relative momentum by the zonally averaged meridional and vertical motions, the horizontal eddy flux divergence, and the Coriolis term. In the lower branches of the Hadley cells the Coriolis force acting on the mean meridional motions serves to decelerate the mean flow. In regions of easterlies this is offset by the transfer of angular momentum from the surface, that is, by an acceleration of the zonal wind, in this case implying the opposition of the easterlies by friction. The opposite situation is found in the lower branches of the Ferrel cells.

In the upper branches of the Hadley cells, the general pattern is of the acceleration of the zonal wind by the action of the Coriolis force on the poleward meridional motions. A balance is maintained here by the strong eddy flux divergence associated with the transport of momentum into the subtropical maximum of the zonal wind. The action of the Coriolis acceleration is also opposed by the advection of relative momentum by the mean meridional motions. In the upper branch of the Ferrel circulation, the opposite sign of the terms is observed.

There must therefore be a cooperative effect between the meridional circulation and the eddy motions which are observed in the atmosphere. Given the observed eddy fluxes of momentum and the value of $[\bar{u}]$, the zonally averaged meridional motions (and hence through continuity the vertical motions) must be as deduced. In fact this was the method used to obtain $[\bar{v}]$ in extratropical latitudes. Alternatively, given the values of $[\bar{u}]$ and of $[\bar{v}]$, the values of the horizontal momentum fluxes may be deduced. These considerations depend on the assumption that the vertical eddy flux of momentum is small everywhere except in the boundary layer, and therefore that the vertical fluxes of momentum are accomplished primarily by the mean motions.

If this assumption were found to be erroneous our picture of the momentum balance could change. Alternatively if independent values of $[\bar{u}]$ and the horizontal eddy fluxes could be obtained, together with an accurate measured value of $[\bar{v}]$, the vertical eddy fluxes of momentum could be derived.

Such improvements are unlikely to drastically change our idea of the circulation. In particular, the tropical region must remain of prime importance in any understanding of the angular momentum balance as a whole.

Table 4.2. Momentum Flux by Standing Eddies $[\bar{u}^*\bar{v}^*]$ (m² sec⁻²)

(a) December–February

p (mb)	1000	850	700	500	400	300	200	150	100	70	50	30	20	10
Lat.														
90°N	0.0	0.0	0.0	0.0	0.0	0.0	0.0	0.0	0.0	0.0	0.0	0.0	0.0	0.0
80	0.3	0.4	0.5	0.6	1.8	2.9	0.0	−1.0	−2.9	−7.5	−12.7	−26.5	−20.0	−10.2
70	0.4	0.5	−0.1	0.1	0.2	0.4	0.0	−3.0	−8.2	−12.0	−11.6	−22.1	−9.5	16.0
60	0.8	1.4	1.0	1.4	2.0	2.5	2.5	−2.0	−9.2	−6.5	−3.1	4.9	23.0	54.7
50	1.0	2.2	2.3	6.0	9.5	11.9	10.3	3.0	0.0	0.2	3.5	14.5	30.0	54.2
40	0.8	2.1	1.9	8.0	10.0	19.4	17.1	12.7	8.7	5.0	4.9	8.0	15.0	25.2
30	0.3	1.2	1.6	2.3	7.5	18.2	22.4	21.0	11.2	5.5	2.2	1.9	3.5	5.6
20	0.2	−0.1	−0.5	2.3	4.5	7.7	4.7	3.4	4.8	4.5	3.6	1.5	0.2	−1.2
10	0.4	−0.3	−0.1	1.2	2.6	2.9	−9.2	−12.8	−0.2					
0	−0.2	−2.0	1.2	−0.1	0.5	1.1	−6.3	−10.6	−1.2					
10	−0.9	0.3	−0.2	−0.4	−0.4	1.6	1.7	0.3	1.7					
20	0.6	−0.1	−0.8	−1.4	−0.5	−0.4	1.7	1.9	2.0					
30	0.5	−0.2	−0.2	−0.7	−0.5	−1.0	−2.7	−1.5	−0.5					
40	0.6	0.6	0.5	−0.4	−1.5	−1.5	−0.8	−0.2	−0.7					
50	1.5	3.1	1.5	0.2	2.8	6.3	3.1	1.1	−0.3					
60	2.0	0.1	0.7	−1.3	−1.0	−0.1	0.2	−0.8	−0.9					
70	—	0.6	−0.5	−0.9	−2.1	−3.0	−3.0	−2.1	−1.3					
80	—	−0.1	−0.1	0.6	−0.1	−0.9	−0.6	−1.2	−1.0					
90°S	—	—	—	0.0	0.0	0.0	0.0	0.0	0.0					

(b) March–May

p (mb)	1000	850	700	500	400	300	200	150	100	70	50	30	20	10
Lat.														
90°N	0.0	0.0	0.0	0.0	0.0	0.0	0.0	0.0	0.0	0.0	0.0	0.0	0.0	0.0
80	0.2	0.2	0.0	−0.2	0.5	1.3	0.8	0.1	0.9	0.1	−0.2	0.3	3.0	6.8
70	0.1	0.1	−0.7	−1.8	−2.1	−2.6	1.0	0.9	1.4	2.9	3.8	6.9	10.5	17.5
60	−0.3	−0.6	−1.5	−2.2	−2.5	−2.6	−1.1	−1.3	−2.1	1.0	3.7	9.9	15.0	22.6
50	−0.1	0.1	1.0	1.5	1.8	2.6	1.6	−0.4	−2.9	−0.3	1.9	6.7	10.0	15.7
40	0.4	1.0	1.9	4.9	3.5	6.0	2.8	1.5	−1.6	−0.8	0.8	3.1	4.6	6.6
30	−0.2	0.1	0.8	2.3	3.0	5.3	5.5	6.6	1.1	0.1	−0.9	−0.2	0.3	1.0
20	1.0	−0.6	0.2	0.5	0.6	0.2	−0.3	−1.2	2.9	1.3	0.2	−1.3	−2.5	−3.7
10	1.8	−0.6	0.2	−0.1	−0.1	−1.1	−4.7	−4.2	0.0					
0	0.8	0.0	0.4	0.4	0.3	−0.2	−1.7	−0.1	0.4					
10	−0.5	0.0	0.1	0.0	0.1	0.3	0.6	4.8	1.9					
20	−0.1	−0.4	−0.5	−0.4	0.1	2.0	−1.5	−0.5	−0.4					
30	0.2	−0.3	0.0	−0.3	−0.4	0.9	−0.6	−3.2	−0.2					
40	0.4	0.3	0.2	−0.7	−0.5	−0.2	−0.2	−0.2	0.5					
50	2.0	2.8	1.1	0.1	−0.2	0.2	0.3	1.3	−0.1					
60	0.6	0.9	0.4	−2.1	−1.7	−1.2	−1.2	−1.9	−2.1					
70	—	−0.7	−0.3	−3.0	−2.9	−2.8	−2.9	−2.7	−3.6					
80	—	0.3	−0.3	−1.2	−2.8	−3.3	−2.8	−1.3	−3.5					
90°S	—	—	—	0.0	0.0	0.0	0.0	0.0	0.0					

Table 4.2. Momentum Flux by Standing Eddies [$\bar{u}^*\bar{v}^*$] (m² sec⁻²) (continued)

(c) June–August

p (mb)	1000	850	700	500	400	300	200	150	100	70	50	30	20	10
Lat.														
90°N	0.0	0.0	0.0	0.0	0.0	0.0	0.0	0.0	0.0	0.0	0.0	0.0	0.0	0.0
80	0.1	0.1	0.3	0.1	−0.2	−0.6	−0.5	−0.3	−0.2	−0.2	−0.2	−0.2	−0.3	−0.4
70	0.3	0.6	0.2	−0.7	−0.5	−0.2	−0.4	0.0	0.4	0.3	0.2	0.0	0.0	0.0
60	−0.3	−0.5	−0.1	0.0	0.3	0.7	0.3	0.4	0.8	0.9	1.0	1.1	0.9	0.8
50	0.5	0.9	1.7	1.6	1.8	2.2	0.8	0.3	−1.2	−0.6	0.0	0.2	0.1	−0.1
40	0.0	1.9	0.5	−0.2	0.0	0.0	2.0	2.8	2.9	−1.1	−0.8	−0.3	−0.1	0.1
30	−0.5	−1.8	0.0	−0.6	−0.3	3.1	10.5	12.9	13.1	2.5	−0.4	−0.6	−0.5	−0.2
20	2.2	−2.2	1.0	1.7	0.9	2.8	5.5	1.5	5.8	3.5	1.7	−0.2	−0.6	−1.3
10	2.5	−2.2	−1.2	−0.9	−0.2	1.8	6.7	6.5	4.4					
0	0.4	−0.7	−1.1	−0.8	−0.3	2.5	12.1	17.6	0.4					
10	−0.1	−1.3	0.5	0.1	−0.3	1.5	11.7	18.1	−1.5					
20	0.3	−1.2	−1.0	−1.4	−0.9	0.5	6.0	8.9	1.4					
30	0.2	0.4	2.2	1.6	−0.1	0.8	0.2	0.6	0.1					
40	0.2	0.9	−0.1	0.3	0.7	3.3	−0.7	0.4	0.9					
50	2.0	3.5	1.7	1.0	−0.1	0.7	0.8	0.2	−2.1					
60	1.0	2.1	1.1	−2.3	−2.6	−2.6	−2.9	−3.0	−10.5					
70	—	−0.3	0.0	−2.1	−2.5	−2.7	−4.3	−5.6	−12.0					
80	—	0.3	−0.5	0.2	−0.1	−0.3	−1.6	−1.3	−4.8					
90°S	—	—	—	0.0	0.0	0.0	0.0	0.0	0.0					

(d) September–November

p (mb)	1000	850	700	500	400	300	200	150	100	70	50	30	20	10
Lat.														
90°N	0.0	0.0	0.0	0.0	0.0	0.0	0.0	0.0	0.0	0.0	0.0	0.0	0.0	0.0
80	0.8	0.6	−0.1	0.1	1.3	1.0	−0.8	0.3	0.2	0.4	0.6	−0.3	0.7	2.5
70	1.0	1.0	−0.6	−0.6	−0.5	−0.5	−1.5	−1.8	−1.5	0.0	1.2	2.7	8.0	17.9
60	0.0	−0.8	−1.0	−1.2	−0.9	−1.5	0.5	0.0	−1.5	−0.4	1.3	6.7	14.5	27.9
50	0.1	0.4	1.4	4.1	7.5	12.5	10.8	6.5	2.5	0.4	2.4	7.0	13.0	21.8
40	−0.1	0.4	0.8	2.8	4.0	8.2	8.6	7.0	5.6	3.0	2.0	4.1	6.5	11.5
30	−0.3	−0.2	0.6	1.5	2.1	6.1	12.0	10.6	8.0	3.0	1.2	1.4	2.4	3.9
20	0.0	−1.2	−0.5	−0.1	0.5	2.2	4.0	6.2	3.5	4.5	4.2	3.0	1.8	1.3
10	0.0	−0.8	−0.5	−0.1	0.1	−0.5	−1.8	0.4	0.4					
0	0.3	−1.1	0.3	0.1	0.3	0.6	4.7	6.9	0.4					
10	−0.5	−1.4	−1.0	−1.2	−0.8	1.2	5.9	10.0	0.7					
20	0.0	−0.5	−1.3	−1.0	−0.6	1.8	5.8	10.5	1.8					
30	0.2	−0.2	0.2	0.6	−0.5	1.7	0.6	0.5	0.9					
40	0.1	0.9	0.7	0.9	−1.9	−2.9	−0.4	0.2	0.3					
50	0.6	1.4	0.5	−1.0	−0.2	0.6	0.4	0.7	0.6					
60	0.0	0.1	1.1	−3.0	−1.9	−0.6	−1.9	−3.0	−3.7					
70	—	0.1	0.1	1.2	0.0	−0.9	−2.6	−4.3	−6.9					
80	—	0.0	0.0	−1.0	−0.2	0.3	−0.2	−0.6	−2.2					
90°S	—	—	—	0.0	0.0	0.0	0.0	0.0	0.0					

Table 4.3. Momentum Flux by Transient Eddies $[\overline{u'v'}]$ (m² sec⁻²)

(a) December–February

p (mb)	1000	850	700	500	400	300	200	150	100	70	50	30	20	10
Lat.														
90°N	0.0	0.0	0.0	0.0	0.0	0.0	0.0	0.0	0.0	0.0	0.0	0.0	0.0	0.0
80	0.8	2.1	2.5	7.0	7.5	6.5	3.0	0.0	−5.0	−3.0	−0.3	−1.2	1.5	4.5
70	0.6	0.8	0.4	4.8	2.0	−3.7	−5.2	−7.0	−8.5	−2.5	0.6	3.3	7.0	16.7
60	−1.0	−3.9	−2.2	−1.7	−4.0	−8.5	−3.8	−3.5	−4.0	1.0	3.3	9.2	16.0	29.7
50	−1.5	−4.3	−2.4	3.7	5.5	6.8	6.4	5.5	5.0	4.5	5.2	10.2	19.0	29.5
40	−1.0	2.0	6.1	16.7	20.0	35.6	25.5	27.3	12.5	9.8	6.0	6.4	10.1	21.7
30	0.0	2.9	7.5	18.4	23.5	38.5	36.9	29.8	14.9	8.5	4.8	6.3	10.0	21.0
20	0.5	1.4	4.8	7.4	10.8	15.6	14.3	11.1	7.1	1.3	1.8	1.8	3.2	10.0
10	0.6	1.8	1.3	0.6	2.5	4.3	−7.6	−8.3	−1.6	−0.3	−0.1	0.2	1.0	1.0
0	0.0	0.7	−1.7	−0.3	−0.6	−0.8	−10.3	−17.3	−3.8	−1.2	0.1	−3.7	−0.2	
10	−1.3	−1.5	−3.4	−2.6	−2.4	−6.1	−8.6	−12.8	−2.0	−0.2	0.6	2.6	0.1	
20	−1.3	−2.7	−3.1	−3.9	−6.0	−12.5	−19.4	−16.6	−6.6	−1.4	−0.3	0.8	1.9	
30	1.0	−2.6	−4.0	−8.0	−13.7	−34.8	−45.3	−30.8	−10.8	1.3	2.3	−1.2	−1.2	
40	1.6	−6.4	−5.6	−10.8	−15.6	−36.8	−45.9	−28.1	−13.5					
50	1.6	0.1	−4.2	−1.0	−3.5	−6.8	−17.7	−7.9	−4.9					
60	0.8	2.8	1.9	7.8	11.0	13.1	2.2	1.2	0.1					
70	—	3.8	4.5	10.3	14.5	19.1	10.7	4.4	3.1					
80	—	0.7	2.8	1.9	0.2	−1.1	1.4	2.4	2.5					
90°S	—	—	—	0.0	0.0	0.0	0.0	0.0	0.0					

(b) March–May

p (mb)	1000	850	700	500	400	300	200	150	100	70	50	30	20	10
Lat.														
90°N	0.0	0.0	0.0	0.0	0.0	0.0	0.0	0.0	0.0	0.0	0.0	0.0	0.0	0.0
80	0.3	0.6	3.0	3.8	4.5	5.5	3.0	1.5	−0.4	−0.5	−0.3	2.1	3.0	5.1
70	0.8	1.1	0.1	−0.6	−2.1	−3.6	−2.4	−2.0	0.0	0.5	0.9	4.4	7.0	10.9
60	−0.1	−0.6	−2.6	−1.3	−5.0	−10.5	−4.3	−2.5	1.9	3.0	3.9	7.0	10.8	15.7
50	−1.2	−1.0	−0.7	0.4	3.0	5.8	2.5	2.5	3.6	3.3	4.1	6.0	10.0	13.5
40	−0.9	1.1	3.4	6.1	12.3	20.5	24.1	17.3	7.4	1.3	2.2	4.7	8.4	11.6
30	−0.8	2.9	4.9	13.1	23.4	32.1	38.1	35.8	13.5	3.1	2.2	2.1	2.7	6.5
20	0.1	1.3	2.7	7.0	10.9	21.2	27.4	19.4	5.6	1.1	0.7	1.0	0.8	−0.2
10	0.1	2.1	2.4	2.7	4.2	7.5	4.8	−2.6	−0.7	−0.6	0.2	−0.4	1.1	0.5
0	−0.7	0.0	0.3	0.6	−0.1	0.9	−2.3	−4.5	−2.3	−1.5	−1.1	−2.3	−2.3	
10	−1.2	−2.2	−3.1	−3.4	−5.8	−5.3	−7.8	−4.4	−3.5	0.5	0.3	2.9	0.0	
20	−1.1	−3.1	−5.2	−7.8	−13.9	−18.6	−26.0	−19.3	−7.8	−1.9	−0.4	0.9	0.1	
30	0.9	−5.0	−6.8	−14.3	−28.8	−36.8	−48.2	−30.2	−12.0	−3.3	1.6	−2.0	−3.1	
40	1.8	−6.0	−5.5	−10.0	−22.5	−31.2	−43.9	−24.4	−15.9					
50	2.0	−0.8	−4.5	−1.2	−7.5	−20.3	−21.5	−10.5	−3.4					
60	2.5	4.6	4.4	11.5	10.5	8.0	−1.1	1.5	3.3					
70	—	5.2	8.6	18.0	21.0	24.8	12.9	8.4	7.3					
80	—	1.1	3.7	5.8	5.0	3.9	4.7	4.9	1.9					
90°S	—	—	—	0.0	0.0	0.0	0.0	0.0	0.0					

Table 4.3. Momentum Flux by Transient Eddies $[\overline{u'v'}]$ (m² sec⁻²) (continued)

(c) June–August

p (mb)	1000	850	700	500	400	300	200	150	100	70	50	30	20	10
Lat.														
90°N	0.0	0.0	0.0	0.0	0.0	0.0	0.0	0.0	0.0	0.0	0.0	0.0	0.0	0.0
80	−0.8	−1.1	−0.5	3.2	4.5	6.0	4.0	0.6	−0.2	−0.1	0.0	0.1	0.1	0.1
70	−0.8	−0.8	−0.7	−1.1	0.0	1.3	−1.0	−0.7	−0.4	−0.2	0.0	0.1	0.2	0.2
60	−1.0	−1.1	−0.2	0.3	0.1	−0.2	1.5	1.3	0.8	0.3	0.2	0.2	0.2	0.2
50	0.5	1.2	2.2	4.5	7.0	11.9	19.2	12.5	3.5	1.0	0.1	0.1	0.1	0.1
40	1.1	0.2	2.3	6.9	11.3	24.9	31.9	17.8	5.4	1.8	0.3	0.0	−0.3	0.1
30	1.1	1.3	2.3	5.1	7.2	14.7	20.3	11.4	0.8	0.2	0.4	−0.3	−0.8	−0.4
20	1.0	1.2	2.2	2.9	3.6	6.2	10.3	4.4	−3.1	−1.5	−0.4	0.4	−0.6	−1.0
10	0.4	1.1	1.9	1.3	1.6	2.3	5.1	4.6	−0.5	−1.5	0.0	−0.3	0.0	
0	−0.1	−0.6	−0.9	−1.2	−1.1	2.4	8.8	9.3	2.5	−3.6	−4.0	−1.0	−1.7	
10	−0.9	−2.7	−3.6	−3.3	−5.3	−1.8	6.3	5.2	0.9	0.2	−0.2	−3.9	−1.1	
20	−0.7	−3.7	−6.6	−10.6	−17.3	−22.4	−16.7	−12.6	−3.7	−1.7	−2.4	−2.2	−6.6	
30	1.0	−5.8	−11.2	−23.6	−34.2	−50.0	−45.9	−28.9	−8.5	−2.0	−2.0	−7.3	−8.4	
40	0.7	−6.2	−13.2	−22.3	−29.6	−40.4	−33.3	−22.6	−11.5					
50	0.5	0.8	−6.8	−11.2	−15.0	−18.0	−10.0	−10.3	−2.5					
60	1.0	4.9	2.4	7.1	7.0	9.3	9.7	1.4	5.2					
70	—	7.6	9.8	19.6	24.5	31.4	22.2	10.8	8.6					
80	—	1.7	5.0	11.7	11.5	11.0	8.3	−2.7	2.2					
90°S	—	—	—	0.0	0.0	0.0	0.0	0.0	0.0					

(d) September–November

p (mb)	1000	850	700	500	400	300	200	150	100	70	50	30	20	10
Lat.														
90°N	0.0	0.0	0.0	0.0	0.0	0.0	0.0	0.0	0.0	0.0	0.0	0.0	0.0	0.0
80	0.0	0.3	1.7	3.5	1.0	−4.1	−2.3	−1.5	0.6	0.8	1.0	0.8	1.5	2.1
70	−1.0	−1.0	0.0	1.3	−2.8	−8.3	−4.5	−1.5	0.5	1.0	1.7	2.2	3.2	6.8
60	−1.8	−2.7	−1.0	−0.1	−0.2	−0.4	1.7	1.6	1.5	1.7	1.8	3.4	5.8	9.1
50	−0.8	−1.1	0.4	5.2	10.0	15.0	18.6	12.0	4.9	2.5	1.8	3.7	6.3	10.3
40	−0.4	0.4	3.7	11.1	20.0	30.0	35.0	24.2	7.1	2.5	2.0	3.5	4.3	7.5
30	1.0	2.8	2.7	8.4	12.9	25.2	30.5	23.6	10.1	3.3	1.8	1.3	2.1	2.8
20	1.0	2.4	2.6	4.7	5.8	11.4	16.3	13.6	6.0	−0.2	0.0	1.2	1.7	1.0
10	0.3	2.5	2.0	2.5	3.0	5.0	8.5	8.3	1.1	0.0	−0.5	0.0	0.1	0.5
0	−0.1	0.8	1.4	−0.1	−0.1	2.1	6.9	6.8	0.1	−1.6	−2.5	−0.4		
10	−1.0	−2.4	−2.6	−2.8	−3.7	−3.3	0.0	−0.4	−2.6	−2.0	0.5	1.1		
20	−1.2	−3.2	−5.9	−8.1	−13.8	−18.4	−19.9	−16.6	−5.0	−2.8	−1.0	−1.1		
30	0.9	−4.9	−8.9	−20.5	−30.7	−41.0	−42.2	−25.2	−6.3	−3.4	−10.0	−12.9		
40	2.0	−4.2	−6.7	−26.6	−35.9	−53.3	−44.6	−21.4	−7.0					
50	1.5	1.0	−5.9	−11.8	−16.0	−20.8	−24.2	−11.8	1.7					
60	1.5	2.6	2.8	2.6	6.5	10.4	1.5	4.9	8.2					
70	—	4.5	6.8	13.6	19.0	27.7	20.4	15.5	11.4					
80	—	1.5	3.1	6.0	6.5	7.3	10.2	6.0	2.9					
90°S	—	—	—	0.0	0.0	0.0	0.0	0.0	0.0					

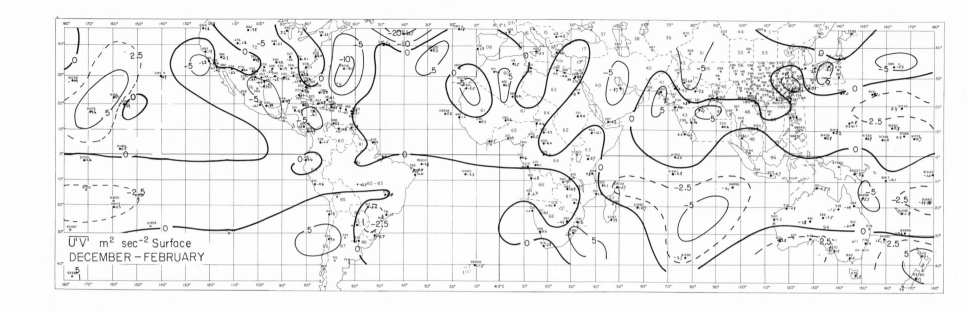

U'V' m² sec⁻² Surface
DECEMBER–FEBRUARY

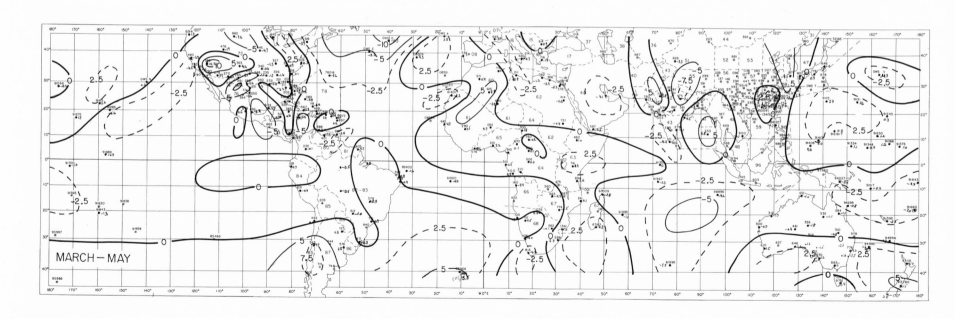

MARCH–MAY

Plate 4.1

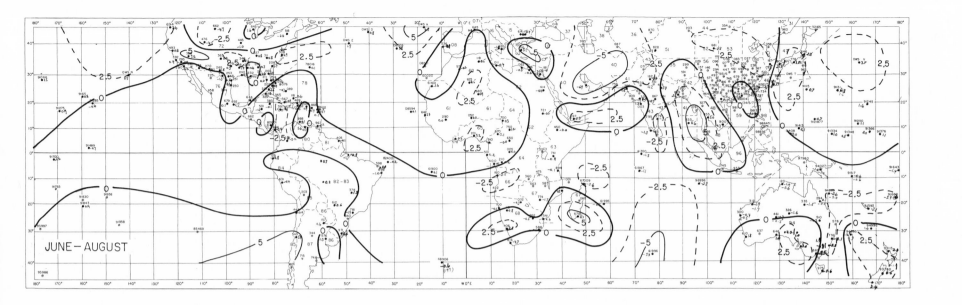

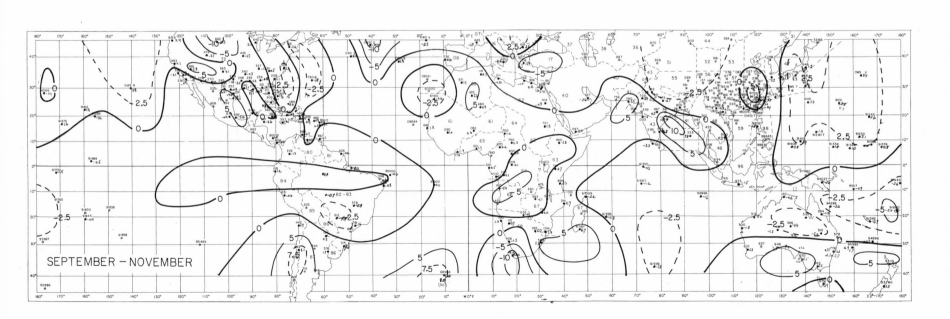

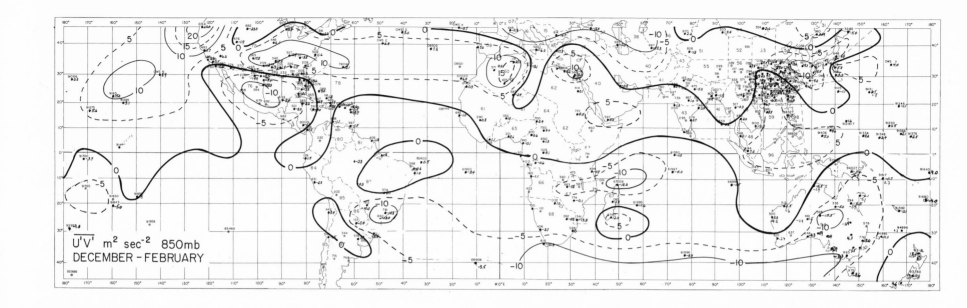

$\overline{U'V'}$ m² sec⁻² 850mb
DECEMBER-FEBRUARY

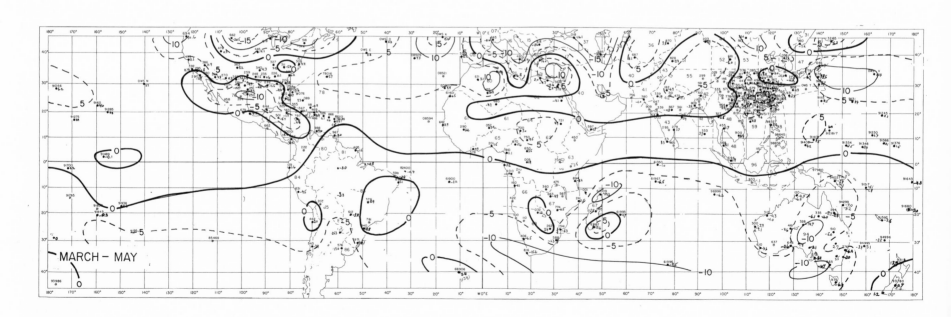

MARCH - MAY

Plate 4.2

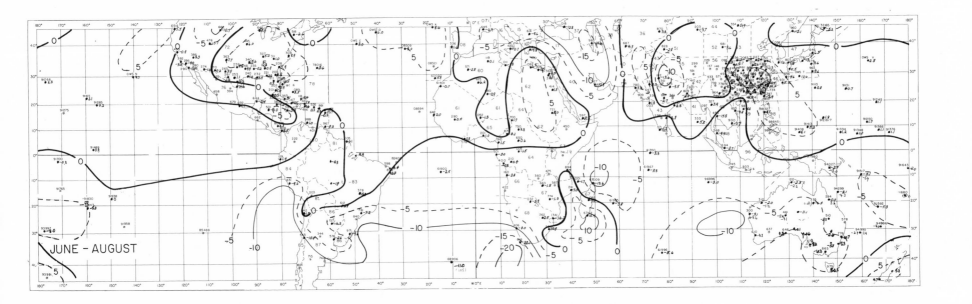

JUNE – AUGUST

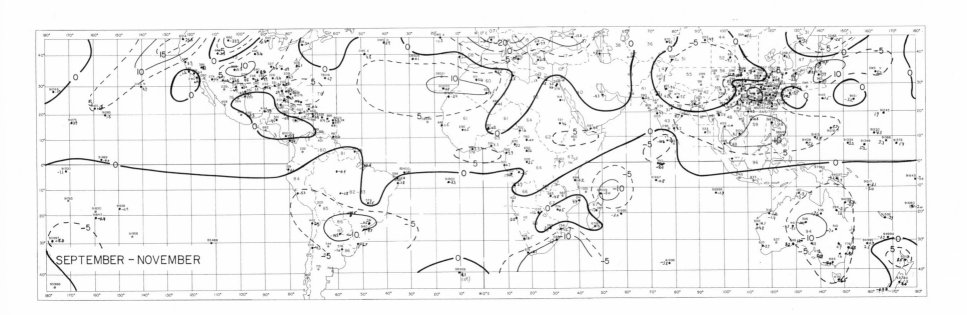

SEPTEMBER – NOVEMBER

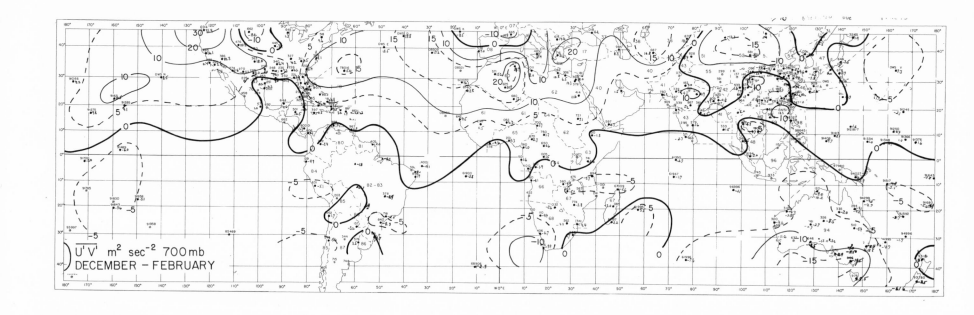

$\overline{U'V'}$ m^2 sec^{-2} 700mb
DECEMBER – FEBRUARY

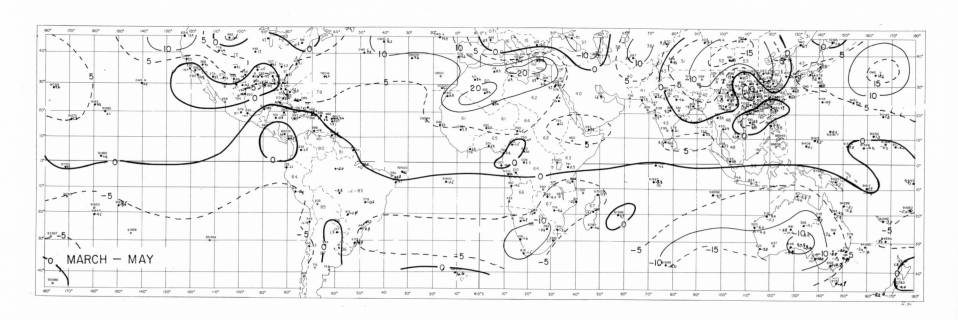

MARCH – MAY

Plate 4.3

174   ANGULAR MOMENTUM BUDGET AND MAINTENANCE OF WINDS

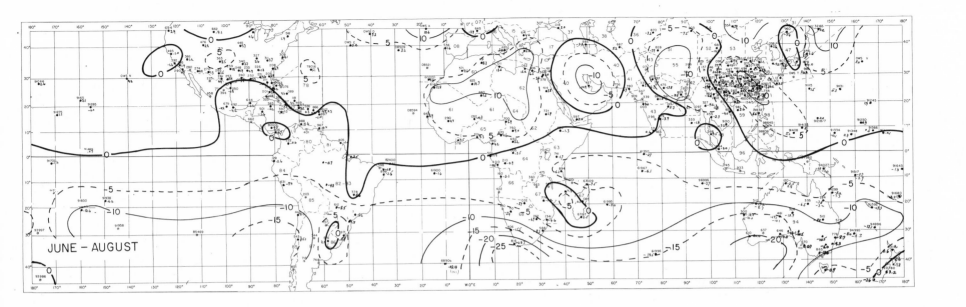

JUNE – AUGUST

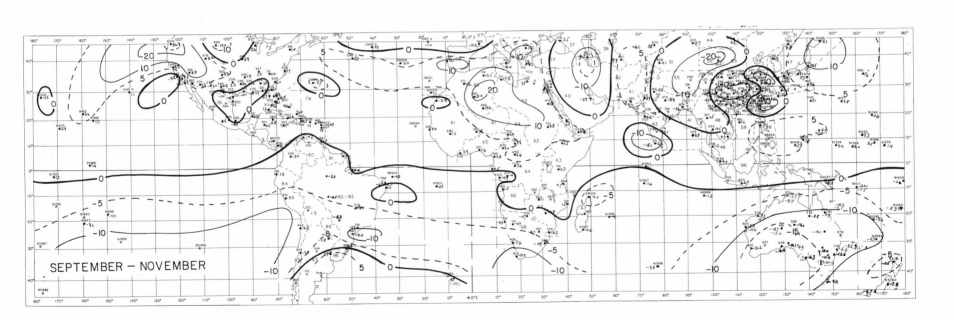

SEPTEMBER – NOVEMBER

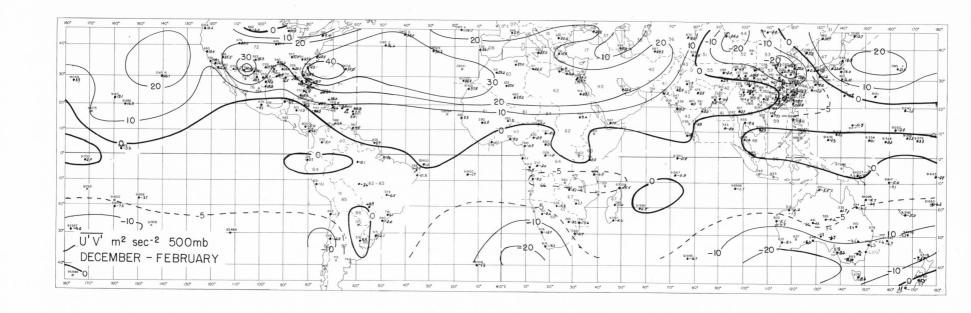

U'V' m² sec⁻² 500mb
DECEMBER – FEBRUARY

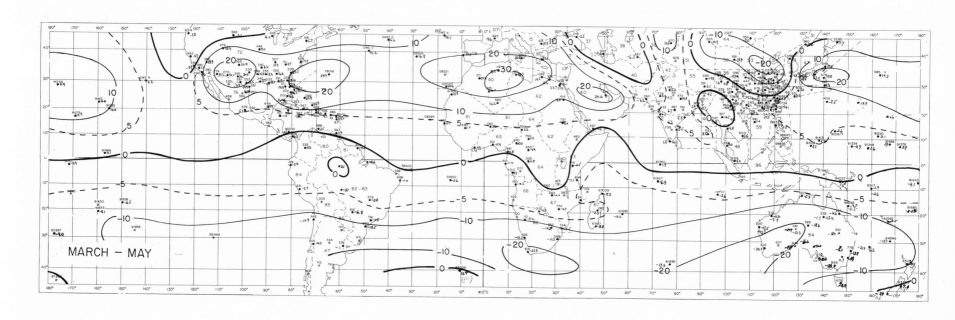

MARCH – MAY

Plate 4.4

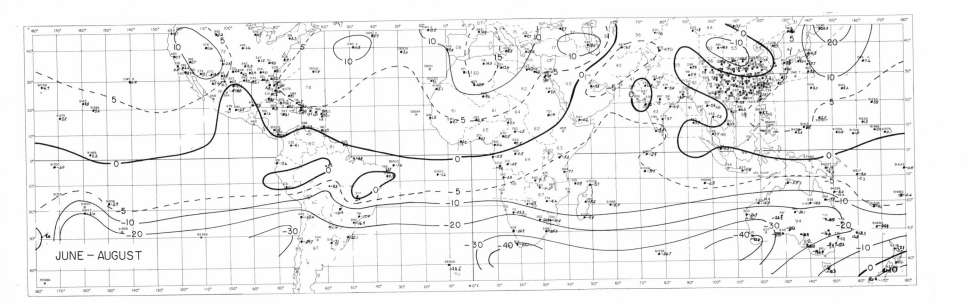

JUNE – AUGUST

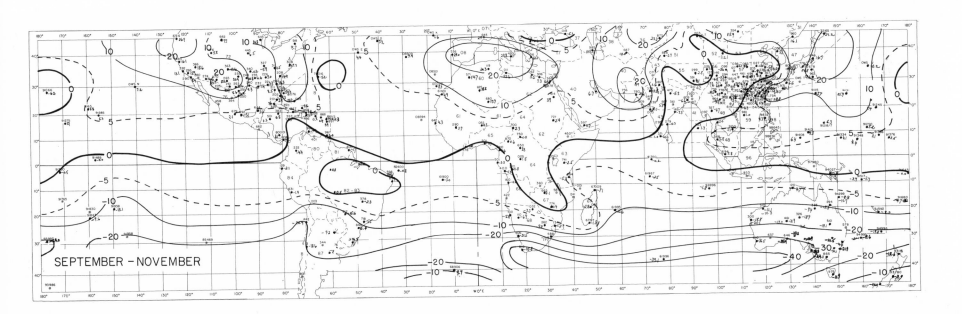

SEPTEMBER – NOVEMBER

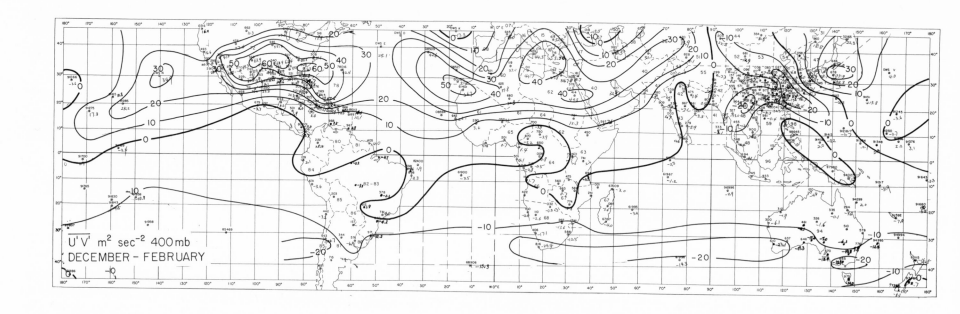

$\overline{U'V'}$ m² sec⁻² 400 mb
DECEMBER - FEBRUARY

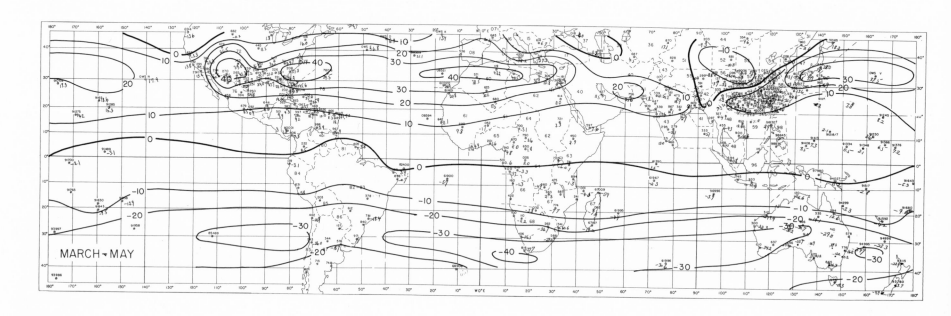

MARCH - MAY

Plate 4.5

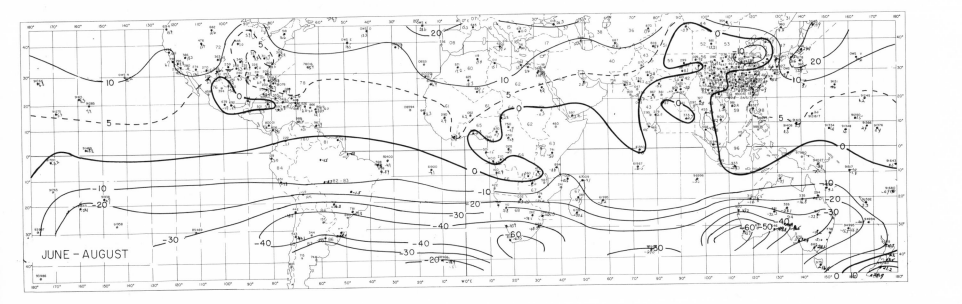

JUNE–AUGUST

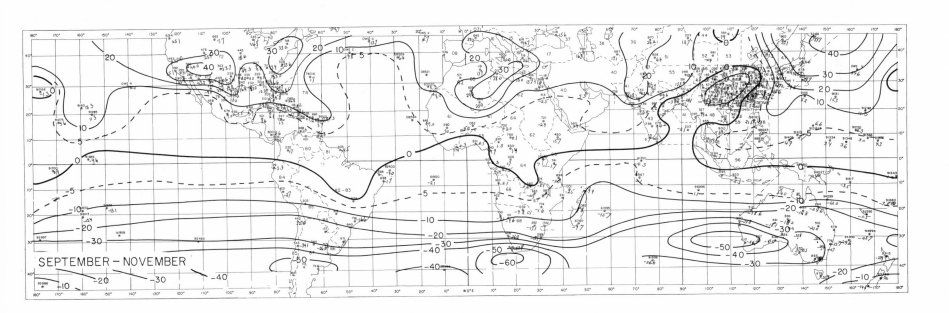

SEPTEMBER–NOVEMBER

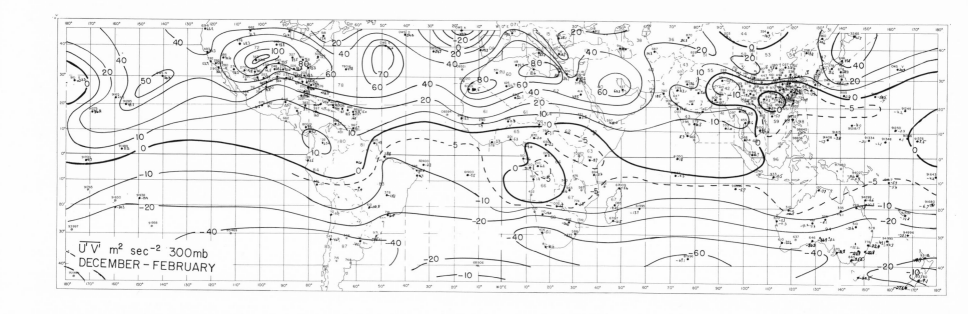

$\overline{U'V'}$ m² sec⁻² 300mb
DECEMBER – FEBRUARY

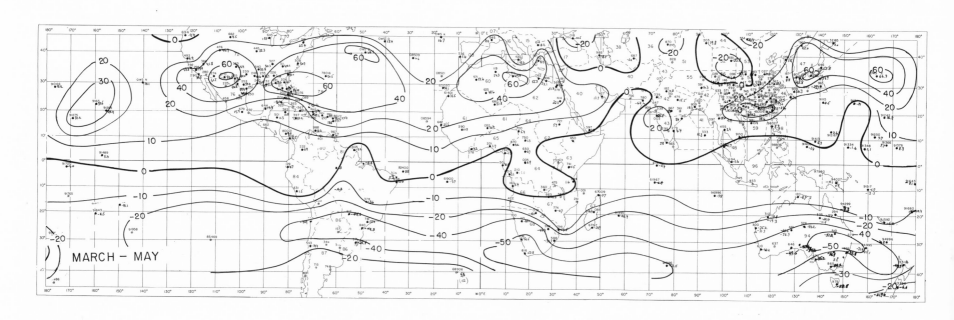

MARCH – MAY

Plate 4.6

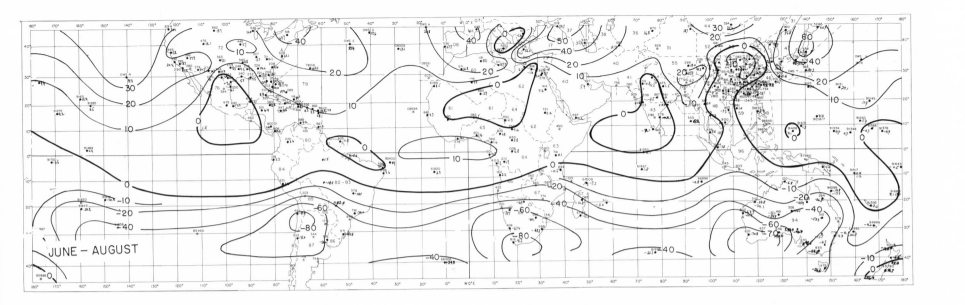

JUNE — AUGUST

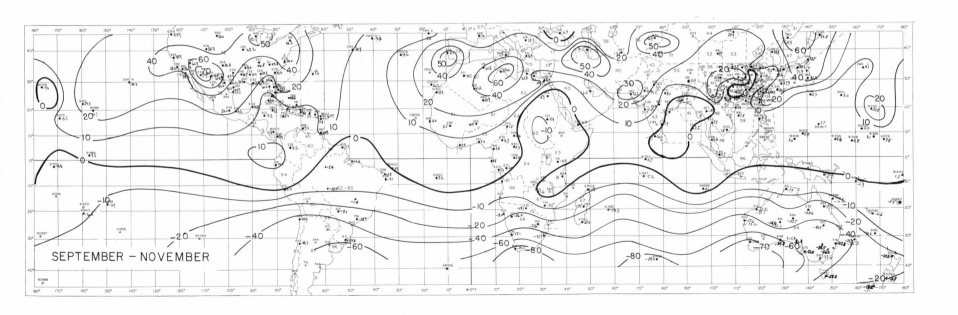

SEPTEMBER — NOVEMBER

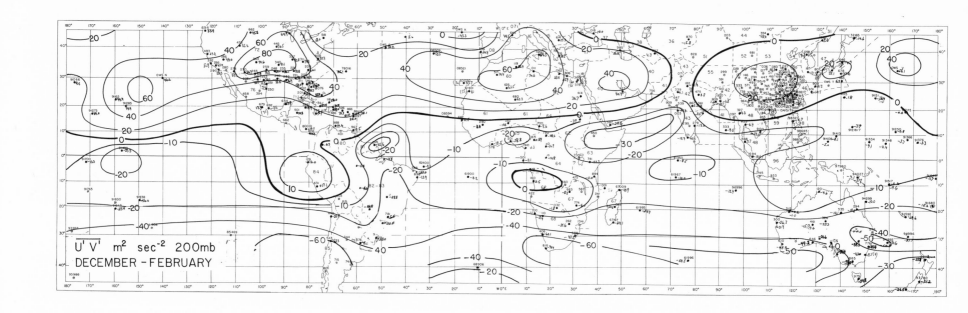

U'V' m² sec⁻² 200mb
DECEMBER – FEBRUARY

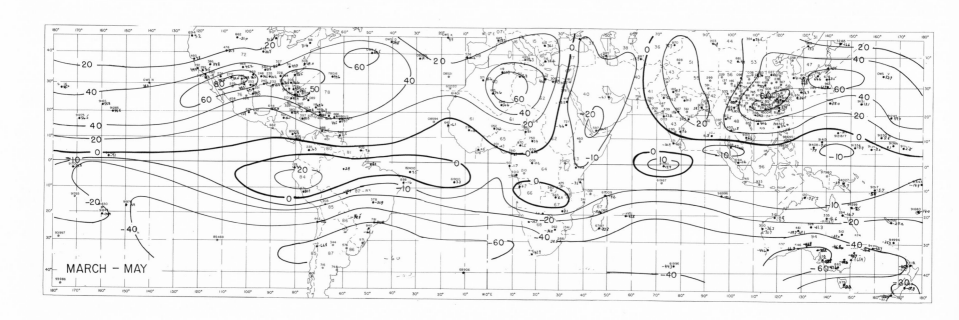

MARCH – MAY

Plate 4.7

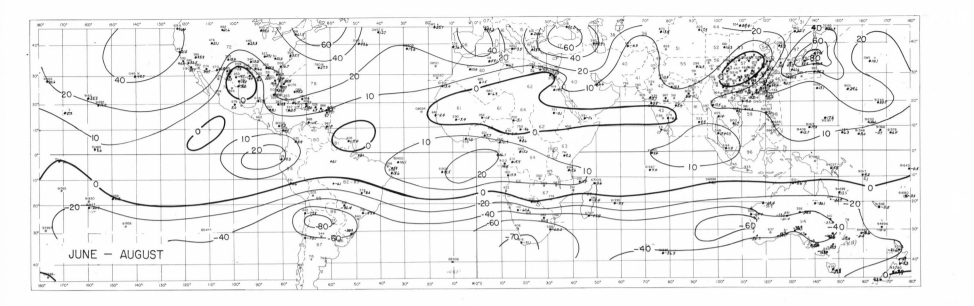

JUNE – AUGUST

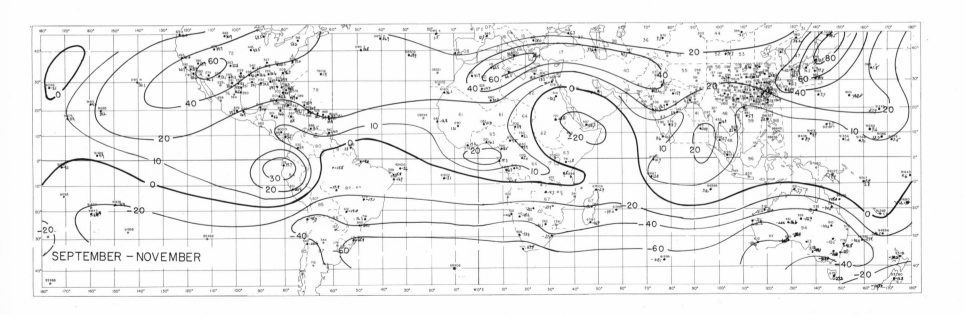

SEPTEMBER – NOVEMBER

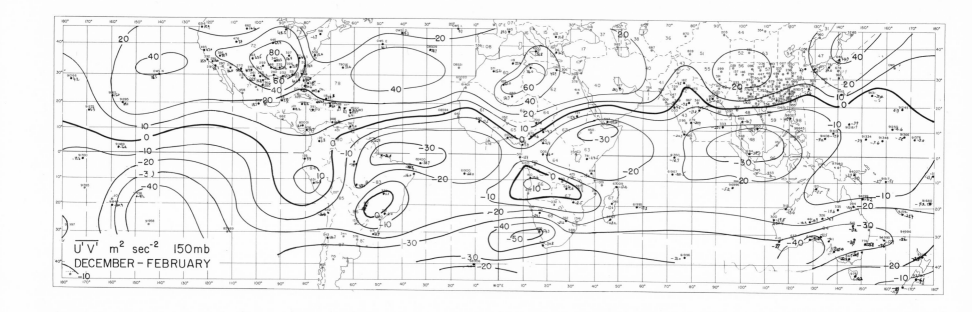

$\overline{U'V'}$ m² sec⁻² 150mb
DECEMBER – FEBRUARY

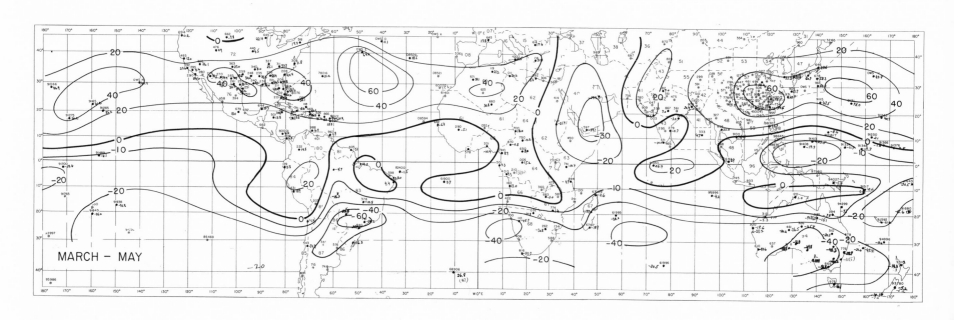

MARCH – MAY

Plate 4.8

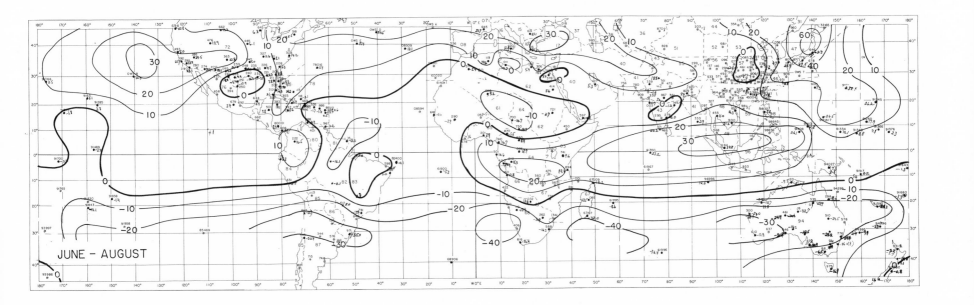

JUNE - AUGUST

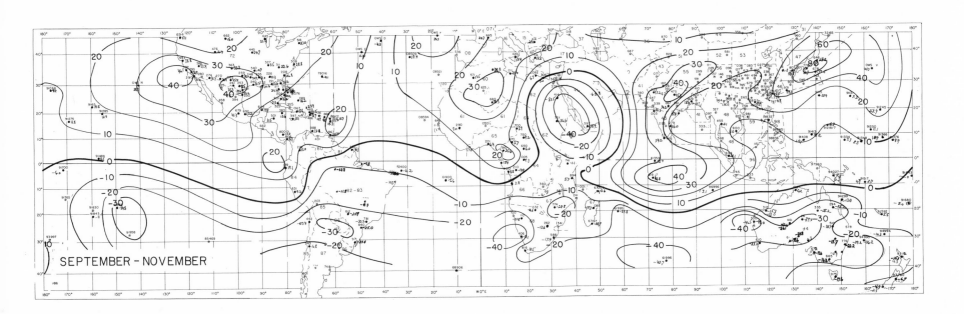

SEPTEMBER - NOVEMBER

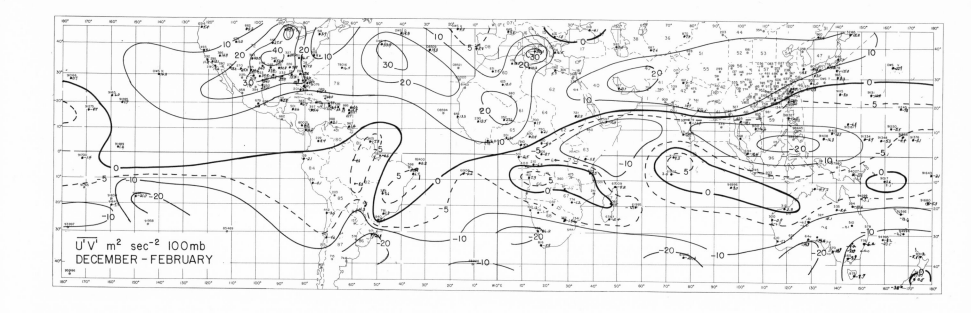

$\overline{U'V'}$ m² sec⁻² 100mb
DECEMBER-FEBRUARY

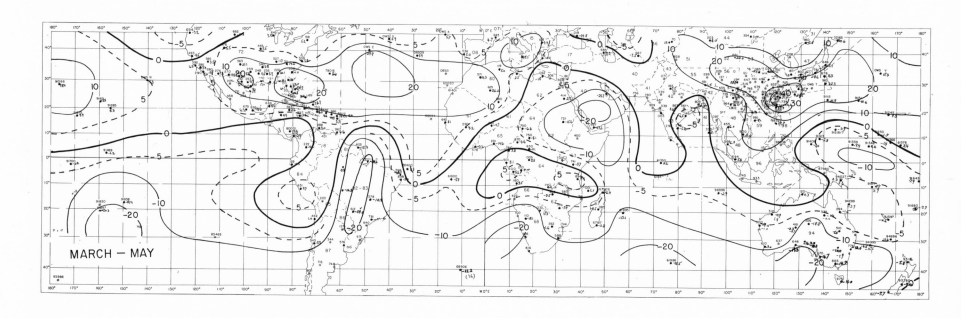

MARCH-MAY

Plate 4.9

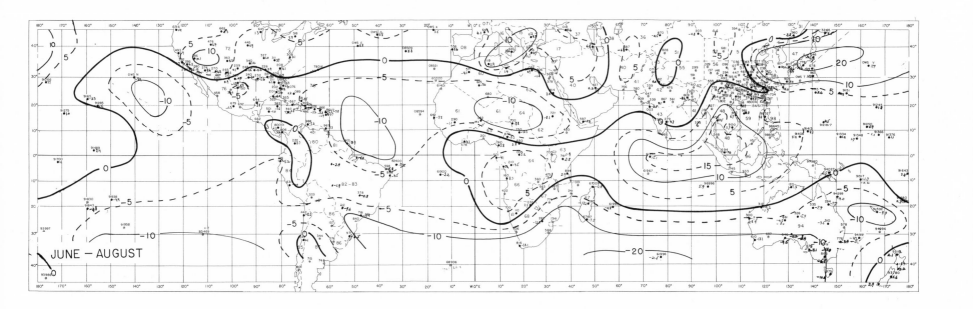

JUNE – AUGUST

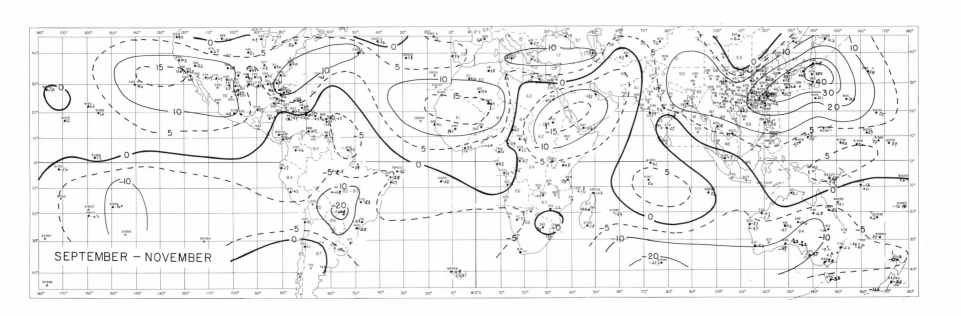

SEPTEMBER – NOVEMBER

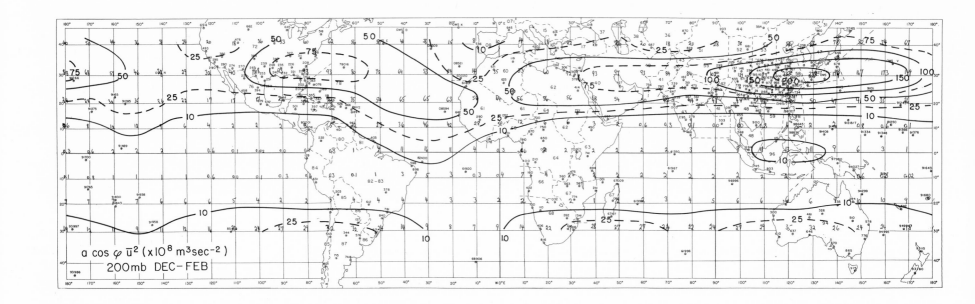

$a \cos \varphi \, \bar{u}^2 \, (\times 10^8 \, m^3 sec^{-2})$
200mb DEC–FEB

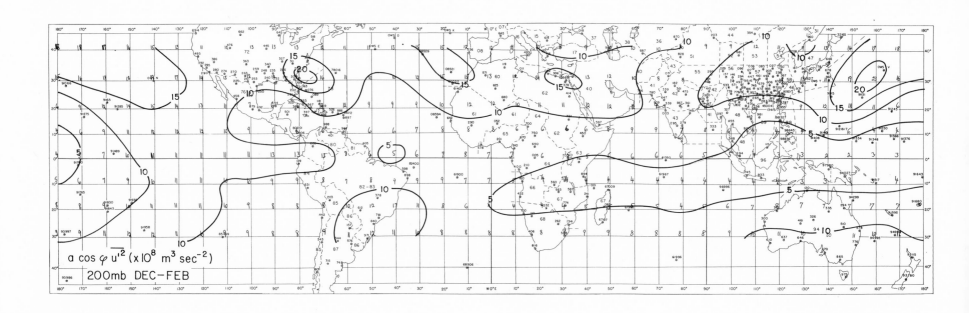

$a \cos \varphi \, \overline{u'^2} \, (\times 10^8 \, m^3 sec^{-2})$
200mb DEC–FEB

Plate 4.10

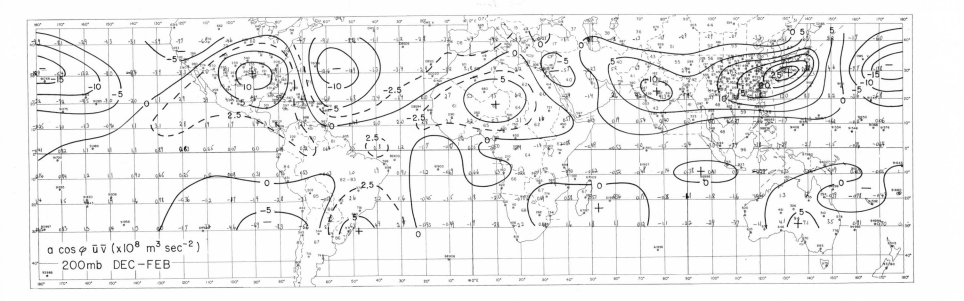

$a \cos \varphi \, \bar{u}\bar{v} \ (\times 10^8 \ m^3 \ sec^{-2})$
200mb DEC—FEB

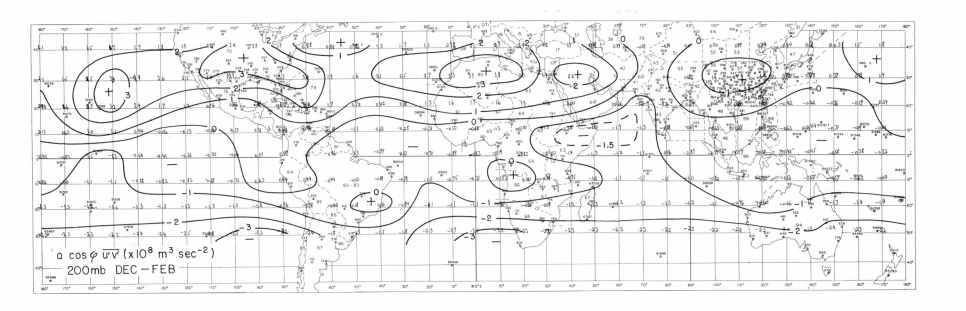

$a \cos \varphi \, \overline{u'v'} \ (\times 10^8 \ m^3 \ sec^{-2})$
200mb DEC—FEB

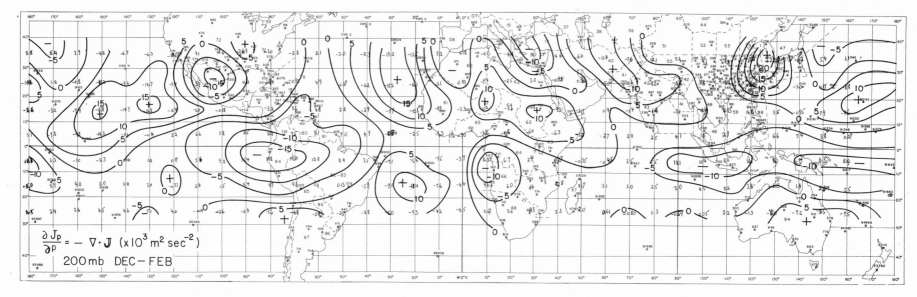

$$\frac{\partial J_p}{\partial p} = - \nabla \cdot \mathbf{J} \ (\times 10^3 \, m^2 \, sec^{-2})$$
200 mb DEC—FEB

Plate 4.11

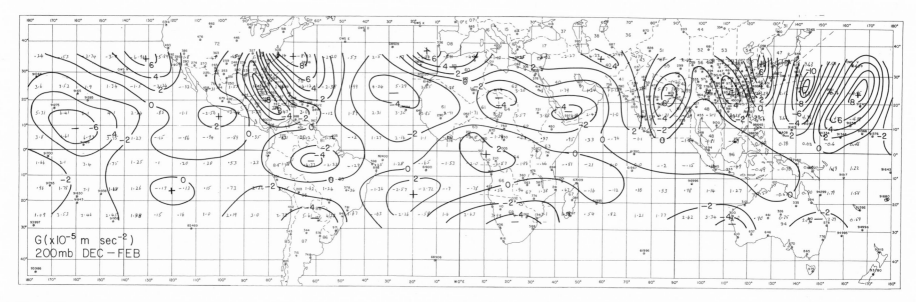

$$G(\times 10^{-5} \, m \, sec^{-2})$$
200 mb DEC—FEB

Plate 4.12

References

Bjerknes, J. 1948. Practical applications of H. Jeffrey's Theory of the general circulation. *Réunion d'Oslo, Programme et Résumé des Memoires*, International Union of Geodesy and Geophysics, Meteorological Association, pp. 13–14.

————. 1955. *Preliminary Study of the Day-to-Day Changes of Angular Momentum during January–February and July–August 1949*. Final Report. Los Angeles: University of California, Department of Meteorology, General Circulation Project, contract no. AF19(122)–48. 13 pp.

Buch, H. 1954. *Hemispheric Wind Conditions during the Year 1950*. Final Report, part 2. Cambridge: Massachusetts Institute of Technology, Department of Meteorology, Planetary Circulations Project, contract no. AF19(122)–153. 126 pp.

Crutcher, H. L. 1959. *Upper Wind Statistics Charts of the Northern Hemisphere (850, 700, and 500 mb levels)*. Office of the Chief of Naval Operations, Navair 50–1C–535, vols. I and II. Washington, D.C.: Government Printing Office.

Eady, E. T. 1950. The cause of the general circulation of the atmosphere. *Centenary Proceedings, Roy. Meteorol. Soc.* pp. 156–172.

Henning, D. 1968. Investigations into regional distribution of transfer of atmospheric parameters over the Equator. Part I, The transfer of relative angular momentum. *Beit. Phys. Atmos.* 41:289–335.

Hutchings, J. W., and W. J. Thompson. 1962. The torque exerted on the atmosphere by the Southern Alps. *New Zealand J. Geol. Geophys.* 5:18–28.

Jeffreys, H. 1926. On the dynamics of geostrophic winds. *Quart. J. Roy. Meteorol. Soc.* 52:85–104.

Kidson, J. W., and R. E. Newell. 1969. Exchange of atmospheric angular momentum between the hemispheres. *Nature* 221:352–353.

Lamb, H. H. 1959. The southern westerlies: A preliminary survey; main characteristics and apparent associations. *Quart. J. Roy. Meteorol. Soc.* 85:1–23.

Lorenz, E. N. 1954. *The Basis for a Theory of the General Circulation*. Final Report, part I. Cambridge: Massachusetts Institute of Technology, Department of Meteorology, General Circulation Project, contract no. AF19–122–153, pp. 522–534.

————. 1967. *The Nature and Theory of the General Circulation of the Atmosphere*. Geneva: W.M.O. 161 pp.

Mintz, Y. 1951. The geostrophic poleward flux of angular momentum in the month of January 1949. *Tellus* 3:195–200.

Molla, A. C., and C. J. Loisel. 1962. On the hemispheric correlations of vertical and meridional wind components, *Geof. Pura e Appl.* 51:166–170.

Munk, W. H., and G. J. F. MacDonald. 1960. *The Rotation of the Earth*. London: Cambridge University Press.

Newell, R. E. 1959. *Some Radar Observations of Tropospheric Cellular Convection*. Weather Radar Research Technical Report no. 33. Massachusetts Institute of Technology, Department of Meteorology, contract no. AF19(604)–2291. 53 pp.

Newell, R. E., D. G. Vincent, T. G. Dopplick, D. Ferruzza, and J. W. Kidson. 1970. The energy balance of the global atmosphere. In *The Global Circulation of the Atmosphere*, G. A. Corby, ed. London: Royal Meteorological Society, pp. 42–90.

Obasi, G. O. P. 1963a. *Atmospheric Momentum and Energy Calculations for the Southern Hemisphere during the IGY*. Report no. 6. Cambridge: Massachusetts Institute of Technology, Department of Meteorology, Planetary Circulations Project. 353 pp.

————. 1963b. Poleward flux of atmospheric angular momentum in the Southern Hemisphere. *J. Atmos. Sci.* 20:516–528.

————. 1965. On the maintenance of the kinetic energy of mean zonal flow in the Southern Hemisphere. *Tellus* 17:95–105.

Priestley, C. H. B. 1948. Heat transport and zonal stress between latitudes. *Réunion d'Oslo, Programme et Résumé des Memoires*, International Union of Geodesy and Geophysics, Meteorological Association, pp. 38–40.

————. 1949. Heat transport and zonal stress between latitudes. *Quart. J. Roy. Meteorol. Soc.* 75:28–40.

————. 1950. On the dynamics of the general atmospheric circulation. *Aust. J. Sci. Res.*, Series A, Physical Sciences, 3:1–18.

————. 1951a. Physical interactions between tropical and temperate latitudes. *Quart. J. Roy. Meteorol. Soc.* 77:200–214.

————. 1951b. A survey of the stress between the ocean and atmosphere. *Aust. J. Sci. Res.*, A, 4, pp. 315–328.

————. 1967. On the importance of variability in the planetary boundary layer. Report of the Study Conference held at Stockholm 28 June–11 July 1967 on the Global Atmospheric Research Programme, sponsored by ICSU/IUGG Committee on Atmospheric Science, COSPAR. World Meteorological Organization. Appendix VI, 5 pp.

Riehl, H. 1954. *Tropical Meteorology*. New York: McGraw-Hill. 392 pp.

Sheppard, P. A. 1953. Momentum flux and meridional motion in the general circulation. *Proceedings of the Toronto Meteorological Conference 1953*, the American Meteorological Society and the Royal Meteorological Society, pp. 103–108.

————. 1958. Transfer across the earth's surface and through the air above. *Quart. J. Roy. Meteorol. Soc.* 84:205–224.

Starr, V. P. 1948. An essay on the general circulation of the earth's atmosphere. *J. Meteorol.* 5:39–43.

————. 1968. *Physics of Negative Viscosity Phenomena.* New York: McGraw-Hill. 256 pp.

Starr, V. P., and R. E. Dickinson. 1963. Large-scale vertical eddies in the atmosphere and the energy of the mean zonal flow, *Geof. Pura e Appl.* 55:133–136.

Starr, V. P., J. P. Peixoto, and J. E. Sims. 1970. A method for the study of the zonal kinetic energy balance in the atmosphere. *Pure and Appl. Geophys.* 80:346–358.

Starr, V. P., and R. M. White. 1952. Note on the seasonal variation of the meridional flux of angular momentum. *Quart. J. Roy. Meteorol. Soc.* 78:62–69.

————. 1954. Balance requirement of the general circulation. Geophysical Research papers, No. 35. Cambridge, Mass.: Geophysical Research Directorate, AFCRL. 57 pp.

Tucker, G. B. 1965. The equatorial troposphere wind regime. *Quart. J. Roy. Meteorol. Soc.* 91:140–150.

U.S. Department of Commerce, 1967. *World Weather Records, 1951–1960.* Washington, D.C.: Government Printing Office. Vols. 1–6.

White, R. M. 1949. The role of mountains in the angular momentum balance of the atmosphere. *J. Meteorol.* 6:353–355.

Widger, W. K. 1949. A study of the flow of angular momentum in the atmosphere. *J. Meteorol.* 6:291–299.

Yeh, T. C., and P. C. Chu. 1958. *Some Fundamental Problems of the General Circulation of the Atmosphere* (in Chinese). Nanking: Academia Sinica, Institute of Geophysics and Meteorology.

Yoshida, K. 1967. An assessment of the transport of momentum in the Australian sector of the Southern Hemisphere. Technical Report No. 7. Melbourne, Australia: International Antarctic Meteorological Research Center, Commonwealth Bureau of Meteorology. 32 pp.

5

Seasonal Variation of Tropical Humidity Parameters
by E. M. Rasmusson

5.1 Introduction

The difficulties inherent in any attempt to describe the features of the climatology and circulation of the tropics from standard synoptic observations have been reviewed by a number of authors, notably Riehl (1954) and Palmer et al. (1955). In common with those for almost all parameters, efforts to define the humidity conditions suffer from the great gaps in the rawinsonde and surface networks which still exist in the region. This present lack of data precludes the possibility of definitive regional water-balance investigations over much of the tropics and leads to some uncertainty in zonally averaged quantities.

In the tropics, the bulk of the horizontal transfer of water vapor takes place below 800 mb. Hence the central analysis problems are connected with the evaluation of the time-mean low-level wind fields, and, to a lesser extent, with the evaluation of the low-level humidity fields. The surface and low-level winds of the continental and island stations of the tropics show a marked diurnal cycle, and this, together with the effects of topography, often overshadows other effects in the lowest 1–2 kilometers. The pronounced influence of local topography and land-sea contrast on the mean low-level wind field was apparent in our data from closely spaced pibal stations over Africa, India, Southeast Asia, and Indonesia. It was not uncommon to find vector differences with magnitudes of several meters per second which were consistent from month to month over distances of 100–200 km or less. The puzzling behavior of the wind, as measured by the radiosonde stations of the Hawaiian Islands and certain Caribbean stations, may well be accounted for primarily in terms of local orographical effects associated with the mountains near the various observing stations (Palmer et al. 1955; Riehl 1954; Rasmusson 1968). All but the smallest atolls exhibit some land-sea breeze effect. For regions with marked topographic relief these diurnal changes become extremely pronounced, and drastic diurnal changes in the wind direction and speed take place (Palmer et al. 1955). Diurnal wind variations associated with large-scale topographic features such as major mountain ranges, large islands, and continents may extend throughout the troposphere (Rasmusson 1966, 1968; Wallace and Hartranft 1969).

These topographically induced features of the wind field have a twofold effect on the analysis. The smaller-scale features may, in areas of sparse data, be interpreted as larger-scale features of the flow, that is, they may lead to serious aliasing. Second, the observed value of the mean wind will depend on the time or times of day at which the observations are taken. Rasmusson (1967b) has pointed out the large diurnal flux variations which occur at many stations surrounding the Caribbean and the Gulf of Mexico, and the necessity of having more than daily observations in order to adequately evaluate the flux divergence over these regions.

Because of the rapid variation of wind and moisture content with height through the planetary boundary layer, one cannot be confident that a linear interpolation between the mandatory levels of 1000 mb and 850 mb will provide an adequate evaluation of the moisture flux in that layer (Palmén 1967). Indeed, examination of rawin data of high vertical resolution from several areas within the tropics clearly indicates the need for at least 50-mb resolution in order to properly evaluate the low-level flux. In addition, the flux between 1000 mb and the surface is of more importance in the tropics than at high latitudes, and it cannot generally be ignored in water-balance computations.

There are other problems which are probably of lesser importance in large-scale water-balance studies but which nevertheless should be kept in mind. Missing humidity observations due to the inability of the sensor to respond to small water-vapor concentrations (i.e., motorboating) will introduce some error, primarily in the middle and upper troposphere. The transfer of water in liquid and solid form is neglected here, as it is in large-scale water-balance computations at higher latitudes.

In spite of these difficulties, a number of aerological investigations have been carried out during the past 20 years which have clarified certain aspects of the water balance of the tropics. Regional studies include those for the northeast trades of the Pacific (Riehl et al. 1951), the Caribbean Sea (Colón 1963), the Gulf of Mexico and Caribbean Sea (Hastenrath 1966; Rasmusson 1966), and the African Continent (Peixoto and Obasi 1965). Riehl and Malkus (1958) have investigated the heat balance of the Equatorial Trough Zone and included an indirect estimate of the water balance.

The first general circulation statistics for water vapor were computed by White (1951), but his computations extended southward only to 25°N and consisted of an evaluation of the transient eddy flux component of the water-vapor transfer. The first attempt to evaluate the total transfer was made by Starr and White (1955) using data from the year 1950. Subsequently, a series of investigations were made of humidity conditions over the Northern Hemisphere during 1950, the final summary of which is presented in a monograph by Peixoto (1958). These investigations were followed by similar studies using the 1958 IGY Northern Hemisphere data (Starr, Peixoto, and Crisi 1965; Peixoto and Crisi 1965). More recently, Starr, Peixoto and McKean (1969) completed a global analysis of the IGY data which provides considerably better definition of mean annual humidity conditions in the vicinity of the equator, and which also provides information on the humidity conditions of the Southern Hemisphere tropics. Lorenz (1967) has reviewed certain results from the 1958 IGY data in his monograph on the general circulation. Since statistics from these data are based on conditions during a single year, only annual and 6-month means were computed by these investigators.

The vapor transport by the Northern Hemispheric mean meridional circulations has been estimated by Palmén and Vuorela (1963) for the 3-month winter season (December–February) and by Vuorela and Tuominen (1964) for the 3-month summer season. Maps of atmospheric water-vapor content have been prepared by Bannon and Steele (1960) and more recently by Tuller (1968). Iida (1968) has made an evaluation of the transequatorial vapor flux using data from 14 stations situated primarily in the western Pacific.

The processing of 5 years of Northern Hemisphere aerological data (May 1958–April 1963) by V. P. Starr's group and of tropical data from the period July 1957–December 1964 (see Kidson et al. 1969) opens the way for a new generation of observational studies of humidity conditions over the Northern Hemisphere and the Southern Hemisphere tropics based upon mean monthly statistics. Even with these data, an initial pilot study (Rasmusson 1967a) indicated the need for additional low-level wind data in order to overcome some of the problems inherent in the evaluation of the low-level vapor flux in the tropics. The basic data have therefore been augmented with the following additional data:

(1) Grid point values of the 1000-mb mean monthly wind components for the Northern Hemisphere from the maps of Crutcher, Wagner, and Arnett (1966). Surface or 1000-mb wind observations from the rawinsonde network alone are only marginally adequate for a proper evaluation of the mean component of the surface flux in the tropics, particularly over ocean areas, as was previously discussed. The maps of Crutcher et al. are based on more data (merchant ships, 3-hourly data, etc.) than are available from the upper air stations alone.

(2) Surface data and pibal data at 950, 900, and 850 mb for more than 150 tropical stations, obtained from wind summaries on file at USAF Environmental Technical Applications Center. These data were included in an effort to improve the vertical resolution of the low-level wind field and to more accurately evaluate the mean wind in those cases where diurnal variations are significant. These data apply generally to Indonesia, India, Pakistan, Southeast Asia, and Africa, but they also provide valuable additions to the basic daily rawinsonde data over South America and the Pacific.

Small but important amounts of data were provided by W. Schallert of the Environmental Data Service, ESSA (processed flux data from 3 Mexican stations). Some additional data were also obtained from "Monthly Climatic Data of the World." Figure 5.1 shows the distribution of stations which provided data at 850 mb.

There is probably little one can do to significantly improve on these data over the Northern Hemisphere until additional observations are available in the areas of sparse coverage. However, somewhat more definitive results can be obtained over the Southern Hemisphere tropics and in the vicinity of the equator by the inclusion of Southern Hemisphere surface wind data from the U.S. Navy Hydrographic Office Pilot Charts and Marine Climatic Atlas of the World. These data have now been processed in connection with studies currently in progress and preliminary analyses have been completed. These analyses will be referred to in the course of this discussion insofar as they clarify or modify our Southern Hemisphere results, which are based on rawinsonde and pilot balloon data only.

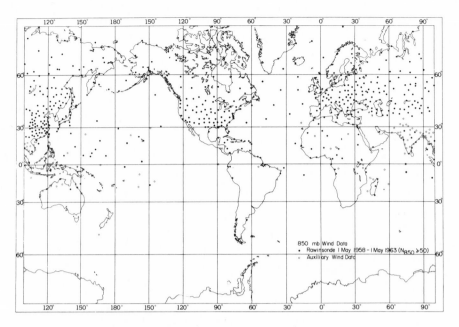

Figure 5.1 Example of data distribution used in this study. Wind data available at 850 mb for the January analyses.

Machine analysis of the various scalar fields was performed using an analysis system developed by J. Welsh of Travelers Research Center (Frazier et al. 1968) and run on the Univac 1108 computer of the Geophysical Fluid Dynamics Laboratory, Princeton, New Jersey. Further details of the analysis scheme are published elsewhere (Oort and Rasmusson 1970).

Although subjective comparison of these analyses with careful hand analyses led to the conclusion that the machine analyses were of good quality, it is undoubtedly true that an experienced analyst, familiar with regional climatology, can improve on the machine analyses.

Notation will be basically that of Starr and Peixoto (1964), and of Rasmusson (1968). In addition to the notation defined in Chapter 1 we define the following quantities:

$$[\overline{W}] = \frac{1}{g} \int_{p_u}^{p_s} [\bar{q}]\, dp, \tag{5.1}$$

water vapor content of a unit column of air (also termed *precipitable water*).

$$[\overline{Q_\lambda}] = \frac{1}{g} \int_{p_u}^{p_s} [\overline{qu}]\, dp, \tag{5.2}$$

total zonal flux of water vapor.

$$[\overline{Q_\varphi}] = \frac{1}{g} \int_{p_u}^{p_s} [\overline{qv}]\, dp, \tag{5.3}$$

total meridional flux of water vapor.

$$M = \frac{1}{g} \int_{p_u}^{p_s} [\bar{q}][\bar{v}]\, dp, \tag{5.4}$$

mean meridional flux of water vapor.

$$SE = \frac{1}{g} \int_{p_u}^{p_s} [\bar{q}^*\bar{v}^*]\, dp, \tag{5.5}$$

meridional standing eddy flux of water vapor.

$$TE = \frac{1}{g} \int_{p_u}^{p_s} [\overline{q'v'}]\, dp, \tag{5.6}$$

meridional transient eddy flux of water vapor.

Here p_s is the pressure at the ground and p_u is the pressure above

which the flux and flux divergence become negligible (it is taken to be 300 mb in this study). It follows from the definitions that

$$[\overline{Q_\varphi}] = M + SE + TE. \tag{5.7}$$

A highly smoothed value of the elevation of the earth's surface was obtained for each grid point, and p_s was estimated from these values. In the computation of the vertically integrated quantities, values of $\bar{q}$, $\overline{qu}$, and $\overline{qv}$ were assumed to decrease linearly from their observed value at 400 mb to zero at 300 mb (where $p = p_u$).

5.2. Low-Level Wind Fields in the Tropics

For purposes of discussing the tropical circulation, it is convenient to visualize the tropical boundaries in terms of some easily identifiable features of the circulation. Since the bulk of the vapor transport takes place in the lower troposphere and is largely determined by the character of the low-level wind field, we have chosen to think of the tropics as the latitude band within which $[\bar{u}] < 0$ at 1000 mb, i.e., where the wind is easterly. The mean northern boundary of the easterlies, as determined from the maps of Crutcher et al. (1966), varies from around 28°N in January–February to around 35°N in August–September (Figure 5.2). Averaged for the year, the boundary lies near 31°N, and thus divides the surface of the hemisphere into two almost equal areas. Our analyses of marine data indicate a seasonal migration of the Southern Hemisphere boundary between approximately the corresponding latitudes. The values of $[\bar{u}]$ and $[\bar{v}]$ at 1000 mb, given in Figures 5.2 and 5.3, are obtained by averaging around the entire latitude circle. Values at grid points for which the 1000-mb surface is below ground level (as given by highly smoothed values of topography) are set to zero. "Typical" values of the wind components for that portion of the latitude circle above ground may be obtained by dividing the values on the diagram by the values given in Table 5.1 at the end of this chapter.

One quickly recognizes the well-known features of the low-level tropical circulation in the mean values of Figures 5.2 and 5.3. The light winds of the "horse latitudes" are found around 30°N. Equatorward of these latitudes, the strong northeast trades of the northern winter and their counterparts, the southeast trades of the southern winter are apparent. At lower latitudes we find the region of light winds (doldrums) associated with the Equatorial Trough. It shifts from one side of the equator to the other with the season in much of the Eastern Hemisphere, but remains in the Northern Hemisphere throughout the year in much of the Western Hemisphere.

Southern Hemisphere marine data indicate that zonally averaged southward flow penetrates only to around 5°S during the northern winter. The data from upper air stations alone indicate a weak penetration to 10–15°S (Figure 5.3). In addition, the marine data show significantly stronger Southern Hemisphere surface easterlies, up to values in excess of 5 m sec⁻¹ during July and August

Perhaps the most significant difference between our averages and those of the earlier studies of Riehl and Yeh (1950) and Tucker (1957), is in the position of the mid-latitude boundary between the zonally averaged surface northerly and southerly winds of the Northern Hemisphere. The maps of Crutcher et al. (1966), from which our data were derived, generally locate this boundary between 30 to 35°N during winter, spring, and summer, and around 35 to 40°N during autumn. Riehl and Yeh show northerlies over the oceans to latitudes higher than 40°N in winter and to around 35°N in summer. Tucker (1957), who computed what corresponds approximately to the mean 1000-mb wind field averaged over both land and sea, found northerlies to around 45°N during winter and to around 40°N in summer.

Since the low-level branches of the mean meridional circulations play a dominant role in the water balance of the tropics, we might briefly note the character of these circulations in the zonally averaged flow at 1000 mb (Figure 5.3). A prominent feature of the flow is the Hadley circulation, which was noted by Riehl in his classic text on tropical meteorology (Riehl 1954), and which has been recently discussed by Kidson et al. (1969). The behavior of the Hadley cells in both hemispheres is discussed in Chapter 3 (see Figure 3.19).

Examination of low-level wind charts (Crutcher et al. 1966; U.S. Weather Bureau 1938) reveals that the monsoon circulations of the Eastern Hemisphere are the large-scale circulation features most responsible for the seasonal variations in the low-level branches of the mean

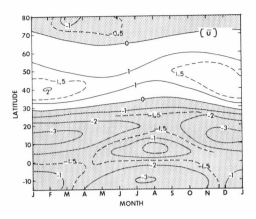

Figure 5.2   Longitudinally averaged zonal wind component [ū] at 1000 mb. Units: m sec⁻¹.

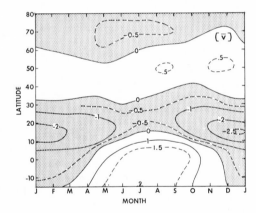

Figure 5.3   Longitudinally averaged meridional wind component [v̄] at 1000 mb. Units: m sec⁻¹.

meridional cells. Winter hemisphere winds exhibit an equatorward component at almost all longitudes. In contrast, the low-level summer hemisphere wind field exhibits a poleward component over most of the Eastern Hemisphere tropics (see Figures 3.14–3.15). Thus we find a summertime situation in which the primary low-level equatorward flow in the latitude belt of the Hadley cell is limited to the eastern portions of the Atlantic and Pacific subtropical highs.

As a matter of interest, one finds support in Figure 5.3 for a weak Ferrel cell between roughly 35 and 55°N, as well as some indication of an arctic Hadley circulation, particularly during spring and summer.

5.3 The Humidity Field

5.3.1 Specific Humidity

The spatial distributions of the semiannually and annually averaged fields of the specific humidity q have been given in the atlas of Peixoto and Crisi (1965) and in the paper of Starr, Peixoto, and McKean (1969).* Maps of mean monthly values of precipitable water content have been constructed by Bannon and Steele (1960) and by Tuller (1968).

Maps of $\bar{q}$ at 1000, 850, 700, 500, and 400 mb, for the 4 mid-season months, are shown in Plates 5.1–5.5. The maps show that:

(a) At 1000 mb in January, the general pattern has a maximum slightly south of the equator. This is especially noticeable in the Pacific. There is a general relation to sea-surface temperatures (see Plate 3.11), and to 1000-mb temperatures as would be expected. Maximum values occur over the continental regions in South America and Africa which correspond to the headwaters of the Amazon and Niger rivers. The absolute maximum of the pattern occurs off the eastern coast of Africa.

(b) In April, the 1000-mb pattern is similar to that of January but shows a slight northward shift and generally slightly higher values in the equatorial region of maximum specific humidity. In July the pattern shows a further northward shift with maximum values of $\bar{q}$ appearing in the region 10–20°N. A marked effect associated with the monsoon circulation is apparent. In October the pattern no longer shows the large values of $\bar{q}$ that were present in July. The broad region of maximum $\bar{q}$ is still present north of the equator.

(c) At 850 mb the overall pattern is similar to that at 1000 mb, although the magnitude of $\bar{q}$ has decreased. There is some evidence of a 3-wave pattern in the structure of the $\bar{q}$ field, with more marked land-sea contrasts apparent at this level than at 1000 mb. Low values tend to occur over continental regions, especially the Sahara and Australia in the Southern Hemisphere winter. The seasonal variation of the pattern is similar to that at 1000 mb. Again the monsoon effect is clearly seen.

(d) At the remaining levels, the overall pattern is similar to that found at lower levels with decreased magnitudes of $\bar{q}$. The seasonal variation of the patterns is similar also, and in general a 3-wave pattern is apparent in the $\bar{q}$ field. This pattern is less obvious in July because of the prominent monsoon effect.

Values of $[\bar{q}]$ for "mid-season" months are given in Table 5.2 at the end of this chapter. Graphs of $[\bar{q}]$ for January and July are given in Figure 5.4. The values for that portion of the latitude circle where the surface of the earth is above the respective pressure surface were set to zero when calculating the zonal average. Humidity values typical of measured values around the latitude circle may be obtained by dividing $[\bar{q}]$ by the appropriate entry in Table 5.1.

The zonally averaged values reflect the distribution of $[\bar{q}]$ which was discussed with respect to the maps. Thus maximum values of $[\bar{q}]$ occur in equatorial low-level regions and values decrease both vertically and poleward. The pattern shifts northward in the Northern Hemisphere summer and southward in Southern Hemisphere summer, with maximum values clearly appearing in the Northern Hemisphere in July. The shift of maximum values into the Southern Hemisphere in January is not as well marked at low levels. The "knees" in the isopleths of

---

* Since the analysis performed in this chapter was completed, two atlases, which include a measure of the water vapor content of the atmosphere through maps of dew point temperature, have been published. They are H. L. Crutcher and J. M. Meserve, 1970: *Selected Level Heights, Temperatures and Dew Points for the Northern Hemisphere*, NAVAIR 50–1C–52, Naval Weather Service Command; and J. J. Taljaard, H. van Loon, H. L. Crutcher, and R. L. Jenne, 1969: *Climate of the Upper Air*, Part 1—*Southern Hemisphere*, Vol. 1: *Temperatures, Dew Points, and Heights at Selected Pressure Levels*, NAVAIR 50–1C–55, Washington, D.C.: Naval Weather Service Command.

$[\bar{q}]$ at middle latitudes are a result of the zonal averaging as mentioned above. The "typical" values of $[\bar{q}]$ which may be obtained using Table 5.1 would show a uniform decrease of $[\bar{q}]$ with height.

Figure 5.5 gives zonally averaged values of $\sigma(q)$ for January and July. The figure shows minimum values of $\sigma(q)$ in equatorial regions and maximum values at more poleward latitudes. The seasonal variation in $\sigma(q)$ shows a northward shifting of the maxima in July in the Northern Hemisphere in the same sense as the seasonal variation of $[\bar{q}]$. The graphs of $\sigma(q)$ should be compared with those of $\sigma(u)$ and $\sigma(v)$ (Figures 3.20, 3.21), which have a minimum in equatorial region and a maximum at about 50°N at 300 mb. Qualitative considerations suggest that $\sigma(q)$ should be related to the standard deviation of the wind in a manner dependent on the gradient of $\bar{q}$, and in a crude sense this is reflected in the patterns of Figure 5.5.

In contrast to wind data, only a small amount of independent specific humidity data were available at 950 and 900 mb, and most of the values at these levels were interpolated from reports at 1000 and 850 mb. Statistical estimates of q were used for a number of stations when direct measurements were unavailable because of "motorboating." In general, missing data from a station will probably be most often associated with motorboating and consequently with low-humidity values, and calculated values of $\bar{q}$ may tend to be slightly higher than their true values, particularly in the middle and upper troposphere.

Table 5.3 and Figure 5.6 reveal some interesting features of the vertically integrated specific humidity distribution, that is, the water vapor content $[\overline{W}]$. Poleward of 10°, the water vapor content falls to a minimum around early February and reaches a maximum around early August. Equatorward of these latitudes, a band of maximum water vapor content, which migrates from around 5 to 10°S during the southern summer to near 10°N during the northern summer, can be clearly identified. Values of $[\overline{W}]$ associated with this maximum range between 4.5 and 5.0 gm cm^{-2}.

Table 5.3 shows the magnitude and time of occurrence of the mean monthly absolute maximum and minimum of $[\overline{W}]$. In the latitude band exhibiting a double maximum, the vapor content is found to be highest during the period when the maximum migrates northward,

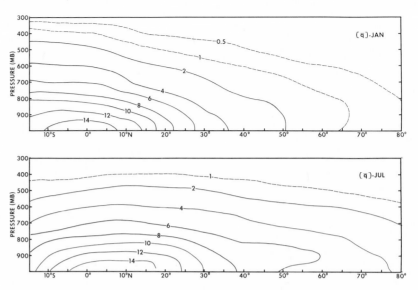

Figure 5.4 Longitudinally averaged specific humidity $[\bar{q}]$ for January and July. Units: gm kg^{-1}.

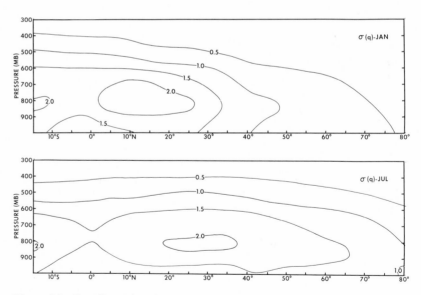

Figure 5.5 Zonally averaged standard deviation of specific humidity $\sigma(q)$ for January and July. Units: gm kg^{-1}.

Table 5.3. Ranges in Mean Monthly Water Vapor Content $[\overline{W}]$. Units: gm cm^{-2}

	Latitude 15°S		10°S		5°S		EQ		5°N		10°N		15°N		20°N		25°N		30°N	
sfc–850 mb																				
Max	Feb.	4.5	Feb.	4.6	Mar.	4.8	Apr.	4.9	May	4.8	Aug.	4.7	Aug.	4.6	Aug.	4.4	Aug.	4.0	Aug.	3.5
Min	Aug.	2.8	Aug.	3.5	Aug.	4.0	Aug.	4.4	Feb.	4.2	Feb.	3.5	Feb.	3.0	Feb.	2.4	Feb.	2.0	Feb.	1.5
Max-Min		1.7		1.1		0.8		0.5		0.6		1.2		1.6		2.0		2.0		2.0
850–300 mb																				
Max	Feb.	2.4	Feb.	2.4	Apr.	2.6	May	2.6	May	2.5	Aug.	2.4	Aug.	2.3	Aug.	2.3	Aug.	2.2	Aug.	1.9
Min	Aug.	1.5	Aug.	1.7	Aug.	2.0	Aug.	2.2	Feb.	2.0	Feb.	1.6	Feb.	1.4	Feb.	1.1	Feb.	1.0	Feb.	0.8
Max-Min		0.9		0.7		0.6		0.4		0.5		0.8		0.9		1.2		1.2		1.1
% of total water vapor above 850 mb		54		51		51		51		51		49		49		49		51		50
% of total seasonal change occurring above 850 mb		53		64		75		80		83		67		56		60		60		55

that is, during the northern spring. Minimum values occur during the respective winters in this latitude band. Minimum values in the vicinity of the equator are almost the same during the northern and southern winters, differing by only 0.1 gm cm^{-2}. The maximum seasonal changes in water vapor content are found in the latitude band 20–30°N, with smaller changes occurring both poleward and equatorward of this region.

About 50 percent of the total water vapor content of the tropical atmosphere is found above 850 mb. Outside of the latitude band 10°S–10°N, this layer also accounts for a comparable percentage of the seasonal change in total water vapor content (50–60%). However, in the vicinity of the equator the seasonal changes above 850 mb become relatively more important (Table 5.3) due to the small seasonal change in the water content of the 1000–850 mb layer.

The seasonal migration of the specific humidity maximum is considerably more restricted than that of the temperature maximum. According to Kidson et al. (1969), the temperature maximum moves from 5°S during the southern summer to 20–25°N during the northern summer. Failure of the humidity maximum to move so far northward appears to be due largely to the difference in the influence of the Northern Hemisphere desert regions on the values of $[\overline{T}]$ and $[\bar{q}]$. Winter-to-summer changes in the temperature of the lower troposphere are extremely pronounced over these regions (see Crutcher and Meserve 1966) when compared with other longitudes and with lower latitudes, and these changes exert a strong influence on the seasonal change in $[\overline{T}]$. Seasonal humidity changes over the deserts exert a weaker influence on the zonally averaged values of $\bar{q}$.

5.3.2 Relative Humidity

Figure 5.7 shows values of the zonally averaged relative humidity for the months of January and July calculated from

$$[\overline{RH(p)}] = \left[\frac{\bar{q}(p)}{q_s(p, \overline{T})}\right], \tag{5.8}$$

where $\bar{q}$, $\overline{T}$ are mean monthly values of specific humidity and tempera-

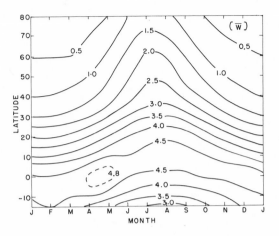

Figure 5.6 Longitudinally averaged, vertically integrated water vapor $[\overline{W}]$. Units: gm cm^{-2}.

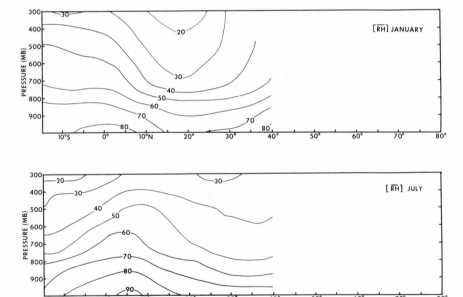

Figure 5.7 Zonally averaged relative humidity for January and July in percent.

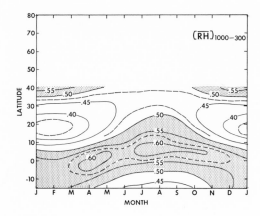

Figure 5.8   Zonally averaged mean relative humidity for the layer 1000 mb–300 mb.

ture. Figure 5.8 gives the zonally averaged mean relative humidity for the 1000–300-mb layer,

$$[\overline{RH}]_{1000-300} = \frac{1}{700} \int_{300}^{1000} [\overline{RH(p)}]\, dp.$$

These values were obtained assuming constant temperatures equal to the monthly mean, and this procedure may introduce errors in the computed values of relative humidity. However, it is unlikely that these errors exceed a few percent for the zonally averaged values, and the seasonal variations in relative humidity are depicted quite adequately.

The lowest relative humidities occur during winter in the subtropics, as expected. The northward migration and decrease in intensity of the minimum from winter to summer are broadly consistent with the movement and change in intensity of the downward-motion branch of the Northern Hemisphere Hadley cell (see Chapter 3, and Oort and Rasmusson 1970). Similarly, the location and seasonal migration of the axis of maximum relative humidity broadly coincides with the position of the ascending branch of the Hadley cells.

Maximum seasonal relative humidity changes in excess of 20 percent occur in the mid-troposphere over the subtropics.

It is worth noting here that it is the relative humidity that is sensed by the instruments normally carried aloft on radiosondes. The instruments do not function at the low temperatures and humidities found above about 300 mb, and as a result there is no routine humidity data available for the upper troposphere and stratosphere. This situation is in contrast to that for the other main parameters discussed in this monograph, wind and temperature, which are routinely measured to heights of about 30 km by the standard balloon network. Basically the deficiency is a technological one.

While the amount of moisture above 300 mb may be small compared with that below, it is of utmost importance in governing the formation of clouds in the upper troposphere and stratosphere. It is also of importance in the radiation budget, as will be seen in Chapter 6 (Volume 2).

Another region where data are inadequate is over the deserts, and here the reason is basically the lack of stations. In view of the immense practical importance of the water budget of desert regions, it seems desirable to increase the station coverage.

5.4 Horizontal Flux

5.4.1 The Zonal Flux

The zonal vapor flux $[\overline{qu}] = [\bar{q}][\bar{u}] + [\bar{q}^*\bar{u}^*] + [\overline{q'u'}]$ consists primarily of the contribution $[\bar{q}][\bar{u}]$ by the longitudinally averaged zonal wind. Zonal transient eddies $[\overline{q'u'}]$ and standing eddies $[\bar{q}^*\bar{u}^*]$ have some interesting characteristics, but their contribution to the total vertically integrated zonal transport is typically 1 to 2 orders of magnitude less than the mean term (Peixoto and Crisi 1965).

Values of $[\overline{qu}]/g$ at selected pressure surfaces for the "mid-season" months are given in Table 5.4 at the end of this chapter. This term $[\overline{qu}]/g$ is somewhat more convenient than $[\overline{qu}]$. The numbers are of a convenient size and the units are appropriate when integrating vertically in pressure coordinates. Values for January and July are illustrated in Figure 5.9. In the tropics, the major contribution to the flux occurs in the 1000–850-mb layer. This is very different from the situation in middle latitudes, where the rapid increase of west winds with height results in a zonal flux maximum which is usually found around the 700-mb level (see Figure 5.9). The line of zero zonal transport, above which the mean zonal transfer in the tropics is westerly, corresponds closely with the line $[\bar{u}] = 0$ as would be expected. The slope of this line is a reflection of the westerly thermal wind in this region, which increases from summer to winter in response to the increase in the north–south thermal gradient in the troposphere.

Values of $[\overline{Q_\lambda}]$, the vertically integrated zonal flux, are shown in Figure 5.10. The pattern is generally similar to that of the 1000-mb zonal wind field; but as the boundary between easterlies and westerlies slopes southward with height, and as higher humidity values are found toward the equator, the maximum vertically integrated westward transport in winter is displaced about 4 to 6° equatorward of the trade-wind maximum. Similarly, the dividing line between easterly and westerly vapor transport is displaced equatorward of the boundary

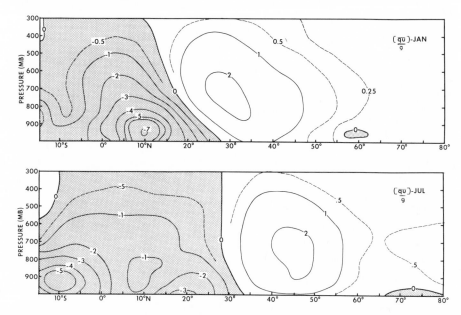

Figure 5.9   Longitudinally averaged zonal flux of water vapor $[\overline{qu}]/g$ for January and July. Units: gm (cm mb sec)$^{-1}$.

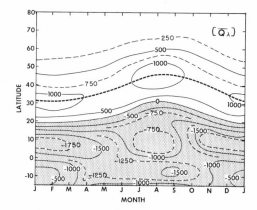

Figure 5.10   Longitudinally averaged, vertically integrated zonal flux of water vapor $[\overline{Q_\lambda}]$. Units: gm (cm sec)$^{-1}$.

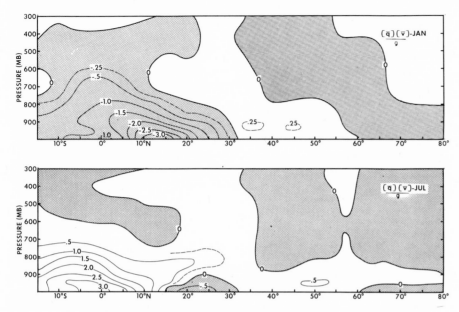

Figure 5.11 Horizontal flux of water vapor by the mean meridional circulation $[\bar{q}][\bar{v}]/g$ for the months of January and July. Units: gm (cm mb sec)$^{-1}$.

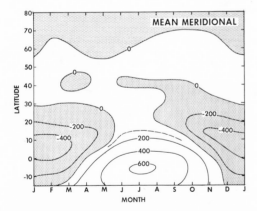

Figure 5.12 Vertically integrated flux of water vapor by the mean meridional circulation. Units: gm (cm sec)$^{-1}$.

between surface easterlies and westerlies by about 8° in winter and by about 4 to 5° in summer. As has been previously pointed out, the data used in this study probably lead to an underestimate of the strength of the low-level easterly flow in the Southern Hemisphere tropics. Consequently, south of the equator, values of the low-level zonal flux (Figure 5.9; Table 5.4), and the vertically integrated zonal flux (Figure 5.10) probably represent underestimates of the actual easterly vapor transport.

5.4.2 The Meridional Flux

Of particular interest, from the standpoint of general circulation studies, are the magnitude and characteristics of the longitudinally averaged meridional flux, $[\overline{qv}]/g$. Values of the total meridional flux for the 4 mid-season months are given in Table 5.5 at the end of this chapter. Vertically integrated values are presented in Figure 5.19. The meridional flux is even more heavily weighted toward the layer 1000–850 mb than is the zonal flux. The maximum value is usually found below 850 mb; most often near 1000 mb in the region of strongest meridional flow. The necessity for a good 1000-mb analysis is apparent.

1 MEAN MERIDIONAL FLUX   The meridional flux $[\bar{q}][\bar{v}]/g$ due to mean motions is illustrated in Figure 5.11 for the months of January and July. The dominant role of the lower branch of the winter-hemisphere Hadley cell is immediately apparent. The Hadley circulation is such as to transport moisture from subtropical to more tropical latitudes so as to maintain the gradient of $[\bar{q}]$ and to provide a source for the precipitation processes which occur in the region.

Since the upper branch of the cell is concentrated just under the tropopause, where water vapor content is low, it contributes little to the total flux. The small values of the flux in the mid-troposphere are a consequence of the small values of both $[\bar{q}]$ and $[\bar{v}]$ in this region. The vertically integrated mean meridional flux is shown in Figure 5.12. Note the large vapor transfer across the equator which occurs during all months, except April and November, with maxima around the time of the solstices.

Conditions during the northern summer are unclear between latitudes 15–30°N. Figure 5.3 suggests a weak Hadley cell circulation in this latitude band, and therefore a weak southward flux of water vapor by the mean meridional circulation. However, our analysis shows the low-level branch of this circulation to be only 150 mb deep (see Figure 5.11), and so weak that it is offset by northward transport aloft. It should be noted that no attempt was made to adjust $[\bar{v}]$ so as to satisfy the condition of no net mass flux through latitude walls. It was felt that for the case of water vapor transfer, which takes place primarily in the lower troposphere, where wind reports are most plentiful, any attempt at satisfying the mass flux condition could well do more harm than good. Even without adjusting $[\bar{v}]$, the net mass flux across latitudes in the tropics was usually found to be surprisingly small (Oort and Rasmusson 1970). This was less the case at higher latitudes, however. The general features of the mean meridional circulations are in satisfactory agreement with the computations of Kidson et al. (1969).

2 STANDING EDDY FLUX Values of the meridional transfer of water vapor by the standing eddies, $[\bar{q}^*\bar{v}^*]/g$, are presented in Table 5.6 at the end of the chapter for the 4 mid-season months. The transfer during January and July is illustrated in Figure 5.13, and the vertically integrated values are shown in Figure 5.14. The standing eddy flux has received relatively little attention in previous studies. As noted by Kidson et al. (1969), there is some uncertainty in the evaluation of this component at low latitudes, where data are sparse. However, one would expect most of the contribution to come from the longer wavelengths, which are evaluated most accurately.

The Northern Hemisphere standing eddy flux exhibits two axes of maximum values. The northern maximum is most strongly developed at about 50°N during winter. This maximum migrates northward and all but disappears during summer. In contrast, the southern maximum is difficult to identify during winter and reaches its maximum intensity during summer around 25°N. Examination of low-level mean monthly winds (Crutcher et al. 1966) and maps of mean monthly sea-level pressure (Byers 1959, p. 266) suggests an association between the northern maximum and the semipermanent oceanic cyclones which are strongly developed during winter and which are weak during summer. So far as the tropics are concerned, the southerly maximum, which appears to be associated with the subtropical high pressure belt and the monsoon circulations, is of primary interest.

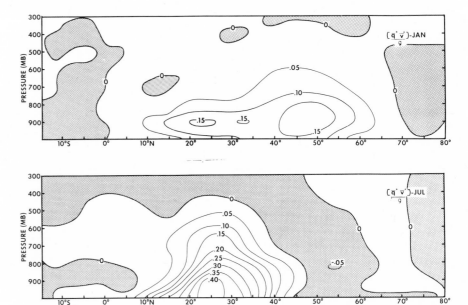

Figure 5.13 Longitudinally averaged meridional flux of water vapor by standing eddies $[\bar{q}^*\bar{v}^*]/g$ for the months of January and July. Units: gm (cm mb sec)$^{-1}$.

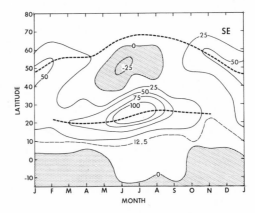

Figure 5.14 Longitudinally averaged vertically integrated meridional flux of water vapor by standing eddies. Units: gm (cm sec)$^{-1}$.

Table 5.7. Standing-Eddy Parameters. Units: $\sigma(\bar{q})$: gm kg^{-1}; $\sigma(v)$: m sec^{-1}

		10°N			20°N			30°N			35°N		
		$\sigma(\bar{q})$	$\sigma(\bar{v})$	$r_{\bar{q},\bar{v}}$	$\sigma(\bar{q})$	$\sigma(\bar{v})$	$r_{\bar{q},\bar{v}}$	$\sigma(\bar{q})$	$\sigma(\bar{v})$	$r_{\bar{q},\bar{v}}$	$\sigma(\bar{q})$	$\sigma(\bar{v})$	$r_{\bar{q},\bar{v}}$
1000 mb	Jan.	1.3	1.1	.16	1.9	1.2	.33	1.6	1.4	.61	1.7	1.5	.63
	Jul.	1.2	2.4	.52	2.2	2.6	.77	2.6	2.6	.85	2.6	2.4	.71
850 mb	Jan.	1.4	1.1	.19	1.8	1.3	.43	1.2	1.7	.54	0.9	2.3	.67
	Jul.	0.8	1.0	.13	2.1	1.6	.48	2.6	2.4	.56	2.3	2.1	.56

The standing eddy transfer may be discussed in terms of the q and v fields through the relationship

$$[\bar{q}^*\bar{v}^*] = r_{\bar{q},\bar{v}}\,\sigma(\bar{q})\,\sigma(\bar{v}),$$

where $\sigma(\bar{q})$, $\sigma(\bar{v})$ are the *spatial standard deviations* of $\bar{q}$ and $\bar{v}$ and $r_{\bar{q},\bar{v}}$ is the linear correlation coefficient of $\bar{q}$ and $\bar{v}$ for *points around a latitude circle*. Values of these three terms at 1000 and 850 mb, for January and July, are given in Table 5.7.

During winter the mean sea-level pressure field over the subtropical Atlantic and Pacific is characterized by semicontinuous east-west ridges of high pressure, as opposed to a highly cellular pattern. Northerly flow dominates over the Indian Ocean, as well as over the Atlantic and Pacific south of the ridge line (northeast trades). Longitudinal variations in the $\bar{v}$ component of the low-level tropical wind field are smallest at this time of year (cf. Plates 3.22–3.23). This contributes to a relatively strong vapor flux by the mean meridional component $[\bar{q}]\,[\bar{v}]$ and a relatively weak flux by the standing eddies $[\bar{q}^*\bar{v}^*]$.

On the other hand, in the summer the mean low-level wind field between 10°N and 40°N is dominated by three major features: (1) the flow around the Atlantic anticyclone centered in the vicinity of the Azores; (2) the flow around the Pacific anticyclone, centered around 35°N, 145°W; and (3) the Asian monsoon circulations. In this latitude band, the low-level wind fields over the Atlantic and Pacific exhibit a markedly more cellular character in summer than in winter. This is reflected in Table 5.7 by an increase in $\sigma(\bar{v})$ from January to July, particularly at 20°N and 30°N.

The increase in the longitudinal variability of $\bar{v}$ at 20 and 30°N is accompanied by an increase in the longitudinal variability of $\bar{q}$ (Table 5.7). If the fields of $\bar{v}$ and $\bar{q}$ were decomposed into Fourier components around a latitude circle,

$$\bar{q} = \sum_{n=-\infty}^{\infty} \overline{Q}_n\, e^{in\lambda}$$

$$\bar{v} = \sum_{n=-\infty}^{\infty} \overline{V}_n\, e^{in\lambda},$$

the standing eddy flux would be given by

$$[\bar{q}^*\bar{v}^*] = \sum_{n=1}^{\infty} 2\,|\overline{Q}_n|\,|\overline{V}_n|\cos\theta_n.$$

Thus an increase in the standing eddy flux could be due either to an increase in the amplitudes of the Fourier coefficients or to smaller phase differences. If the fields of $\bar{q}$ and $\bar{v}$ were composed of a single Fourier coefficient, the correlation coefficient would be a measure of this phase difference, that is:

$$r = [\bar{q}^*\bar{v}^*]/\sigma(\bar{q})\,\sigma(\bar{v}) = \cos\theta.$$

The increase in the calculated value of r together with the increase of $\sigma(\bar{q})$ and $\sigma(\bar{v})$ suggests that both an increase in amplitude of $\bar{q}^*$ and $\bar{v}^*$ and a more favorable phase relationship contributes to the enhanced summertime standing eddy flux.

The main regional characteristics of the summer time-mean meridional water vapor flux (standing eddy plus mean meridional) between 10°N and 40°N, may be summarized as follows. South of 15°N, the bulk of the northward vapor transfer takes place over the Arabian Sea, the Bay of Bengal, and the South China Sea. Poleward of 15°N, the northward transfer over the Asian sector diminishes, and the Atlantic and Pacific anticyclones assume the role of chief northward transporters of water vapor. Here relatively dry air is transferred southward over the eastern Atlantic and Pacific, while relatively moist air is transferred northward over the western Atlantic and Pacific as well as over the Caribbean Sea, the Gulf of Mexico, and the eastern United States.

The standing eddy flux is predominantly poleward (Figure 5.14), and thus in the direction of decreasing values of longitudinally averaged water vapor content. There are, however, exceptions to this pattern. Some equatorward transfer is indicated south of 10°N, but the small values leave the analysis in doubt. A weak equatorward transfer also appears between 40°N and 60°N during the summer months. The direction of transfer appears to be reliable here. If so, the flux is countergradient in the sense that the standing eddies are transporting water vapor in the direction of increasing values of $[\bar{q}]$.

3 TRANSIENT EDDY FLUX   Before turning our attention to the transient eddies it might be well to emphasize a point which the author feels is sometimes overlooked. The usual method of partitioning the flux, as used in this study, gives rise to a transient eddy term which represents contributions from the entire spectrum of eddy motions. Thus it reflects variations on all spatial scales up to and including wave number zero, the transient meridional circulation. Furthermore, the transient eddy term represents not only variations associated with traveling disturbances, but also variations more closely associated with slow changes in the semipermanent features of the atmosphere (the standing eddies), that is, variations in position, shape, and intensity of the Icelandic and Aleutian lows and the semipermanent subtropical highs. Thus, for annual averages, the seasonal march alone may contribute a sizable portion of the covariance of the transient eddy term. For monthly means, the contribution due to the seasonal trend will be considerably less, perhaps negligible. On the other hand, year-to-year variations in the monthly mean values during the 5-year period used for this study can give rise to a transient eddy contribution. While the transient eddy terms which are calculated in this way are not in any sense incorrect, these facts must be kept in mind when interpreting their physical meaning.

Values of the meridional transfer of water vapor by the transient eddy flux $[\overline{q'v'}]/g$ are given, for the mid-season months, in Table 5.8 at the end of this chapter. Values for January and July are illustrated in Figure 5.15.

Figure 5.16 shows the vertically integrated values of the transient eddy flux. A rather clear seasonal migration of the dividing line between northward and southward flux is apparent near the equator. The position of this line agrees closely with that of the line of maximum atmospheric water vapor content (see Figure 5.6), and differences may well lie within the errors of the computations (gradients of both quantities are very flat near their respective maxima and minima). The implication is, of course, that the vertically integrated transient eddy flux is always down the gradient. Furthermore, the transient eddy flux in tropical latitudes is, for the most part, opposed to the mean meridional flux, as can be seen from a comparison of Figures 5.12 and 5.16. Thus the transient eddies tend to erode the maximum of $[\bar{q}]$ in the

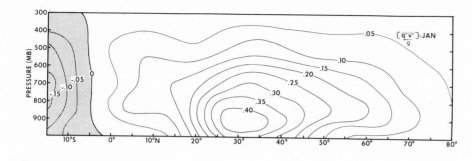

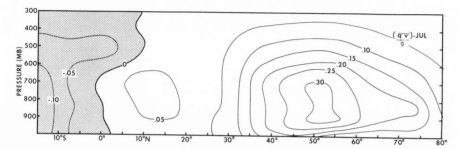

Figure 5.15 Longitudinally averaged meridional flux of water vapor by transient eddies $[\overline{q'v'}]/g$ for the months of January and July. Units: gm (cm mb sec)$^{-1}$.

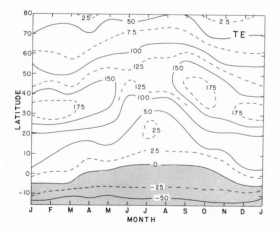

Figure 5.16 Longitudinally averaged, vertically integrated meridional flux of water vapor by transient eddies. Units: gm (cm sec)$^{-1}$.

tropical regions, while the mean meridional motions tend to maintain it. The likelihood of such a situation was first noted by Sutcliffe (1956).

The strength of the transient eddy transfer in low latitudes increases with distance from the latitude of maximum water vapor content, in general, indicating an essentially divergent flux throughout the tropics and throughout the year. The only exception to this pattern is the small region of flux convergence in summer between about 10 and 20°N. The transient eddies act, in general, to remove water vapor from the tropics. This outflow reaches a maximum during winter and a minimum during summer. This is the reverse of what was found in the case of the standing eddy flux which gave a maximum outflow of water vapor in summer and a minimum in winter.

As was noted previously, the subtropical high-pressure belts of the Northern Hemisphere are centered around 25–30°N during winter and appear on mean monthly maps as more or less continuous east-west ridges of high pressure. These ridge lines are, however, often fractured by disturbances on individual daily maps. Such a situation would be expected to result in a modest standing eddy but a sizable northward transient eddy flux from the subtropics during the winter season (Figures 5.14, 5.16). In summer, on the other hand, large anticyclonic cells dominate the Atlantic and Pacific regions on the mean monthly maps and are prominent features on most daily maps as well. Such a pattern of mean flow would be expected to give an enhanced standing eddy flux in summer. Furthermore, most migratory cyclones and anticyclones, which give rise to the bulk of the transient eddy flux, follow paths well to the north of their normal winter tracks and give rise to the northward displacement in summer of the maximum transient eddy flux.

Most of the northward transient eddy flux takes place over the oceans (see Peixoto and Crisi 1965), in association with the oceanic cyclone belts (Rasmusson 1967a). Most of the summertime transient outflow from the tropics is concentrated in the extreme western portions of the oceans and over eastern North America. The more intense outflow of winter is distributed over a broader region of the western Atlantic and Pacific.

Figure 5.17 shows the distribution of the temporal correlation coefficient r_{qv}:

$$r_{qv} = \left[\frac{\overline{q'v'}}{\sigma(q)\ \sigma(v)} \right].$$

Figure 5.18 shows the seasonal variation of this quantity at the equator and at 30°N. These figures, taken together with Figure 5.15, point up some interesting differences in the character of the wintertime transient eddy flux at low and middle latitudes.

During the northern winter, north of about 20°N, the transient eddy flux is strongest in the layer 1000–700 mb, with maximum values at about 900 mb. On the other hand, the transient eddy flux at low latitudes exhibits no maxima below 850 mb, but instead shows a tendency for higher values in the 850–400-mb layer. This distribution arises as a consequence of the fact that the magnitude of the low-level flux diminishes rapidly south of 25°N, while the flux in the middle troposphere decreases slowly as one moves from middle to low latitudes. A somewhat similar situation apparently exists in the Southern Hemisphere during the southern winter, accounting for the July mid-tropospheric transient eddy maximum at the equator.

Figure 5.17 reveals relatively high correlations associated with the mid-latitude transient maximum, both in January and July. However, the figure also shows a second maximum at low latitudes in the mid-troposphere of the winter hemisphere. This relatively high correlation between the variation of q and v in both the low-latitude middle troposphere and the middle-latitude regions suggests, as in the discussion of spatial correlation, that the eddies of q and v maintain a "favorable" phase relationship in these regions. The enhanced transient eddy flux at low latitudes is a reflection of this fact, although the magnitude of the flux is smaller than that at middle latitudes as the amplitudes of the disturbances are much smaller. Note from Figure 5.18 how these mid-tropospheric correlation maxima appear in the vicinity of the equator during the respective winter seasons.

4 TOTAL MERIDIONAL FLUX   Figure 5.19 shows the total, vertically integrated meridional flux,

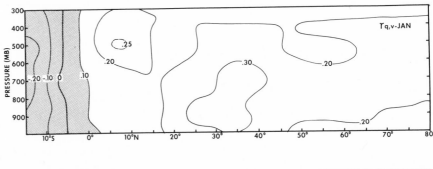

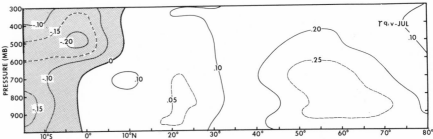

Figure 5.17   Longitudinally averaged correlation coefficient between q and v for January and July.

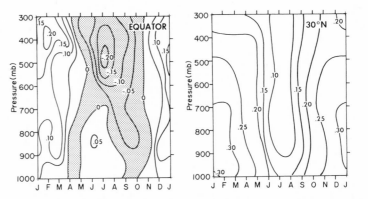

Figure 5.18   Longitudinally averaged correlation coefficient between q and v in the vicinity of the equator and at 30°N.

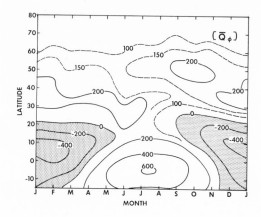

Figure 5.19   Longitudinally averaged vertically integrated meridional flux of water vapor $[\overline{Q}_\varphi]$. Units: gm (cm sec)$^{-1}$.

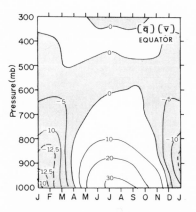

Figure 5.20   Flux of water vapor across the equator by the mean meridional circulation. Units: 10 gm (cm mb sec)$^{-1}$.

$$[\overline{Q}_\varphi] = \int_{p_u}^{p_s} \left\{ \frac{[\overline{q}][\overline{v}]}{g} + \frac{[\overline{q}^*\overline{v}^*]}{g} + \frac{[\overline{q'v'}]}{g} \right\} dp,$$

which is the sum of the mean meridional, standing eddy, and transient eddy components. Figures 5.11, 5.13, and 5.15 show the vertical and latitudinal distributions of the three terms in the integrand, while Figures 5.12, 5.14, and 5.16 give the seasonal variation of the integrals of the terms. These figures clearly show the dominant role of the mean meridional water vapor flux in the tropics and the larger values relative to the flux at middle and high latitudes. At middle latitudes the standing eddy flux dominates in summer, while the transient eddy flux is largest in winter. At high latitudes the transient eddy flux is the dominant term in all seasons.

The mean meridional flux shows not only the largest absolute values but the most marked seasonal variation, with the direction of the flux changing sign with season. The convergence of this flux is such as to maintain the tropical maximum of water vapor content by transporting water vapor from subtropical to more tropical latitudes.

At middle and high latitudes the net flux is poleward in all seasons and is dominated by the fluxes due to the transient and standing eddies. The flux is from subtropical to more poleward latitudes in this case and acts to transport water vapor down the gradient of $[\overline{W}]$.

This pattern of water vapor transport forms the basis for the discussion of the water balance of the tropical atmosphere.

5.4.3. Transequatorial Transfer

The final topic in the discussion of the water vapor flux as such concerns the transport across the equator. As is apparent from the discussion of the previous section, the meridional flux of water vapor varies markedly with season in the vicinity of the equator. Figure 5.20 shows the mean meridional transfer across the equator. As mentioned above, this term dominates the transequatorial flux as transient eddies contribute little to the total transfer (see Figure 5.16), and the standing eddy transfer is even smaller. The seasonal flux variations are almost entirely due to variations in $[\overline{v}]$ as the changes in $[\overline{q}]$ are relatively small, and roughly 180° out of phase with changes in $[\overline{v}]$.

It is apparent from Figure 5.20 and the previous discussion that the strongest transequatorial flow is accomplished by the southern Hadley cell. Our computations show significant differences in the $[\bar{v}]$ profiles typical of the lower branch of each cell near the equator (Figure 5.21). The value of $[\bar{v}]$ associated with the southern Hadley cell is found to be a maximum at or very near the surface, while $[\bar{v}]$ associated with the northern Hadley cell shows a maximum between 700 and 800 mb. Further discussion of these differences will be found in Oort and Rasmusson (1970). Figure 5.21 makes it clear that the greater trans-equatorial flux at the southern Hadley cell is not due simply to higher values of $[\bar{v}]$. It also arises as a consequence of the fact that maximum $[\bar{v}]$ values in the lower branch of the northern Hadley cell occur at higher levels, where values of $[\bar{q}]$ are lower.

The mean annual transequatorial flux is found to be 110 gm cm^{-1} sec^{-1} into the Northern Hemisphere. Rakipova (1966) calculated this quantity from the heat balance maps of Budyko (1963), and arrived at an estimate of around 145 gm cm^{-1} sec^{-1}. Starr, Peixoto, and McKean (1969) obtained a figure of 112 gm cm^{-1} sec^{-1}, a value very close to that which can be obtained from Sellers' (1965) estimate of the global distribution of $\bar{E} - \bar{P}$.

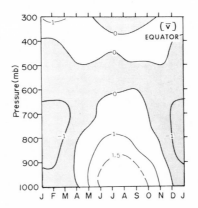

Figure 5.21 Mean meridional component $[\bar{v}]$ in the vicinity of the equator. Units: m sec^{-1}.

5.5 The Water Balance of the Tropical Atmosphere

5.5.1 Atmospheric Vapor Flux Divergence

Assuming hydrostatic equilibrium and no significant divergence of the horizontal flux of water in liquid and solid form, a water-balance equation can be written in the following form for a column of air extending from the ground to a pressure p_u (see Rasmusson 1968);

$$\frac{\partial W}{\partial t} + \mathbf{\nabla} \cdot \mathbf{Q} = E - P, \qquad (5.9)$$

where

$$\mathbf{Q} = \mathbf{i}Q_\lambda + \mathbf{j}Q_\varphi.$$

$\mathbf{i}, \mathbf{j}$ are eastward and northward pointing unit vectors, respectively; P is

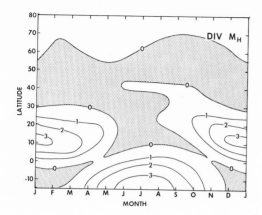

Figure 5.22   The contribution of the horizontal advection term to the vertically integrated vapor flux divergence of the mean meridional circulation. Units: gm (cm² mo)⁻¹.

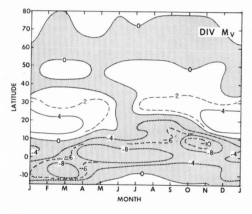

Figure 5.23   The contribution of the vertical advection term to the vertically integrated vapor flux divergence of the mean meridional circulation. Units: gm (cm² mo)⁻¹.

the precipitation rate; and E is the rate of evaporation and transpiration at the earth's surface. E will simply be referred to as evaporation in our subsequent discussion.

For mean monthly and longitudinally averaged quantities, equation (5.9) becomes

$$\left[\overline{\frac{\partial W}{\partial t}}\right] + \frac{1}{a \cos \varphi} \frac{\partial}{\partial \varphi} ([\overline{Q}_\varphi] \cos \varphi) = [\overline{E}] - [\overline{P}]. \qquad (5.10)$$

In words, a difference between the average rates of evaporation and precipitation within a latitude band must be balanced by a change in the quantity of water vapor stored in the overlying atmosphere and/or a net transfer of water vapor to or from the latitude band.

The horizontal vapor flux divergence on a pressure surface may be decomposed in a number of ways. We chose the following:

$$\frac{1}{a \cos \varphi} \frac{\partial}{\partial \varphi} ([\overline{qv}] \cos \varphi) = \frac{[\overline{v}]}{a} \frac{\partial [\overline{q}]}{\partial \varphi} + \frac{[\overline{q}]}{a \cos \varphi} \frac{\partial}{\partial \varphi} ([\overline{v}] \cos \varphi)$$

$$+ \frac{1}{a \cos \varphi} \frac{\partial}{\partial \varphi} ([\overline{q}^* \overline{v}^*] \cos \varphi) + \frac{1}{a \cos \varphi} \frac{\partial}{\partial \varphi} ([\overline{q'v'}] \cos \varphi). \qquad (5.11)$$

Similarly, the vertically integrated divergence may be decomposed into 4 terms:

$$\frac{1}{a \cos \varphi} \frac{\partial}{\partial \varphi} ([\overline{Q}_\varphi] \cos \varphi)$$

$$= \text{Div } M_H + \text{Div } M_V + \text{Div } SE + \text{Div } TE, \qquad (5.12)$$

where the terms on the right-hand side of equation (5.12) represent the vertically integrated values of the respective terms of equation (5.11). 1 Div M_H (FIGURE 5.22)   This term arises as a consequence of the mean meridional circulation and represents the divergence due to the advection of $[\overline{q}]$ by the meridional wind. The largest values of this term are positive (divergence) and are found in the winter hemisphere, where both $[\overline{v}]$ and $\partial [\overline{q}]/a \partial \varphi$ are large and of the same sign. The pattern resembles that of $[\overline{v}]$ (Figure 5.3) quite closely and reflects the advection of $[\overline{q}]$ by the motion in the lower branches of the Hadley cells. 2 Div M_V (FIGURE 5.23)   This term also arises as a consequence of the mean meridional circulation. It may be interpreted in a number of ways. For instance,

$$\frac{[\bar{q}]}{a \cos \varphi} \frac{\partial}{\partial \varphi} ([\bar{v}] \cos \varphi) = -[\bar{q}] \frac{\partial [\bar{\omega}]}{\partial p} = [\bar{\omega}] \frac{\partial [\bar{q}]}{\partial p} - \frac{\partial}{\partial p} [\bar{\omega}][\bar{q}], \qquad (5.13)$$

where the term on the left-hand side may be identified with that part of the divergence which arises as a consequence of mass divergence in the horizontal. The variation of $[\bar{v}]$ with latitude is intimately connected with vertical motions through the continuity equation. For practical purposes the second term on the right drops out in a vertical integration ($[\bar{\omega}] \sim 0$ at p_s; $[\bar{q}] \sim 0$ at p_u). The term Div M_V, therefore, is related to the integrated vertical advection of $[\bar{q}]$ by $[\bar{\omega}]$. As $[\bar{q}]$ always decreases with height, the contribution of the term will be a weighted average of the vertical motion. A negative contribution (convergence) will imply "predominant" ascending motion at levels where $\partial [\bar{q}]/\partial p$ is largest (i.e., at low levels), and vice versa. This "predominant" ascending motion will imply convergence, through mass continuity, in the horizontal.

This is the dominant term in equation (5.12) in the tropics. It is clearly associated with convergence (negative values) in the ascending motion branches of the Hadley cells and with divergence in the latitudes of the descending motion branches of the cells. Unfortunately, it is also the most difficult term to evaluate due to the difficulty of evaluating $\partial [\bar{v}] \cos \varphi/\partial \varphi$. Errors of a few tenths of a meter per second in $[\bar{v}]$ in the lower levels result in serious error in the flux divergence. The general features of the true pattern are believed to be fairly well represented on Figure 5.23, but it is quite doubtful whether the details are depicted accurately.

3 Div ($M_H + M_V$) (FIGURE 5.24) This figure shows the divergence of the total flux due to the mean meridional motions. It is simply the sum of the terms given in the two previous figures. It clearly shows the presence of strong convergence in the tropical regions and strong divergence at subtropical latitudes. The regions of strong convergence in the tropical zone are related reasonably well, in time and space, to the maximum values of $[\overline{W}]$ (Figure 5.6) and have large values generally in all but the Northern Hemisphere winter months. The divergence at subtropical

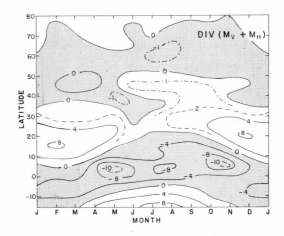

Figure 5.24 The net contribution to the vertically integrated vapor flux divergence of the mean meridional circulation. Units: gm (cm² mo)⁻¹.

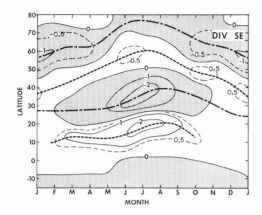

Figure 5.25   The vertically integrated divergence of the standing eddy flux. Units: gm (cm² mo)⁻¹.

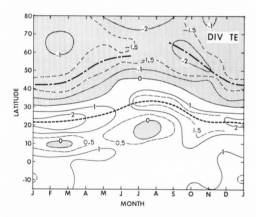

Figure 5.26   The vertically integrated divergence of the transient eddy flux. Units: gm (cm² mo)⁻¹.

latitudes, on the other hand, exhibits maximum values for these same winter months.

4 Div *SE* (FIGURE 5.25)   This term represents the divergence contribution by the standing eddies. During winter, when the semipermanent oceanic low-pressure systems are most strongly developed, its major contribution is found in the middle and high latitudes. During summer, when the monsoon circulations and subtropical highs become the dominant features of the low-level mean monthly wind field, the major contribution shifts to lower latitudes. It is apparent that neglect of this term in summertime water-balance estimates for the Northern Hemisphere will lead to serious error.

5 Div *TE* (FIGURE 5.26)   This term represents the divergence contribution by the transient eddies. Unlike the divergence associated with the mean meridional motions, the pattern is one of general divergence in the tropical regions. The maximum values of the divergence are found in the subtropical regions in winter, as would be expected from the previous comments on the strength and distribution of the eddy flux of moisture. At higher latitudes there is a convergence of the transient eddy flux.

5.5.2 Broadscale Features of the Hydrologic Cycle of the Tropics

Summation of the divergence terms just discussed gives the total vapor flux divergence of the atmosphere, a term of basic importance in the atmospheric water balance (equation 5.9). This total vapor flux divergence term is shown in Figure 5.27.

Neglecting for the moment the small changes in atmospheric storage from month to month, the values on Figure 5.27 represent the difference between mean monthly evaporation and precipitation averaged around a latitude circle. This is the net amount of water vapor which is transferred to or from the atmospheric column during the month. As has been noted by many previous investigators, the subtropical region is the source region from which atmospheric water vapor is exported. The extratropical latitudes and the Equatorial Trough are sink regions into which atmospheric water vapor is imported. The computations show the Northern Hemisphere source regions to be strongest during fall and winter, at which time the longitudinally averaged evaporation exceeds precipitation by as much as 10 gm cm⁻² mo⁻¹. The pattern is not symmetric about the equator, as has been noted previously in terms of the transequatorial flux (section 5.4.3). With regard to the wintertime flow, Malkus (1962) states, "The average trade-flow is vigorous, importing cool, dry air rapidly into the Tropics, so that exchange of all properties is greatest in the winter months. Interaction between the disturbances of the mid-latitude westerlies and the trades is also common from time to time, and the Tropics are frequently invaded by polar troughs and shearlines, leftover remnants of the polar front which cause strong resurgences of the trades in their rear."

When examining the results one must keep in mind the fact that the current aerological network is not sufficient to reliably resolve details of the flux divergence in the vicinity of the Equatorial Trough. Lack of data in the aerologically "silent" eastern Pacific alone precludes this possibility. Thus, particular caution must be exercised in the interpretation of the results between 15°N and 15°S. Nevertheless, the broad-scale features revealed by the existing network of stations, taken together with data obtained from remote satellite sensors, appear to provide information from which rough quantitative estimates of the seasonal variations of zonally averaged precipitation can ultimately be made.

As a preliminary attempt in this direction we consider the following data: (1) the vertically integrated zonally averaged values of water vapor content (Figure 5.6); (2) the values of zonally averaged precipitation (Figure 5.30) which are obtained from equation (5.10) using the flux divergence as shown in Figure 5.27 and the evaporation estimates of Budyko (1963); (3) the zonally averaged brightness values of Taylor and Winston (1968), shown in Figure 5.31. These data were obtained from the ESSA 3 and ESSA 5 satellites and are presented in digitalized form. The data span the 13-month period from February 1967 through February 1968; (4) the cloudiness values of Sadler (1969), shown in Figure 5.32. These data were obtained from operational nephanalyses prepared by the National Environmental Satellite Center of ESSA during the two-year period February 1965–January 1967.

Consider first the conditions at the equator. Figure 5.28 gives the departures of the monthly mean values of $[\bar{q}]$ from the annually averaged specific humidity in the vicinity of the equator. Although departures are small, they clearly exhibit the expected semiannual variation.

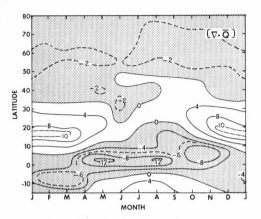

Figure 5.27 The total vertically integrated vapor flux divergence. Units: gm $(cm^2 \ mo)^{-1}$.

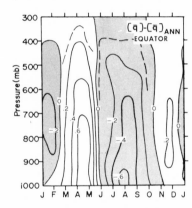

Figure 5.28 Departure from the annual averaged values of specific humidity in the vicinity of the equator. Units: gm kg^{-1}.

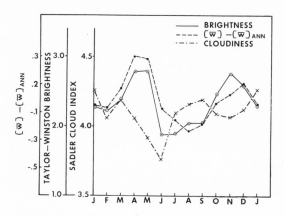

Figure 5.29 Seasonal variations in longitudinally averaged water-vapor content, cloudiness, brightness, and precipitation in the vicinity of the equator.

Table 5.9. Semiannual Differences at the Equator for Various Parameters

	Dec–Jun	Jan–Jul	Feb–Aug	Mar–Sep	Apr–Oct	May–Nov
p (mb)	$[\bar{q}]$: gm kg^{-1}					
300	0.1	0.1	0.1	0.1	0.0	0.0
400	0.2	0.1	0.1	0.2	0.2	0.0
500	0.3	0.3	0.1	0.3	0.4	0.2
700	0.3	0.0	0.3	0.5	0.5	0.5
850	0.4	0.2	0.2	0.7	0.7	0.6
1000	0.1	0.4	0.7	0.7	0.9	0.8
$[\bar{W}]$: gm cm^{-2}						
	0.3	0.2	0.1	0.3	0.3	0.3
Brightness values (Taylor and Winston 1968)						
	.70	.39	.17	.24	.33	.04
Cloudiness Index 2.5°N–2.5°S (Sadler 1969)						
	.41	.16	−.12	−.02	−.03	−.16

The highest moisture content is found in late April and in early December. The spring maximum closely coincides with the time at which the southern Hadley cell first extends into the Northern Hemisphere, while the late fall maximum bears a similar relationship to the development of the northern Hadley cell. Minimum values coincide with the times of maximum development of the respective Hadley circulations and the strongest transequatorial flow of water vapor.

These variations in specific humidity in terms of water vapor content are compared to the Taylor-Winston brightness values and to the Sadler cloud index in the vicinity of the equator in Figure 5.29. The pattern of the seasonal variation of brightness and water vapor content are quite similar. The vapor content and brightness values not only show the same pattern of semiannual variation, but both parameters exhibit a clear annual variation as well. Table 5.9 gives the semiannual differences for each of these quantities.

These differences serve to roughly identify the amplitude and phase of the annual variation (i.e., for a perfect sinusoidal annual variation the maximum difference would occur between values six months apart at the maximum and minimum values of the cycle). The differences here show the annual cycle has its maximum values in the Northern Hemisphere winter and spring.

In contrast to the brightness values, the seasonal variations in Sadler's mean cloudiness between 2.5°S and 2.5°N bear little relationship to either the annual (Table 5.9) or semiannual variations in water vapor content.

Thus a question remains as to the relationship between seasonal variations in zonal average water content and zonal average precipitation in the vicinity of the equator. The brightness data strongly suggest such a relationship, but the cloud data offer no evidence in support of this conclusion.

On a broader scale, the fields of $[\bar{P}]$, brightness, cloudiness, and $[\bar{W}]$ (Figures 5.30, 5.31, 5.32, 5.6) show that

(1) A cloudiness maximum is found between 2.5°N and 10.0°N during every month, with a maximum value in July. A less sharply defined maximum is found between 2.5°S and 12.5°S from September through

May. Only during January, when it exhibits its highest value, does this maximum exceed the value of the concurrent Northern Hemisphere maximum. A cloudiness minimum is located near the equator, which merges with the Southern Hemisphere subtropical minimum during June–August. Gradients of cloudiness between the subtropics and equator are strongest in the Northern Hemisphere.

(2) There are differences in the details of the cloudiness and brightness fields near the equator. Perhaps the most significant of these is the difference in the relative intensity of the Northern Hemisphere and Southern Hemisphere maxima. From December through April the maximum brightness values are clearly found in the Southern Hemisphere. In contrast, the cloudiness values show Northern Hemisphere maxima which are comparable to, or greater than, Southern Hemisphere values for all months, except January.

(3) The axis of maximum $[\overline{W}]$ follows a simple seasonal cycle, which is similar to what would be obtained from a heavy smoothing of the brightness field.

(4) Only in their gross aspects do the seasonal changes in the $[\overline{P}]$ maxima bear a resemblance to the seasonal behavior of the cloudiness and brightness maxima. $[\overline{P}]$ maxima are seen in both hemispheres only in April, December, and January. The absence of a Northern Hemisphere maximum in February and March may be due to smoothing of the divergence features in the region of strong gradient between the equator and 15°N. During Northern Hemisphere winter, the axis of maximum computed precipitation corresponds well with the axis of maximum brightness and with the cloudiness maximum of the Southern Hemisphere. During May–September, however, the $[\overline{P}]$ axis appears to be located 4–8 degrees too far south, judging from the features of the 3 other fields. Inspection of the machine-analyzed wind fields indicates that this apparent inconsistency can be largely eliminated by reasonable modifications of the machine analyses over regions of sparse data, such as the Indian Ocean and eastern Pacific. However, the fact that such modifications can be made within the framework of the existing data network merely illustrates the degree of uncertainty in the analyses. Finally, a good correspondence exists between brightness, cloudiness,

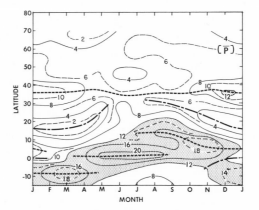

Figure 5.30 The estimate of longitudinally averaged precipitation obtained by using Budyko's estimates of evaporation and the computed values of vapor flux divergence. Units: gm (cm² mo)⁻¹.

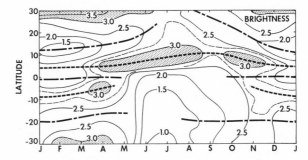

Figure 5.31 Longitudinally averaged brightness values. From Taylor and Winston (1968). The units are relative on a scale of 10.

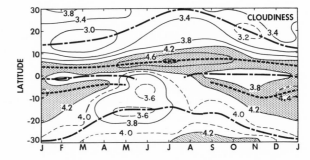

Figure 5.32 Longitudinally averaged cloudiness. From Sadler (1969). The units are approximately eighths of total sky cover.

and precipitation values in the subtropical region where all values have a relative minimum.

The following tentative conclusions can be drawn from a synthesis of these data.

(1) During the northern summer there is little doubt of the existence of a single maximum of zonally averaged precipitation in the tropics. This maximum is apparently centered between 5°N and 10°N, or possibly even a little farther north during late summer.

(2) There is evidence for the existence of double maxima, one on either side of the equator, during the northern winter and spring.

(3) The analyses indicate a rather rapid shift in the precipitation maximum from one hemisphere to the other in the equinoctial seasons. There is substantial, though not conclusive, evidence that these seasons mark the time of maximum zonally averaged precipitation at the equator. However, even during these seasons, precipitation appears to exhibit a relative minimum at the equator, when considered as a function of latitude.

Somewhat more reliable estimates can be obtained for average flux divergence over rather broad latitude bands. Thus, in order to increase the reliability of precipitation estimates, and to bring out the broad-scale features of the tropical hydrologic cycle, the mean water-budget components have been computed for selected latitude bands.

In dealing with averages over broad latitude bands, the divergence term may be more appropriately expressed, through application of the Gauss theorem as $(\overline{F}_2 - \overline{F}_1)/A$, where

$$\overline{F}_2 = a \oint \overline{Q}_\varphi(\varphi_2) \cos \varphi_2 \, d\lambda$$

is the total northward flux across the northern bounding latitude φ_2;

$$\overline{F}_1 = a \oint \overline{Q}_\varphi(\varphi_1) \cos \varphi_1 \, d\lambda$$

is the total northward flux across the southern bounding latitude φ_1; and

$$A = 2\pi a^2 (\sin \varphi_2 - \sin \varphi_1)$$

is the area between φ_2 and φ_1.

The water-balance equation (5.10) may then be written as

$$\overline{E} - \overline{P} = \Delta\overline{W} + (\overline{F}_2 - \overline{F}_1)/A. \tag{5.14}$$

That is, the excess or deficit of evaporation over precipitation in the region must be balanced by a change in the amount of water vapor and/or a net flux of water vapor through the lateral boundaries of the region. The values of the fluxes across boundaries and of the change in water vapor content are evaluated from the data. The values of evaporation are those given by Budyko (1963). The precipitation is obtained as a residual in the equation.

Table 5.10a gives values of these hydrological components averaged over the entire Northern Hemisphere. The seasonal variations in the transequatorial flux imply an excess of precipitation over evaporation in the summer hemisphere, and the reverse in the winter hemisphere. In fact, the computations show that the Southern Hemisphere supplies close to 25 percent of the moisture which falls as precipitation over the Northern Hemisphere during the months of June through August, while around 20 percent of the water which is evaporated in the Northern Hemisphere during January–March is apparently transferred to the Southern Hemisphere. There is, as previously noted, a mean annual transfer of water vapor into the Northern Hemisphere. This transfer apparently amounts to around 5 percent of the total precipitation of the hemisphere. On balance, then, precipitation exceeds evaporation in the Northern Hemisphere, while the reverse is true in the Southern Hemisphere. Balance considerations then require a mean annual southward transfer across the equator by rivers and ocean currents of around 14×10^5 kg yr^{-1}. It should be noted that except for the transfer across the equator, the mean annual precipitation is entirely determined by the estimate of mean annual evaporation.

Estimates of the hydrological components for the extratropical region 30–90°N are given in Table 5.10b. There is inflow of moisture from the tropics across 30°N during every month, and this inflow apparently supplies the moisture for better than 20 percent of the precipitation in these latitudes.

The water balance of the tropical latitude band 0–30°N is given in

Table 5.10c. Note that the values associated with the transequatorial flux are different than those given in Table 5.10a. This is a consequence of the different areas for which the water balance is calculated in the two cases (equation 5.14) and does not imply a difference in the net transfer of water vapor across this boundary.

Water vapor is transported from the tropics to higher latitudes in all months. From May through October precipitation exceeds evaporation and serves to remove water vapor from the region, but the deficit is made up by enhanced transfer across the equator by the southern Hadley circulation.

A comparison of the values of $\bar{E}$ and $\bar{P}$ in the tropical region with those for the extratropical region vividly illustrates the greater intensity of the tropical hydrological cycle. On the annual average, however, even though the tropical values of $\bar{E}$ and $\bar{P}$ are much larger than extratropical values, they differ by a smaller amount. This near balance of $\bar{E}$ and $\bar{P}$ in tropical regions implies, in effect, that the net annual transequatorial flux goes toward balancing the vapor deficit at higher latitudes.

In order to contrast the hydrological regimes of subtropical and equatorial latitudes, averages for the latitude bands 15°N–30°N, and 15°S–15°N were also computed. Values for the 15°N–30°N latitude band are presented in Table 5.10d. This region acts as a source of water vapor for the atmosphere. Precipitation exceeds evaporation only in the month of September, and then only by a small amount. Since the average annual position of the Equatorial Trough is slightly north of the equator, it seems reasonable to expect the corresponding latitude band of the Southern Hemisphere to serve as an even stronger source of water vapor. The mean annual $\bar{E}$ and $\bar{P}$ charts of Budyko indicate this to be the case.

Table 5.10e gives the hydrological components for the latitude band 15°S–15°N. Precipitation exceeds evaporation in all months of the year. The inflow of water vapor which is required to balance this loss is supplied from both hemispheres on the annual average. The largest net fluxes into the region occur in June and November. Fluxes into the region from both hemispheres occur simultaneously in the spring and fall transition periods. During the winter and summer seasons there is a flux out of the region into the summer hemisphere. The net flux is into

the region in all months, however. Vuorela and Tuominen (1964) have previously commented on this feature of the circulation.

A value of 26.6 gm cm^{-2} was calculated for the annual flux of water vapor into this region. By comparison, a value for $\bar{P} - \bar{E}$ of around 36 gm cm^{-2} was obtained using the charts of Budyko (1963).

Local evaporation, as given by Budyko (1963), amounts to 83 percent of the mean annual precipitation, while inflow of moisture accounts for 17 percent. During April and November, however, inflow amounts to 25 percent of the precipitation.

5.6 Concluding Remarks

Our discussion has dealt primarily with the zonal average monthly variations in humidity, horizontal vapor transfer, vertically integrated flux divergence, and water balance of the total atmospheric column in the tropics. We have attempted to bring out the broad-scale features of the seasonal cycle, which, it is hoped, can serve as a climatological framework for future regional water and heat balance studies.

Hydrologically speaking, there is a great deal of detail in the low-latitude tropics. Mean monthly composite satellite pictures for the year 1967 (Kornfield et al. 1967) show a relatively narrow zone of cloudiness (Intertropical Convergence Zone) a few degrees north of the equator over much of the Atlantic and Pacific oceans. There is, in addition, evidence of a comparable zone of cloudiness in the South Pacific during the late northern winter and early northern spring of 1967. Raman (1965) reports the existence of a distinct trough system over the Indian Ocean sector. A trough is found in each hemisphere during all months of the year. Lower and middle troposphere cyclonic vortices are embedded therein.

Ideally, one would like to compute the water vapor flux divergence on a regional basis rather than merely to deal with zonal averages. Accurate data of this type would be invaluable aids in water-balance studies over the tropical land masses; in the opinion of the author they will be an indispensable tool in the eventual accurate evaluation of the precipitation distribution over the tropical oceans. Regional computations for the relatively data-rich region of the Caribbean Sea and Gulf of Mexico (Rasmusson 1966; Hastenrath 1966) amply demonstrate the feasibility

Table 5.10a. Mean Values—Northern Hemisphere. Units: gm cm⁻²

	J	F	M	A	M	J	J	A	S	O	N	D	Ann
Inflow across Equator	−1.6	−1.7	−1.5	0.0	1.3	2.0	2.4	2.4	2.0	1.3	0.0	−1.2	5.4
$\Delta\bar{W}$	−0.1	0.0	0.2	0.3	0.4	0.3	0.1	−0.1	−0.3	−0.3	−0.3	−0.2	0.0
$\bar{E}$	8.5	7.6	7.1	6.4	6.6	7.2	7.5	7.6	7.7	8.0	8.4	8.9	91.5
$\bar{P}$	6.9	5.9	5.5	6.1	7.5	8.9	9.8	10.1	10.0	9.6	8.7	7.9	96.9

Table 5.10b. Mean Values—30°N–90°N. Units: gm cm⁻²

	J	F	M	A	M	J	J	A	S	O	N	D	Ann
Inflow: 30°N	1.6	1.5	1.4	1.3	1.5	1.5	1.6	1.2	0.6	0.7	1.0	1.4	15.3
$\Delta\bar{W}$	0.0	0.0	0.1	0.3	0.4	0.5	0.2	−0.2	−0.5	−0.4	−0.3	−0.1	0.0
$\bar{E}$	5.6	5.0	5.0	4.4	4.5	4.5	4.3	4.4	4.9	5.2	5.6	5.9	59.3
$\bar{P}$	7.2	6.5	6.3	5.4	5.6	5.5	5.7	5.8	6.0	6.3	6.9	7.4	74.6

Table 5.10c. Mean Values—0°–30°N. Units: gm cm⁻²

	J	F	M	A	M	J	J	A	S	O	N	D	Ann
Outflow: 30°N	1.6	1.5	1.4	1.3	1.5	1.5	1.6	1.2	0.6	0.7	1.0	1.4	15.3
Outflow: 0°	3.4	3.4	2.8	0.0	−2.6	−4.0	−4.7	−4.7	−4.0	−2.7	0.0	2.4	−10.7
$\Delta\bar{W}$	−0.1	0.0	0.2	0.3	0.3	0.2	0.1	0.0	−0.2	−0.3	−0.3	−0.2	0.0
$\bar{E}$	11.4	10.2	9.2	8.4	8.7	9.9	10.7	10.8	10.5	10.8	11.2	11.9	123.7
$\bar{P}$	6.5	5.3	4.8	6.8	9.5	12.2	13.7	14.3	14.1	13.1	10.5	8.3	119.1

Table 5.10d. Mean Values—15°N–30°N. Units: gm cm⁻²

	J	F	M	A	M	J	J	A	S	O	N	D	Ann
Outflow: 30°N	3.2	3.0	2.9	2.8	3.2	3.1	3.3	2.6	1.2	1.4	2.1	2.9	31.7
Outflow: 15°N	4.4	4.0	3.4	2.9	1.0	−2.9	−2.7	−2.0	−1.1	2.4	5.9	5.7	21.0
$\Delta\bar{W}$	−0.1	0.0	0.3	0.4	0.5	0.4	0.2	0.0	−0.4	−0.5	−0.5	−0.3	0.0
$\bar{E}$	12.2	10.9	9.7	8.7	8.1	8.8	9.9	10.4	10.4	10.6	11.0	12.4	123.1
$\bar{P}$	4.7	3.9	3.1	2.6	3.4	8.2	9.1	9.8	10.7	7.3	3.5	4.1	70.1

Table 5.10e. Mean Values—15°N–15°S. Units: gm cm⁻²

	J	F	M	A	M	J	J	A	S	O	N	D	Ann
Outflow: 15°S	1.5	0.6	−0.4	−2.0	−2.4	−5.2	−3.0	−2.9	−2.9	−2.2	−1.0	1.0	−16.9
Outflow: 15°N	−2.0	−1.8	−1.6	−1.4	−0.4	1.4	1.2	1.0	0.5	−1.1	−2.8	−2.6	−9.7
$\Delta\bar{W}$	0.0	0.0	0.1	0.1	−0.1	−0.1	0.0	0.0	0.0	0.0	0.0	0.0	0.0
$\bar{E}$	10.9	10.0	10.0	9.5	9.8	12.0	12.2	11.7	11.0	11.1	11.4	11.4	131.0
$\bar{P}$	11.4	11.2	11.9	12.8	12.7	13.9	14.0	13.6	13.4	14.4	15.2	13.0	157.6

of this approach. However, a much denser and more uniform system of aerological observations is required than is now available over most of the tropics. An improved aerological network would allow the strong flux gradients associated with the detailed structure of the divergence field to be properly resolved. Accurate regional water-balance computations for these areas must therefore await significant improvement in the aerological network, together with additional satellite information, or perhaps a more sophisticated application of current satellite observations.

Estimates of the amount of latent heat which is released into the total atmospheric column can be derived from our precipitation estimates. The complete quantitative evaluation of the role of condensation heating in the general circulation of the tropics requires a more detailed knowledge of the temporal and spatial variations of the release of latent heat, however. This problem is directly related to the evaluation of the vertical transfer of water in the atmosphere, a topic not covered in our discussion. Vertical vapor transfer is difficult to investigate with standard rawinsonde data alone. Satellite pictures, and research such as that of Riehl and Malkus (1958) and Gray (1968), indicate that over the oceans the vertical transfer of water vapor and hence the condensation heating of the atmospheric column tend to be concentrated in clusters of convective cells with typical dimensions of 100–600 km. Convection over the continents, where the diurnal heating cycle plays a major role, appears to be less well organized. With advances in satellite technology, and with the augmentation of new observational systems such as the constant-level balloon, it seems reasonable to expect an adequate description of these processes to emerge during the next few years.

Acknowledgments

The unselfish cooperation of a number of people made this study possible. Most of the Northern Hemisphere data became available through the efforts of Prof. V. P. Starr, whose help and cooperation over the past several years are deeply appreciated. Rawinsonde data for the Southern Hemisphere and for a number of key Northern Hemisphere stations were generously provided by Prof. R. E. Newell and Dr. J. W. Kidson. The low-level pibal data were processed by the author in connection with his reserve activities at the U.S. Air Force Environmental Technical Applications Center. The full cooperation and encouragement of the staff of this organization were very much appreciated.

Mr. J. Welsh provided the basic analysis routines used in the study. His cooperation and continuous availability for consultation contributed significantly to the study. A major part of the basic data processing was expertly accomplished by H. Frazier and E. Sweeton.

Most of the procedures and methodology used in the final analysis of the data were developed in cooperation with Dr. A. Oort. His work in this area, and our many hours of fruitful discussion, have made a significant contribution to this study. In addition, the many stimulating discussions and most helpful suggestions from Dr. S. Manabe and Dr. K. Miyakoda, and the continued interest and encouragement of Dr. J. Smagorinsky are greatfully acknowledged.

Thanks are also due to P. Tunison for drafting the figures, and to Mrs. Christine Morgan and Mrs. Yvonne Towns for typing the manuscript.

Table 5.1. Percent of Grid Points for Which Pressure Surface Was Above Smoothed Topography

Latitude	1000 mb	950 mb	900 mb	850 mb	700 mb
15°S	78	86	90	100	100
10	87	92	94	100	100
5	88	93	96	100	100
0	89	94	96	99	100
5	86	94	97	100	100
10	87	97	97	100	100
15	84	96	98	100	100
20	81	91	98	100	100
25	75	86	92	97	100
30	67	78	82	89	100
35	68	74	79	86	100
40	69	74	78	85	100
45	71	80	86	92	100
50	70	81	90	97	100
55	71	85	96	100	100
60	74	87	97	100	100
65	75	88	94	97	100
70	85	93	94	94	100
75	91	91	93	95	100
80°N	90	91	95	99	100

Table 5.2. $[\bar{q}]$. Units: gm kg^{-1}

p (mb)	Latitude −15	−10	−5	0	5	10	15	20	25	30	35
January											
400	1.3	1.2	1.1	1.0	0.8	0.6	0.5	0.4	0.4	0.4	0.3
500	2.5	2.5	2.4	2.3	1.8	1.4	1.1	1.0	0.9	0.8	0.6
700	6.1	6.0	6.0	6.0	5.3	4.2	3.4	2.9	2.5	2.1	1.8
850	11.0	11.0	10.9	10.6	10.2	9.0	7.8	6.3	4.9	3.6	3.0
900	11.6	12.1	12.5	12.5	11.5	10.8	9.3	7.6	5.8	4.4	3.5
950	12.8	13.6	14.2	14.5	13.8	12.7	11.4	8.5	6.6	5.0	4.0
1000	13.0	14.3	15.3	15.4	14.5	13.5	11.5	8.9	6.9	5.2	4.2
April											
400	1.0	1.1	1.2	1.2	1.0	0.8	0.6	0.6	0.5	0.5	0.4
500	2.0	2.4	2.6	2.6	2.3	1.8	1.4	1.3	1.2	1.1	0.8
700	5.4	6.1	6.8	6.8	6.0	4.8	4.0	3.7	3.4	2.8	2.4
850	9.8	10.8	11.2	11.3	11.0	9.7	8.4	7.4	6.0	4.5	3.6
900	10.3	11.9	12.8	13.1	12.8	11.7	10.4	9.1	7.1	5.3	4.2
950	11.3	13.4	14.6	15.1	14.6	13.8	12.4	10.2	8.0	6.0	4.7
1000	11.4	14.1	15.7	16.0	15.2	14.8	13.2	10.7	8.3	6.2	5.0
July											
400	0.7	0.7	0.8	0.9	1.0	1.0	1.0	1.0	1.0	0.9	0.8
500	1.4	1.6	1.8	2.0	2.4	2.0	2.1	2.2	2.1	2.0	1.8
700	4.0	4.6	5.4	6.1	6.3	6.1	5.8	5.6	5.4	5.1	4.6
850	7.6	8.8	9.8	10.4	11.0	11.0	10.8	10.4	9.4	7.9	7.0
900	8.2	9.9	11.3	12.2	12.7	12.7	12.6	12.1	10.6	8.6	7.6
950	9.0	11.3	13.0	14.1	14.3	14.7	14.4	13.2	11.7	9.7	8.4
1000	9.1	12.0	14.1	15.0	14.8	15.1	14.4	13.2	11.7	9.9	8.8
October											
400	0.8	0.9	0.9	1.0	1.1	1.0	0.8	0.8	0.7	0.6	0.5
500	1.6	1.8	2.0	2.2	2.4	2.2	1.8	1.7	1.6	1.4	1.2
700	4.5	4.9	5.6	6.3	6.4	6.0	5.4	4.9	4.2	3.5	2.9
850	8.4	9.4	10.0	10.6	11.0	10.5	9.8	9.2	7.9	6.3	5.2
900	9.0	10.5	11.6	12.4	12.8	12.5	11.8	11.0	9.2	7.2	5.9
950	10.0	11.8	13.2	14.3	14.6	14.6	13.9	12.2	10.3	8.1	6.5
1000	10.1	12.3	14.0	15.1	15.2	15.5	14.4	12.5	10.6	8.3	6.8

Table 5.4. $[\overline{qu}]/g$. Units: gm (cm mb sec)$^{-1}$

p (mb)	Latitude −15	−10	−5	0	5	10	15	20	25	30	35
January
400	.04	−.16	−.24	−.39	−.44	−.18	.27	.80	1.12	1.03	0.74
500	−.04	−.26	−.54	−.95	−1.04	−.52	.37	1.25	1.72	1.56	1.15
700	−.47	−.41	−1.00	−2.05	−2.65	−1.88	−.34	1.26	2.17	2.21	1.84
850	−1.88	−.58	−.85	−2.37	−4.49	−5.47	−3.57	−.64	1.26	2.02	2.02
900	−1.55	−.86	−1.45	−3.16	−5.41	−6.72	−4.98	−1.87	0.47	1.78	1.95
950	−1.47	−1.17	−1.68	−3.16	−5.12	−6.77	−5.43	−2.54	−.27	1.30	1.62
1000	−1.02	−1.05	−1.63	−2.68	−3.53	−4.71	−4.55	−2.53	−1.15	0.36	1.00
April
400	.29	−.18	−.47	−.53	−.35	.00	.35	.69	.86	.89	.74
500	.18	−.55	−1.10	−1.31	−1.06	−.45	.27	1.00	1.53	1.53	1.26
700	−1.01	−1.85	−2.52	−3.09	−3.24	−2.31	−.87	.85	2.05	2.15	1.86
850	−4.48	−3.91	−2.70	−2.76	−4.10	−4.28	−2.88	−.90	.79	1.29	1.57
900	−4.37	−4.08	−3.15	−3.29	−4.47	−5.21	−4.23	−2.15	−.34	.82	1.51
950	−3.89	−3.90	−3.22	−3.36	−4.18	−5.15	−4.59	−2.71	−1.00	.36	1.24
1000	−2.88	−2.98	−2.71	−2.88	−3.02	−4.20	−4.57	−3.29	−1.80	−.21	.75
July
400	.50	−.05	−.35	−.51	−.57	−.48	−.37	−.33	−.18	.20	.71
500	.54	−.34	−.80	−.91	−.90	−.88	−.82	−.72	−.37	.37	1.15
700	−.48	−1.54	−1.84	−1.49	−1.31	−1.57	−1.49	−.87	−.45	.49	1.51
850	−3.92	−4.38	−3.60	−2.35	−1.12	−.72	−1.10	−1.61	−1.13	.08	1.32
900	−4.34	−5.11	−4.34	−2.85	−1.29	−.89	−1.67	−2.24	−1.92	−.43	1.10
950	−4.11	−5.11	−4.37	−2.92	−1.17	−.79	−1.93	−2.64	−2.28	−.72	.84
1000	−3.27	−4.28	−3.98	−2.83	−1.20	−1.46	−2.89	−3.31	−2.41	−.90	.31
October
400	.24	.01	−.16	−.24	−.27	−.19	−.06	.13	.36	.55	.57
500	.27	−.48	−.80	−.92	−1.01	−.89	−.53	.03	.67	1.11	1.22
700	−1.00	−2.03	−2.17	−2.15	−2.43	−2.53	−2.04	−1.06	.04	.94	1.42
850	−3.92	−4.14	−3.11	−2.37	−2.43	−3.00	−3.59	−3.15	−1.76	−.29	.90
900	−3.96	−4.51	−3.71	−2.85	−2.54	−3.26	−4.20	−4.12	−2.79	−.90	.68
950	−3.72	−4.38	−3.67	−2.84	−2.16	−2.94	−4.19	−4.15	−2.94	−1.09	.49
1000	−3.00	−3.63	−3.32	−2.56	−1.59	−2.48	−3.57	−3.55	−2.47	−.70	.42

Table 5.5. $[\overline{qv}]/g$. Units: gm (cm mb sec)$^{-1}$

p (mb)	Latitude −15	−10	−5	0	5	10	15	20	25	30	35
January
400	−.02	−.04	.00	.00	.00	.01	.02	.04	.06	.06	.05
500	−.14	−.10	.00	.05	.06	.07	.07	.06	.09	.13	.12
700	−.04	−.18	−.57	−.69	−.32	.09	.20	.23	.30	.39	.39
850	−.59	−.68	−.92	−1.20	−1.41	−1.16	−.45	.20	.44	.59	.61
900	−.78	−.86	−1.09	−1.33	−1.76	−1.75	−1.02	−.29	.22	.64	.82
950	−.69	−.83	−.92	−1.11	−1.83	−2.59	−2.02	−1.07	−.18	.54	.81
1000	−.63	−1.06	−1.24	−1.00	−1.49	−3.26	−3.34	−2.00	−.65	.31	.57
April
400	.11	.02	.03	.05	.06	.05	.04	.07	.08	.07	.07
500	−.04	−.05	−.04	−.01	.00	.04	.06	.08	.10	.11	.12
700	.10	.00	−.22	−.32	−.35	−.16	.01	.15	.28	.29	.31
850	.73	.75	.65	.31	−.27	−.48	−.10	.40	.58	.56	.55
900	1.08	1.03	.80	.30	−.48	−.79	−.46	.06	.36	.53	.61
950	1.35	1.26	.75	.19	−.98	−1.64	−1.31	−.68	−.03	.54	.63
1000	1.45	1.06	.37	−.11	−1.08	−2.51	−2.63	−1.75	−.79	.12	.34
July
400	−.08	−.05	−.04	−.01	.04	.04	.02	.02	.02	.01	.00
500	−.07	−.12	−.14	−.15	−.08	.01	.01	.05	.12	.17	.12
700	.16	.26	.30	.28	.16	.05	.04	.26	.41	.40	.27
850	1.07	1.77	1.99	1.50	.80	.68	.72	.80	.70	.60	.45
900	1.63	2.25	2.47	2.09	1.22	1.00	.79	.62	.47	.54	.54
950	2.08	2.91	3.14	3.01	1.89	1.32	.84	.37	.23	.58	.64
1000	2.18	2.70	3.07	3.39	2.77	1.74	.20	−.51	−.37	.23	.58
October
400	.00	−.01	.00	.01	.03	.04	.04	.05	.06	.07	.08
500	.06	−.02	−.06	−.06	.00	.07	.10	.10	.13	.16	.14
700	.41	.20	−.01	−.08	−.11	−.07	.02	.16	.28	.39	.37
850	.58	1.02	1.10	.80	.25	−.05	−.10	.02	.00	.15	.33
900	.84	1.30	1.49	1.33	.53	−.14	−.55	−.60	−.39	.01	.33
950	1.15	1.66	1.94	1.98	.98	−.15	−.94	−1.20	−.80	−.16	.27
1000	1.26	1.57	1.94	2.42	1.83	−.43	−2.04	−2.08	−1.43	−.55	−.08

Table 5.6. $[\bar{q}^*\bar{v}^*]/g$. Units: gm (cm mb sec)$^{-1}$

p (mb)	Latitude										
	−15	−10	−5	0	5	10	15	20	25	30	35
January											
400	.00	.00	.00	.00	.00	.01	.01	.01	.00	.00	.00
500	.00	.00	.00	.00	.00	.01	.01	.01	.00	.00	.02
700	.00	.00	−.01	.00	.00	.00	.00	.00	.01	.02	.04
850	.00	−.01	−.01	.01	.01	.03	.05	.10	.11	.10	.12
900	.00	−.03	−.02	.00	.01	.06	.11	.15	.16	.14	.14
950	.00	−.04	−.03	−.01	.00	.03	.08	.14	.14	.14	.14
1000	−.01	−.04	−.04	.00	.02	.02	.02	.06	.08	.09	.11
April											
400											
500	.00	.00	.00	.00	.00	.00	.00	.00	.00	.01	.01
700	.00	−.01	.00	−.01	−.01	−.01	−.02	−.01	.02	.03	.03
850	−.01	−.03	−.01	.00	.00	.04	.13	.17	.13	.07	.04
900	−.02	−.03	−.01	.01	.00	.03	.18	.29	.22	.12	.04
950	−.01	−.02	.00	.02	.01	.03	.16	.26	.20	.10	.04
1000	−.01	.00	.01	.02	.04	.10	.20	.19	.10	.05	.01
July											
400											
500	.00	−.01	.00	.01	.01	.00	−.01	.01	.04	.04	.02
700	.00	.01	.01	.01	.02	.03	.06	.12	.18	.16	.06
850	−.01	−.01	−.01	−.01	.00	.01	.04	.16	.32	.31	.23
900	−.04	−.03	−.02	−.01	−.01	.01	.09	.29	.42	.36	.26
950	−.02	.01	.02	.01	.00	.04	.14	.33	.43	.35	.27
1000	−.02	.03	.05	.03	.03	.13	.26	.36	.43	.38	.30
October											
400											
500	.00	.00	.00	.00	.00	.01	.01	.02	.02	.02	.01
700	.01	.00	.00	−.01	.00	.02	.03	.04	.05	.04	.02
850	−.01	−.01	−.01	−.02	−.01	.04	.08	.09	.08	.05	.03
900	.00	.00	.00	−.01	−.01	.02	.08	.13	.12	.09	.04
950	.00	.01	.01	.00	−.01	.00	.04	.13	.15	.10	.05
1000	.00	−.01	.00	.00	−.01	.01	.05	.10	.10	.06	.04

Table 5.8. $[\overline{q'v'}]/g$. Units: gm (cm mb sec.)$^{-1}$

p (mb)	Latitude										
	−15	−10	−5	0	5	10	15	20	25	30	35
January											
400	−.04	−.02	.00	.02	.04	.04	.03	.03	.05	.06	.06
500	−.12	−.06	.00	.06	.09	.09	.07	.06	.08	.12	.12
700	−.18	−.08	.00	.04	.09	.14	.15	.18	.24	.29	.29
850	−.15	−.08	.00	.06	.12	.12	.14	.26	.38	.41	.38
900	−.12	−.06	.00	.04	.07	.08	.12	.24	.38	.42	.42
950	−.10	−.05	−.01	.02	.04	.06	.12	.23	.38	.43	.41
1000	−.08	−.03	.00	.00	.00	.02	.08	.18	.31	.37	.34
April											
400	−.03	−.02	−.01	.01	.04	.05	.04	.05	.06	.07	.08
500	−.07	−.06	−.04	−.01	.02	.05	.06	.07	.10	.13	.14
700	−.14	−.08	−.04	.00	.04	.05	.04	.05	.16	.25	.29
850	−.11	−.08	−.06	−.03	.01	.05	.08	.15	.28	.34	.37
900	−.10	−.06	−.04	−.02	.00	.03	.08	.18	.28	.33	.36
950	−.08	−.05	−.03	−.02	.00	.03	.10	.21	.30	.34	.33
1000	−.08	−.03	−.02	−.02	−.01	.01	.08	.19	.28	.31	.28
July											
400	−.04	−.03	−.03	−.02	−.01	.01	.01	.01	.02	.03	.06
500	−.07	−.06	−.06	−.07	−.04	.00	.02	.02	.03	.05	.10
700	−.14	−.07	−.02	.01	.06	.07	.06	.04	.04	.07	.16
850	−.14	−.09	−.04	.00	.03	.06	.06	.04	.04	.07	.15
900	−.14	−.08	−.03	.00	.04	.05	.06	.04	.04	.08	.15
950	−.12	−.08	−.04	−.01	.01	.02	.03	.03	.04	.07	.12
1000	−.12	−.08	−.03	−.01	.00	.02	.02	.03	.04	.07	.12
October											
400	−.04	−.03	−.02	−.01	.00	.02	.04	.05	.05	.05	.06
500	−.08	−.04	−.03	−.03	.00	.03	.08	.09	.10	.12	.13
700	−.16	−.09	−.04	−.04	−.10	.03	.09	.15	.22	.30	.34
850	−.06	−.02	−.01	−.01	.00	.04	.11	.17	.24	.30	.34
900	−.07	−.02	.00	.00	.00	.03	.08	.14	.23	.30	.36
950	−.08	−.02	.00	.01	.01	.05	.08	.14	.24	.32	.38
1000	−.09	−.01	.01	.02	.02	.02	.04	.11	.22	.33	.36

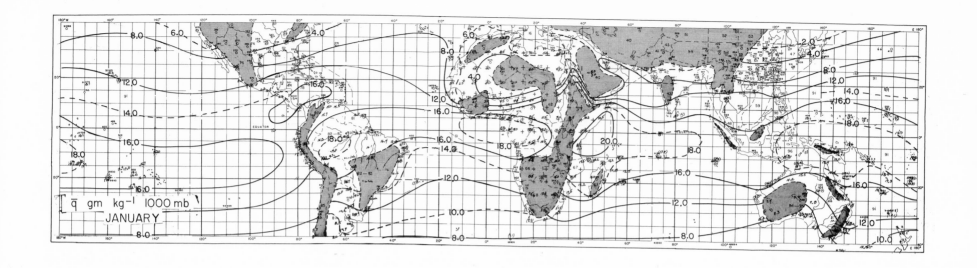

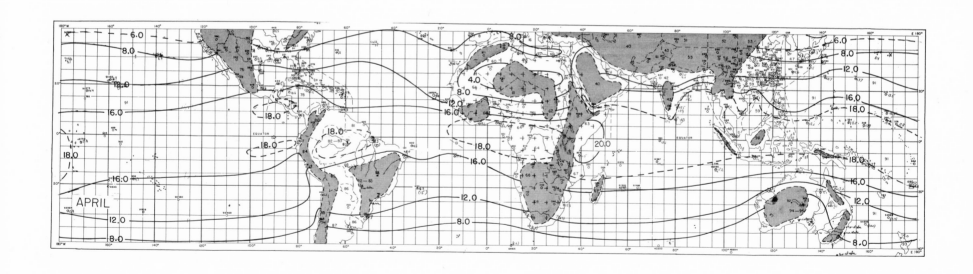

Plate 5.1

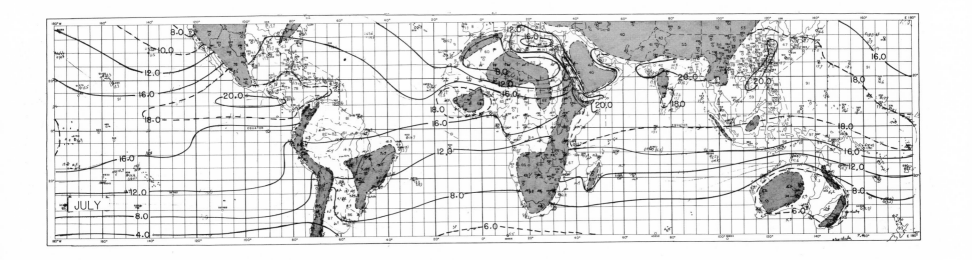

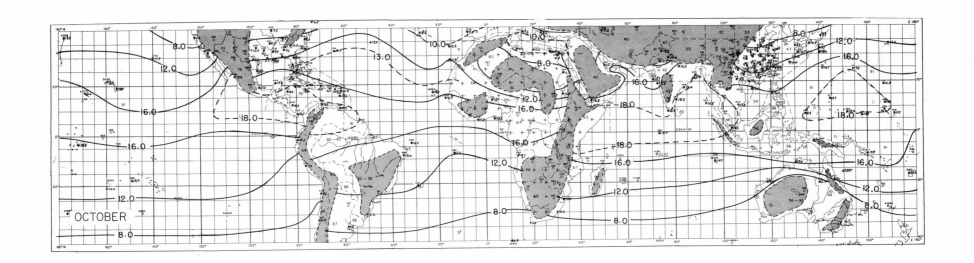

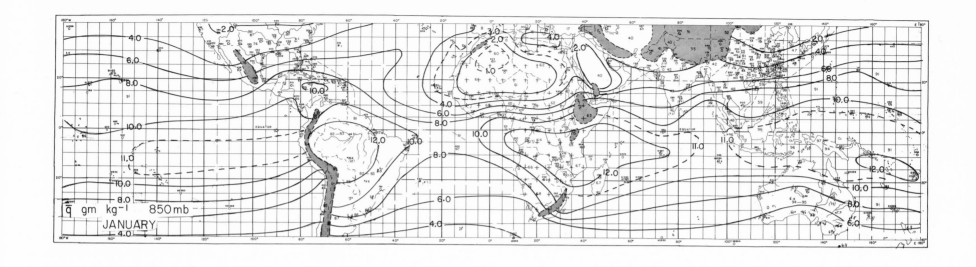

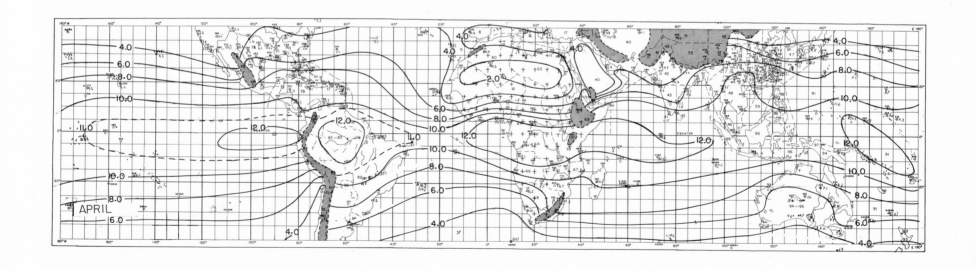

Plate 5.2

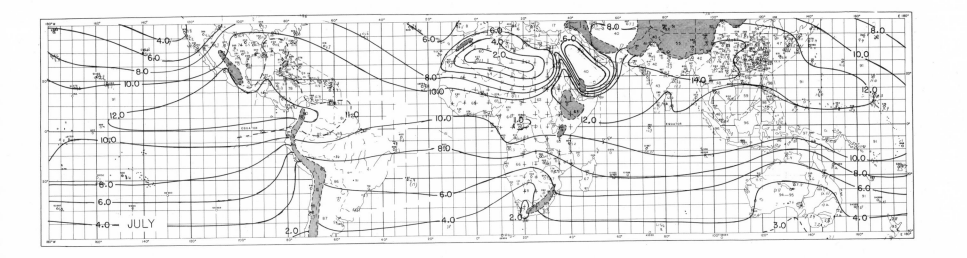

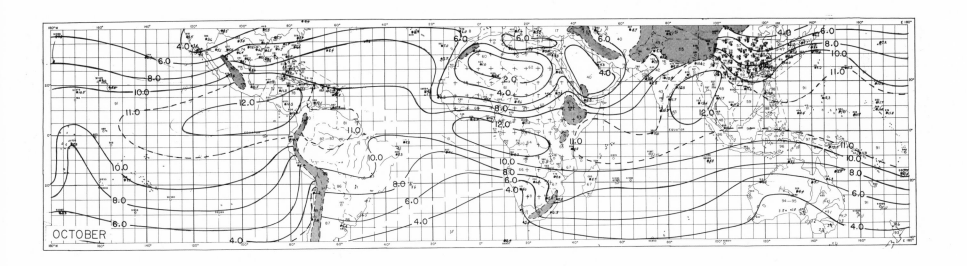

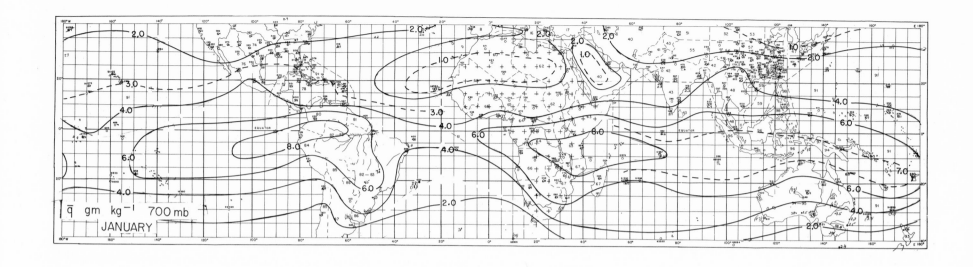

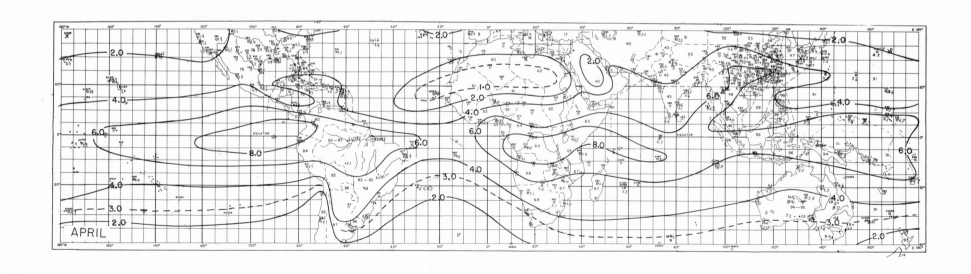

Plate 5.3

230 SEASONAL VARIATION OF TROPICAL HUMIDITY PARAMETERS

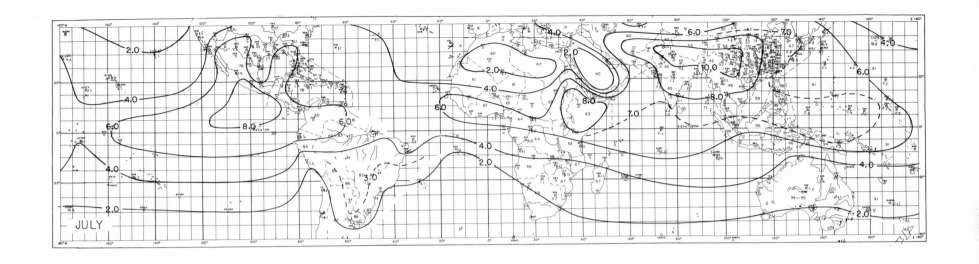

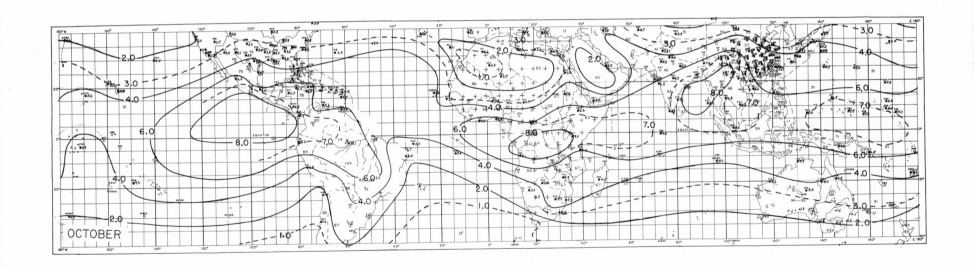

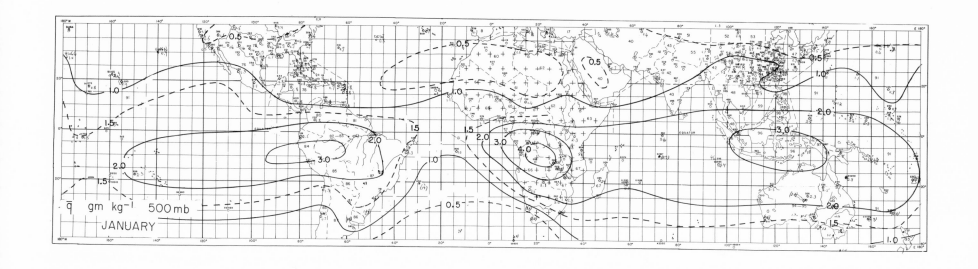

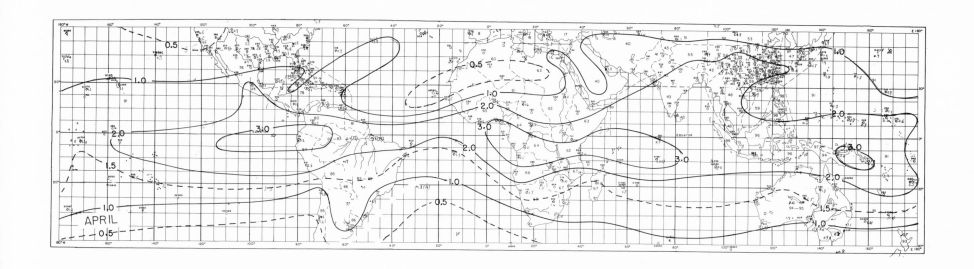

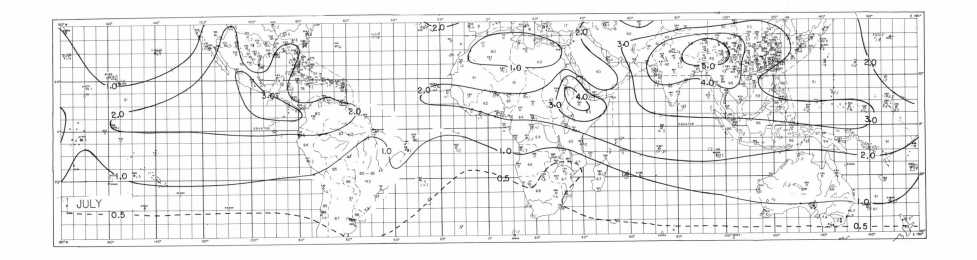

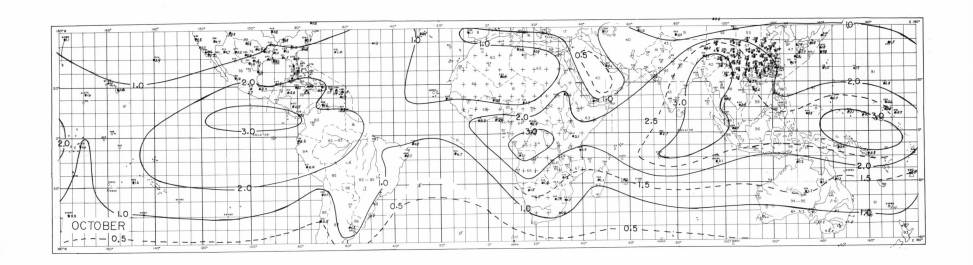

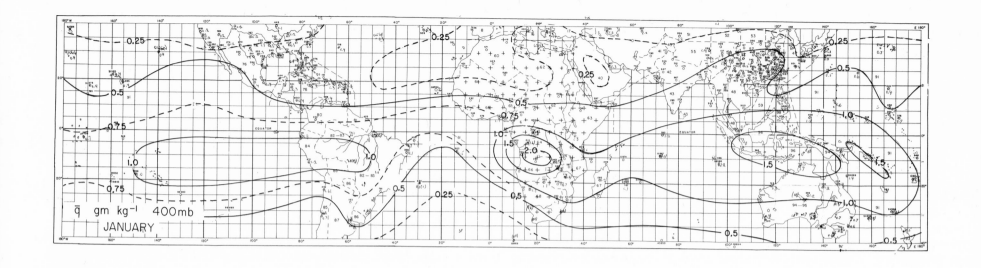

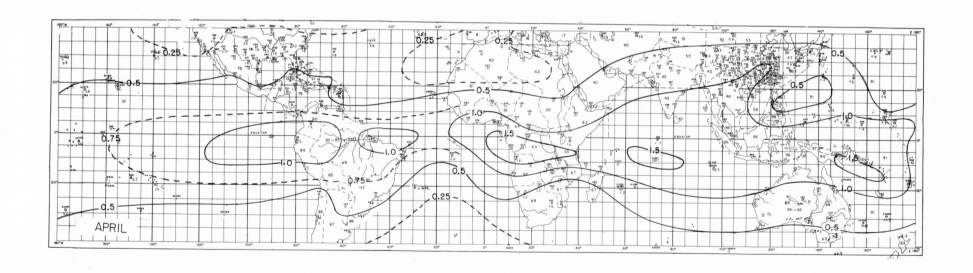

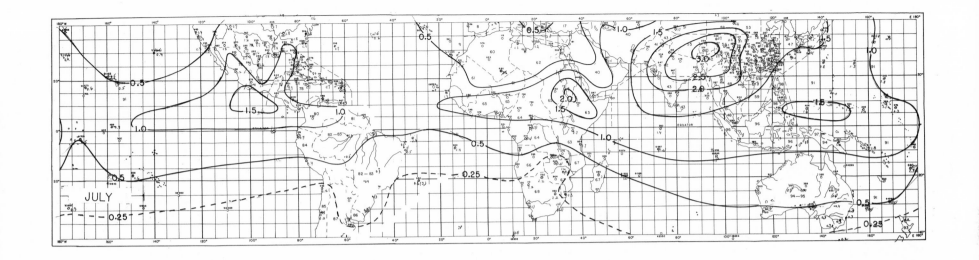

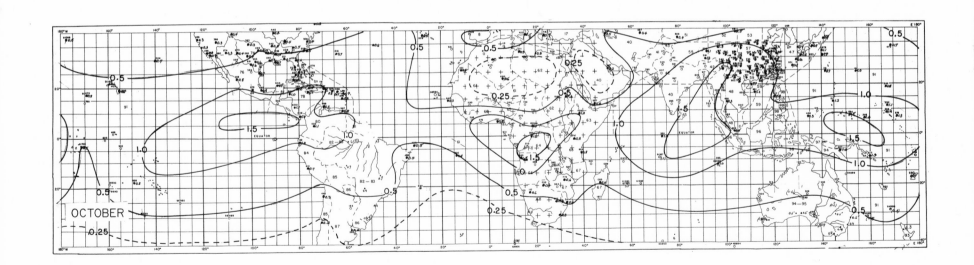

References

Bannon, J. K., and L. P. Steele. 1960. Average water vapor content of the air. Meteorological Office, *Geophys. Mem.* no. 102 (Second number, vol. 13).

Brooks, C. E. P., and T. M. Hunt. 1930. The zonal distribution of rainfall over the earth, *Memoirs, Roy. Meteorol. Soc.* 3:28, 139–158.

Budyko, M. I. 1963. *Atlas Teplovogo Balansa Zemnogo Shara.* Moscow: USSR Glavnaia Geofizicheskaia Observatoria. 69 pp. (I. A. Donehoo, trans.), *Guide to the Atlas of the Heat Balance of the Earth.* WB/T–106. Washington, D.C.: U.S. Weather Bureau, Dec. 1964. 25 pp.

Byers, H. R. 1959. *General Meteorology.* New York: McGraw-Hill. 540 pp.

Colón, J. 1963. Seasonal variations of the heat flux from the sea surface to the atmosphere over the Caribbean Sea. *J. Geophys. Res.* 68:1421–1430.

Crutcher, H. L., and J. M. Meserve. 1966. *Selected Level Temperatures and Dew Points for the Northern Hemisphere.* Navair 50–1C–52. Washington, D.C.: U.S. Environmental Data Service.

Crutcher, H. L., A. C. Wagner, and J. Arnett. 1966. *Components of the 1000 mb Winds.* Navair 50–1C–51. Asheville, N.C.: U.S. National Records Center.

Frazier, H. M., E. R. Sweeton, and J. G. Welsh. 1968. *Data Processing Support to a Program for Observational and Theoretical Studies of Planetary Atmospheres.* Progress Report, Massachusetts Institute of Technology Subcontract SR13730 for NSF Grant CA–1310X. Hartford: Travelers Research Center. 87 pp.

Gray, William M. 1968. Global view of the origin of tropical disturbances and storms. *Mon. Weath. Rev.* 96:669–700.

Hastenrath, S. L. 1966. The flux of atmospheric water vapor over the Caribbean Sea and the Gulf of Mexico. *J. Appl. Meteorol.* 5:778–788.

Iida, M. 1968. Computations of the transports of momentum, sensible and latent heat across the equator. *J. Meteorol. Soc. Japan* 46:1–13.

Kidson, J. W., D. G. Vincent, and R. E. Newell. 1969. Observational studies of the general circulation of the tropics: long term mean values. *Quart. J. Roy. Meteorol. Soc.* 95:258–287.

Kornfield, J., A. F. Hasler, K. J. Hanson, and V. E. Suomi. 1967. Photographic cloud climatology from ESSA III and V computer produced mosaics. *BAMS,* 48:878–889.

Lorenz, E. 1967. *The General Circulation.* Geneva: TR. 115, World Meteorological Organization. 161 pp.

Malkus, Joanne S. 1962. Large scale interactions. In *The Sea,* N. M. Hill, ed., New York: Interscience vol. I, pp. 88–294.

Oort, A. H., and E. M. Rasmusson. 1970. On the annual variation of the monthly mean meridional circulation. *Mon. Weath. Rev.* 98:423–442.

Palmén, E. 1967. *Evaluation of Atmospheric Moisture Transport for Hydrological Purposes.* Report no. 1, Reports on WMO/IHD Projects. Geneva: WMO. 63 pp.

———, and L. A. Vuorela. 1963. On the mean meridional circulations in the northern hemisphere during the winter season. *Quart. J. Roy. Meteorol. Soc.* 89:131–138.

Palmer, C. E., C. W. Wise, L. T. Stempson, and G. H. Duncan. 1955. *The Practical Aspect of Tropical Meteorology.* A. F. Surveys in Geophysics no. 76, GRD, Bedford, Mass.: AFCRL. 195 pp.

Peixoto, J. P. 1958. *Hemispheric Humidity Conditions during the Year 1950.* Report no. 3. Cambridge: Massachusetts Institute of Technology, Department of Meteorology, General Circulation Project. 142 pp.

———, and A. R. Crisi. 1965. *Hemispheric Humidity Conditions during the IGY.* Report no. 6. Cambridge: Massachusetts Institute of Technology, Department of Meteorology, Planetary Circulations Project. 166 pp.

———, and G. O. P. Obasi. 1965. *Humidity Conditions over Africa during the IGY.* Report no. 7. Cambridge: Massachusetts Institute of Technology, Department of Meteorology, Planetary Circulations Project. 143 pp.

Rakipova, L. R. 1966. Heat transfer and general circulation of the atmosphere. *Izv. Atmospheric and Oceanic Physics* 2:983–986.

Raman, C. R. V. 1965. *Cyclonic Vortices on either Side of the Equator and their Implications.* Bombay: Symposium on Meteorological Results of the IIOE.

Rasmusson, E. M. 1966. *Atmospheric Water Vapor Transport and the Hydrology of North America.* Report no. A1. Cambridge: Massachusetts Institute of Technology, Department of Meteorology. Planetary Circulations Project. 170 pp.

———. 1967a. Large scale water vapor transfer in the atmosphere. In *Transport Phenomena in Atmospheric and Ecological Systems.* New York: American Society of Mechanical Engineers, pp. 37–68.

———. 1967b. Atmospheric water vapor transport and the water balance of North America. Part 1, Characteristics of the water vapor flux field. *Mon. Weath. Rev.* 95:403–426.

———. 1968. Atmospheric water vapor transport and the water balance of North

America. Part 2, Large scale water balance investigations. *Mon. Weath. Rev.* 96:720–734.

Riehl, H. 1954. *Tropical Meteorology*. New York: McGraw-Hill. 392 pp.

—————, and J. S. Malkus. 1958. On the heat balance in the equatorial trough zone. *Geophysica* 6:503–538.

—————, and T. C. Yeh. 1950. The intensity of the net meridional circulation. *Quart. J. Roy. Meteorol. Soc.* 76:182–188.

—————, T. C. Yeh, J. S. Malkus, and N. E. La Seur. 1951. The Northeast Trade of the Pacific Ocean. *Quart. J. Roy. Meteorol. Soc.* 77:598–626.

Sadler, James C. 1969. *Average Cloudiness in the Tropics From Satellite Observations.* IIOE Meterological Monograph no. 2. Honolulu: East-West Center Press.

Sellers, William D. 1965. *Physical Climatology*. Chicago: University of Chicago Press, pp. 100–127.

Starr, V. P., and J. P. Peixoto, 1964. The hemispheric eddy flux of water vapor and its implications for the mechanics of the general circulation. *Arch. f. Met., Geoph., u. Biokl.* Ser. A, 14, 111–130.

—————, J. P. Peixoto, and A. R. Crisi. 1965. Hemispheric water balance for the IGY. *Tellus* 17:463–472.

—————, J. P. Peixoto, and R. McKean. 1969. Pole-to-pole moisture conditions for the IGY. *Pure and Appl. Geophys.* 75:300–331.

—————, and R. M. White. 1955. Direct measurement of the hemispheric poleward flux of water vapor. *J. Mar. Res.* 14:217–225.

Sutcliffe, R. C. 1956. Water balance and the general circulation of the atmosphere. *Quart. J. Roy. Meteorol. Soc.* 82:385–395.

Taylor, V. Roy, and Jay S. Winston. 1968. *Monthly and Seasonal Mean Global Charts of Brightness from ESSA 3 and ESSA 5 Digitized Pictures, February 1967–February 1968*. Washington, D.C.: ESSA Technical Report NESC 46. 7 pp., 17 charts.

Tucker, G. B. 1957. Evidence of a mean meridional circulation in the atmosphere from surface wind observations. *Quart. J. Roy. Meteorol. Soc.* 83:290–302.

Tuller, Stanton E. 1968. World distribution of mean monthly and annual precipitable water. *Mon. Weath. Rev.* 96:785–797.

U.S. Weather Bureau. 1938. *Atlas of Climatic Charts of the Oceans*. Weather Bureau no. 1247. Washington, D.C.: Government Printing Office. 130 pp.

Vuorela, L. A., and I. Tuominen. 1964. On the zonal mean and meridional circulations and the flux of moisture in the northern hemisphere during the summer season. *Geof. Pura e Appl.* 57:167–180.

Wallace, J. M., and F. R. Hartranft. 1969. Diurnal wind variations surface to 30 km. *Mon. Weath. Rev.* 97:446–455.

White, R. M. 1951. The meridional eddy flux of energy. *Quart. J. Roy. Meteorol. Soc.* 77:188–199.

Appendix I

Station List for Tropical Study

STATION LIST FOR TROPICAL STUDY

KN	WMO	STATION NAME	LAT	LONG	HT	00	06	12	18	TYPE	SOURCE	COMMENTS
1	07645	NIMES/COURBESSANT	43.9N	4.4E	60	1				WT	TRC TAPES	Type of report
2	08509	LAJES (AZORES)	38.8N	27.1W	54	1				WT	TRC TAPES	
3	08521	FUNCHAL	32.6N	16.9W	110			1		T	ASHEVILLE	W: wind data
4	08536	LISBOA/PORTELA	38.8N	9.1W	110	1				WT	TRC TAPES	T: thermodynamic data
5	08594	SAL	16.7N	23.0W	55			1		T	ASHEVILLE	
6	15614	SOFIA	42.8N	23.4E	589	1				WT	TRC TAPES	Source of data
7	16420	MESSINA	38.2N	15.6E	54	1				WT	TRC TAPES	
8	16716	ATHINAI/HELLINIKON	37.9N	23.7E	9	1				WT	TRC TAPES	TRC: Travelers Research Corporation
9	31960	VLADIVOSTOK	43.1N	131.9E	138	1				WT	TRC TAPES	ASHVL: Asheville weather records center
10	32165	JUZNO-KURIL'SK	44.0N	145.8E	40	1				WT	TRC TAPES	MCDS: Micro-cards
11	36870	ALMA ATA	43.2N	76.9E	347	1				WT	TRC TAPES	FRN.TAB: Foreign tabulated data
12	37018	TUAPSE	44.1N	39.1E	78	1				WT	TRC TAPES	
13	37985	LENKORAN	38.7N	48.8E	-11	1				WT	TRC TAPES	Heights are in meters.
14	38687	CARDZOU	39.1N	63.6E	193	1				WT	TRC TAPES	
15	40181	BEER YAAQOV	31.9N	34.8E	63	1				WT	TRC TAPES	3 MONTHS' DATA ONLY
16	40427	BAHRAIN /MUHARRAQ	26.3N	50.6E	2	1				WT	TRC+ASHVL	
17	40597	ADEN/KHORMAKSAR	12.8N	45.0E	3	1				WT	TRC+ASHVL	
18	41780	KARACHI AP	24.9N	67.1E	22	1		1		WT	ASHEVILLE	1959-60 MISSING
19	41920	DACCA	23.8N	90.4E	8	1				T	TRC TAPES	STARTS 7/62 (MAY BE 41917)
20	42071	AMRITSAR	31.6N	74.9E	234	1				T	TRC TAPES	ENDS 5/60
21	42182	NEW DELHI/SAFDARJUNG	28.6N	77.2E	216	1				T	TRC TAPES	
22	42339	JODHPUR	26.3N	73.0E	224	1		1		WT	TRC+ASHVL	
23	42361	GWALIOR	26.2N	78.2E	207	1		1		WT	TRC+ASHVL	
24	42410	GAUHATI	26.1N	91.7E	54	1				T	TRC TAPES	4 MONTHS' DATA
25	42411	GAUHATI	26.2N	91.8E	55	1				WT	TRC TAPES	
26	42475	ALLAHABAD/BAMHRAULI	25.4N	81.7E	98	1		1		WT	MCDS+ASHVL	
27	42647	AHMADABAD	23.1N	72.6E	55	1				WT	TRC TAPES	STARTS 11/62
28	42809	CALCUTTA/DUM DUM	22.8N	88.4E	10	1		1		WT	TRC TAPES	
29	42867	NAGPUR/SONEGAON	21.1N	79.0E	310	1		1		WT	TRC+ASHVL	
30	42909	VERAVAL	20.9N	70.4E	8	1				T	TRC TAPES	ENDS 7/61
31	43003	BOMBAY/SANTACRUZ AP	19.1N	72.8E	15	1		1		WT	TRC+ASHVL	
32	43149	VISHAKHAPATNAM	17.7N	83.2E	3	1		1		WT	TRC+ASHVL	
33	43279	MADRAS/MINAMBAKKAM	13.0N	80.2E	16	1		1		WT	TRC+ASHVL	
34	43295	BANGALORE	13.0N	77.6E	921	1				T	TRC TAPES	8 MONTHS' DATA
35	43333	PORT BLAIR	11.7N	92.7E	79	1		1		WT	TRC+ASHVL	
36	43371	TRIVANDRUM	8.5N	77.0E	64	1		1		WT	TRC+ASHVL	
37	44354	YUGUDZYV	44.9N	110.1E	914	1				WT	TRC TAPES	
38	45004	HONG KONG/KING'SPARK	22.3N	114.2E	66	1				WT	TRC+ASHVL	
39	46692	TAIPEI	25.0N	121.5E	9	1				WT	TRC+ASHVL	
40	46697	TAOYUAN	25.0N	121.2E	48	1				WT	TRC+ASHVL	ENDS 5/60
41	46734	MAKUNG NO 1	23.5N	119.6E	22	1				WT	TRC+ASHVL	ENDS 1958
42	46747	TUNGKONG	22.5N	120.4E	8	1				WT	TRC+ASHVL	ENDS 5/60
43	47058	PYONGYANG	39.0N	125.8E	29	1				WT	TRC TAPES	
44	47187	MOSULPO AB	33.2N	126.2E	6	1				WT	TRC TAPES	
45	47646	TATENO	36.0N	140.1E	27	1				WT	TRC TAPES	

RN	WMO	STATION NAME	LAT	LONG	HT	00	06	12	18	TYPE	SOURCE	COMMENTS
46	47678	HACHIJOJIMA/OMURE	33.1N	139.8E	156	1				WT	TRC TAPES	
47	47778	SHIONOMISAKI	33.4N	135.7E	75	1				WT	TRC TAPES	
48	47807	FUKUOKA	33.6N	130.4E	6	1				WT	TRC TAPES	
49	47827	KAGOSHIMA/YOSHINO	31.6N	130.6E	283	1				WT	TRC TAPES	
50	47909	NAZE/FUNCHATOGE	28.4N	129.6E	295	1				WT	TRC+ASHVL	
51	47931	KADENA AB	26.4N	127.8E	46	1				WT	TRC TAPES	
52	47963	TORISHIMA	30.5N	140.3E	83	1				WT	TRC TAPES	
53	48327	CHIANGMAI	18.8N	99.0E	313	1				WT	TRC+ASHVL	
54	48455	BANGKOK	13.7N	100.5E	12	1				WT	TRC+ASHVL	
55	48568	SONGKHLA	7.2N	100.6E	10	1				WT	TRC+ASHVL	
56	48694	SINGAPORE AP	1.4N	104.0E	10	1				WT	MCDS+ASHVL	ENDS 1/61
57	48819	LANG	21.0N	105.9E	5	1				WT	TRC TAPES	STARTS 5/62
58	48855	DANANG/TOURANE	16.0N	108.2E	7	1				WT	TRC+ASHVL	
59	48900	SAIGON/TANSONNHUT	10.8N	106.7E	10	1				WT	TRC TAPES	
60	51828	HOTIEN	37.1N	79.9E	1389	1				WT	TRC TAPES	
61	52203	HAMI	42.8N	93.4E	735	1				WT	TRC TAPES	
62	52681	MINCHIN	38.7N	103.1E	1368	1				WT	TRC TAPES	
63	55299	HEIHO	32.0N	92.1E	4000	1				WT	TRC TAPES	
64	55591	LASA	29.7N	91.0E	3658	1				WT	TRC TAPES	
65	56029	YUSHU	33.1N	96.8E	3873	1				WT	TRC TAPES	
66	56096	WOUTOU	33.4N	104.7E	1090	1				WT	TRC TAPES	
67	56137	CHANGTU	31.2N	97.3E	3200	1				WT	TRC TAPES	
68	56146	KANTZE	31.6N	100.0E	3320	1				WT	TRC TAPES	
69	56294	CHENGTU	30.7N	104.1E	498	1				WT	TRC TAPES	
70	56492	YEHPIN	28.8N	104.5E	286	1				WT	TRC TAPES	
71	56533	TZUKEI	27.7N	98.4E	63	1				WT	TRC TAPES	
72	56571	HSIHCHANG	27.9N	102.3E	1599	1				WT	TRC TAPES	STARTS 1/60
73	56691	WEINING	26.8N	104.2E	2235	1				WT	TRC TAPES	
74	56739	TENG CHUNG	25.1N	98.5E	1628	1				WT	TRC TAPES	
75	56778	KUNMING	25.0N	102.7E	1893	1				WT	TRC TAPES	
76	56779	(KUNMING)	25.1N	102.8E	1905	1				WT	TRC+ASHVL	
77	56964	SZEMOA	22.6N	101.0E	1319	1				WT	TRC+ASHVL	
78	56989	HOKOW	24.4N	103.9E	134	1				WT	TRC TAPES	
79	57036	SIAN	34.2N	108.9E	400	1				WT	TRC TAPES	
80	57083	CHENG CHOW	34.7N	113.7E	110	1				WT	TRC TAPES	
81	57127	HANCHUNG	33.1N	107.2E	520	1				WT	TRC TAPES	
82	57447	ENSHIH	30.3N	109.4E	437	1				WT	TRC TAPES	
83	57461	YEHCHANG	30.7N	111.3E	707	1				WT	TRC TAPES	
84	57494	HANKOW	30.4N	114.3E	23	1				WT	TRC TAPES	
85	57515	SAPINPA	29.5N	106.6E	261	1				WT	TRC TAPES	CONTINUATION OF 57516
86	57516	PAISHIH TO	29.5N	106.3E	290	1				WT	TRC TAPES	ENDS 3/61
87	57679	CHANGSHA	28.2N	112.8E	49	1				WT	TRC TAPES	
88	57745	CHIHKIANG	27.4N	109.6E	268	1				WT	TRC TAPES	
89	57816	KWEIYANG	26.6N	106.7E	1071	1				WT	TRC TAPES	
90	57957	KWEILIN	25.2N	110.1E	156	1				WT	TRC TAPES	
91	57972	CHENGHSIEN	25.8N	113.0E	171	1				WT	TRC TAPES	
92	57993	KANCHOW	25.8N	114.8E	110	1				WT	TRC TAPES	
93	58027	HSUCHOW	34.2N	117.2E	45	1				WT	TRC TAPES	

KN	WMO	STATION NAME	LAT	LONG	HT	00	06	12	18	TYPE	SOURCE	COMMENTS
94	58150		33.8N	120.5E	2	1				WT	TRC TAPES	
95	58203	FUYANG	32.9N	115.8E	32	1				WT	TRC TAPES	
96	58238	NANCHING	32.1N	118.8E	72	1				WT	TRC TAPES	
97	58321	HOFEI	31.9N	117.2E	28	1				WT	TRC TAPES	
98	58367	SHANGHAI	31.2N	121.4E	5	1				WT	TRC TAPES	STARTS 1/60
99	58606	NANCHANG	28.7N	116.0E	49	1				WT	TRC TAPES	
100	58633	CHUNCHOW	29.0N	118.9E	80	1				WT	TRC TAPES	
101	58636		28.5N	118.0E								
102	58666	TACHEN TAO	28.4N	121.9E	199	1				WT	TRC TAPES	
103	58725	SHAO WU	27.3N	117.5E	255	1				WT	TRC TAPES	
104	58847	FUCHOW	26.1N	119.3E	88	1				WT	TRC TAPES	
105	58912	CHANGTING	25.9N	116.4E	301	1				WT	TRC TAPES	
106	58965	DAOYUNG	25.1N	121.3E	48	1				WT	TRC TAPES	10 MONTHS' DATA
107	59082	CHUCHIANG	24.8N	113.5E	87	1				WT	TRC TAPES	
108	59134	SHAMEN	24.4N	118.1E	39	1				WT	TRC TAPES	ENDS 3/61
109	59135	CHINMEN	24.4N	118.3E	91	1				WT	TRC+ASHVL	
110	59137		24.3N	118.1E	23	1				WT	TRC TAPES	PATCHY
111	59211	POSEH	23.4N	106.5E	198	1				WT	TRC TAPES	7 MONTHS ONLY
112	59265	WUCHOW	23.5N	111.4E	119	1				WT	TRC+ASHVL	
113	59287	KUANGCHOW	23.2N	113.3E	18	1				WT	TRC TAPES	
114	59316	SHANTOU	23.4N	116.7E	5	1				WT	TRC+ASHVL	
115	59345	MAKUNG	23.5N	119.5E	40	1				WT	TRC+ASHVL	
116	59431	NANNING	22.8N	108.3E	75	1				WT	TRC TAPES	ENDS 9/58
117	59559	HENGCHUNG	22.0N	120.8E	22	1				WT	TRC+ASHVL	
118	59758	HAIKOW	20.0N	110.4E	14	1				WT	TRC+ASHVL	
119	59981	SISHA	16.8N	112.3E	2	1				WT	TRC+ASHVL	
120	60020	SANTA CRUZ	28.5N	16.2W	37	1				WT	TRC+ASHVL	
121	60119	KENITRA/PORT LYAUTEY	34.3N	6.6E	12	1				WT	TRC TAPES	NO WINDS UNTIL 3/61
122	60571	COLOMB BECHAR	31.6N	2.2W	781			1		WT	TRC TAPES	
123	60625	AOULEF	27.0N	1.1E	290			1		WT	ASHEVILLE	(FORMERLY 60570)
124	60680	TAMANRASSET	22.8N	5.5E	1378			1		WT	ASHEVILLE	
125	61052	NIAMEY	13.5N	2.2E	226	1				WT	MCDS+ASHVL	
126	61401	FORT TRINQUET	25.2N	11.6W	360	1				WT	TRC+ASHVL	
127	61642	DAKAR/OUAKAM	14.7N	17.4W	39	1				WT	MICROCARDS	
128	62011	WHEELUS AFB TRIPOLI	32.9N	13.3E	4	1				WT	TRC+ASHVL	
129	62062	TOBRUK	32.1N	24.0E	14	1				WT	TRC TAPES	
130	62306	MERSA MATRUH	31.3N	27.2E	30	1				WT	TRC TAPES	
131	62366	CAIRO	30.1N	31.6E	74	1				WT	TRC TAPES	
132	62378	HELWAN	29.9N	31.2E	116	1				WT	TRC TAPES	ENDS 7/60
133	62414	ASSWAN	24.0N	32.9E	194	1				WT	TRC TAPES	STARTS 8/60
134	62721	KHARTOUM	15.6N	32.6E	380	1				WT	TRC+ASHVL	
135	64400	POINTE-NOIRE	4.8S	11.9E	17	1				W	FOREIGN	FR. STARTS 9/60
136	64650	BANGUI	4.4N	18.6E	381	1				WT	TRC+ASHVL	
137	64700	FORT-LAMY	12.1N	15.0E	295	1				WT	TRC+ASHVL	
138	64750	FORT ARCHAMBAULT	9.1N	18.4E	365	1				W	FOREIGN	FR. STARTS 11/61
139	65201	LAGOS/IKEJA	6.6N	3.3E	40	1				WT	TRC+ASHVL	
140	65578	ABIDJAN	5.2N	3.9W	16	1				WT	TRC+ASHVL	
141	72201	KEY WEST	24.6N	81.8W	6	1				WT	TRC+ASHVL	

RN	WMO	STATION NAME	LAT	LONG	HT	00	06	12	18	TYPE	SOURCE	COMMENTS
142	72202	MIAMI	25.8N	80.3W	4	1				WT	TRC TAPES	
143	72206	JACKSONVILLE	30.4N	81.6W	9	1				WT	TRC TAPES	
144	72208	CHARLESTON	32.9N	80.0W	15	1				WT	TRC TAPES	
145	72211	TAMPA	28.0N	82.5W	3	1				WT	TRC TAPES	
146	72221	VALPARISO	30.5N	86.5W	29	1				WT	TRC TAPES	
147	72226	MONTGOMERY	32.3N	86.4W	62	1				WT	TRC TAPES	
148	72232	BURRWOOD(BOOTHVILLE)	29.0N	89.4W	5	1				WT	TRC TAPES	
149	72235	JACKSON	32.3N	90.1W	101	1				WT	TRC TAPES	
150	72240	LAKE CHARLES	30.2N	93.2W	10	1				WT	TRC TAPES	
151	72248	SHREVEPORT	32.5N	93.8W	79	1				WT	TRC TAPES	
152	72250	BROWNSVILLE	25.9N	97.4W	6	1				WT	TRC TAPES	
153	72251	CORPUS CHRISTI	27.8N	97.5W	13	1				WT	TRC TAPES	
154	72253	SAN ANTONIO	29.5N	98.5W	242	1				WT	TRC TAPES	
155	72259	FT WORTH	32.8N	97.1W	176	1				WT	TRC TAPES	
156	72261	DEL RIO	29.4N	100.9W	313	1				WT	TRC TAPES	
157	72265	MIDLAND	31.9N	102.2W	872	1				WT	TRC TAPES	
158	72270	EL PASO	31.8N	106.4W	1194	1				WT	TRC TAPES	
159	72274	TUCSON	32.1N	110.9W	779	1				WT	TRC TAPES	
160	72280	YUMA NEVA	32.7N	114.6W	134							
161	72290	SAN DIEGO	32.7N	117.2W	124	1				WT	TRC TAPES	
162	72295	LOS ANGELES	33.9N	118.4W	38	1				WT	TRC TAPES	
163	72308	NORFOLK	36.9N	76.2W	9	1				WT	TRC TAPES	
164	72311	ATHENS	34.0N	83.3W	247	1				WT	TRC TAPES	
165	72327	NASHVILLE	36.1N	86.7W	184	1				WT	TRC TAPES	
166	72340	LITTLE ROCK	34.7N	92.2W	81	1				WT	TRC TAPES	
167	72363	AMARILLO	35.2N	101.7W	1099	1				WT	TRC TAPES	
168	72386	LAS VEGAS	36.1N	115.2W	664	1				WT	TRC TAPES	
169	72392	PT ARGUELLO	34.6N	120.6W	22	1				WT	TRC TAPES	
170	72445	COLUMBIA	39.0N	92.4W	239	1				WT	TRC TAPES	
171	72476	GRAND JUNCTION	39.1N	108.5W	1475	1				WT	TRC TAPES	
172	72493	OAKLAND	37.7N	122.2W	2	1				WT	TRC TAPES	
173	72518	ALBANY	42.8N	73.8W	89	1				WT	TRC TAPES	
174	72645	GREEN BAY	44.5N	88.1W	214	1				WT	TRC TAPES	
175	72662	RAPID CITY	44.1N	103.1W	966	1				WT	TRC TAPES	
176	72694	SALEM	44.9N	123.0W	61	1				WT	TRC TAPES	
177	74794	CAPE KENNEDY	28.5N	80.6W	3	1				WT	TRC TAPES	
178	76458	MAZATLAN	23.2N	106.4W	78	1				WT	TRC+ASHVL	
179	76644	MERIDA	21.0N	89.7W	22	1				WT	TRC+ASHVL	NO WINDS UNTIL 1/61
180	76679	TACUBAYA	19.4N	99.2W	2306	1				WT	TRC+ASHVL	NO WINDS UNTIL 5/62
181	76692	VERA CRUZ	19.2N	96.1W	16	1				WT	TRC+ASHVL	
182	78016	KINDLEY FIELD	32.4N	64.7W	6	1				WT	TRC TAPES	
183	78063	GOLD ROCK CREEK AFB	26.6N	78.3W	6	1				WT	TRC TAPES	
184	78076	COFFIN HILLS AFB	25.3N	76.3W	9	1				WT	TRC+ASHVL	
185	78089	BONEFISH BAY AFB	24.1N	74.5W	9	1				WT	TRC+ASHVL	
186	78118	TURKS ISLAND AFB	21.5N	71.1W	7	1				WT	TRC+ASHVL	
187	78325	HAVANA	23.2N	82.4W	49	1				WT	TRC+ASHVL	
188	78355	CAMAGUEY	21.4N	77.9W	120	1				WT	TRC TAPES	
189	78367	GUANTANAMO	19.9N	75.2W	23	1				WT	TRC+ASHVL	

KN	WMO	STATION NAME	LAT	LONG	HT	00	06	12	18	TYPE	SOURCE	COMMENTS
190	78383	GRAND CAYMAN	19.2N	81.4W	6	1				WT	TRC+ASHVL	
191	78397	KINGSTON/PALISADOES	17.9N	76.8W	12	1				WT	TRC+ASHVL	
192	78467	SABANA DELAMAR	19.1N	69.4W	11	1				WT	TRC TAPES	
193	78486	SAN DOMINGO	18.5N	69.9W	14	1				WT	TRC+ASHVL	STARTS 8/62
194	78501	SWAN ISLAND	17.4N	83.9W	11	1				WT	TRC+ASHVL	
195	78526	SAN JUAN	18.4N	66.0W	19	1				WT	TRC+ASHVL	
196	78806	ALBROOK FIELD	9.0N	79.6W	66	1				WT	TRC+ASHVL	
197	78861	COOLIDGE AFB ANTIGUA	17.1N	61.8W	10	1				WT	TRC TAPES	STARTS 4/61
198	78862	COOLIDGE FLD ANTIGUA	17.1N	61.8W	10	1				WT	TRC+ASHVL	ENDS 8/60
199	78866	JULIANA AP/STMAARTEN	18.0N	63.1W	0	1				WT	TRC+ASHVL	
200	78897	RAIZET GUADELOUPE	16.3N	61.5W	8	1				WT	TRC+ASHVL	
201	78967	CHAGUARAMAS TRINIDAD	10.7N	61.4W	13	1				WT	TRC+ASHVL	
202	78988	PLESMAN AP,CURACAO	12.2N	69.0W	0	1				WT	TRC+ASHVL	
203	80001	SAN ANDRES	12.8N	81.7W	6	1				WT	TRC+ASHVL	
204	80222	BOGOTA/ELDORADO	4.6N	74.1W	2547	1				WT	TRC+ASHVL	STARTS 8/60
205	91066	MIDWAY ISLAND	28.2N	177.3W	13	1				WT	TRC+ASHVL	
206	91115	IWO JIMA AFB	24.8N	141.3E	106	1				WT	TRC+ASHVL	
207	91131	MARCUS ISLAND	24.3N	154.0E	17	1				WT	TRC+ASHVL	
208	91165	LIHUE	22.0N	159.4W	45	1				WT	TRC+ASHVL	
209	91217	TAGUAC GUAM	15.0N	144.8E	111	1				WT	TRC+ASHVL	
210	91218	ANDERSON APT	13.6N	144.9E	162							
211	91245	WAKE ISLAND	19.3N	166.6E	4	1				WT	TRC+ASHVL	
212	91250	ENIWETCK ATOLL	11.3N	162.3E	6	1				WT	TRC+ASHVL	
213	91275	JOHNSTON ISLAND	16.7N	169.5W	5	1				WT	TRC+ASHVL	
214	91285	HILO	19.7N	155.1W	11	1				WT	TRC+ASHVL	
215	91334	TRUK	7.4N	151.8E	2	1		1		WT	ASHEVILLE	
216	91348	PONAPE	7.0N	158.2E	46	1		1		WT	ASHEVILLE	
217	91366	KWAJALEIN	8.7N	167.7E	8	1				WT	TRC+ASHVL	
218	91376	MAJURO	7.1N	171.4E	3	1		1		WT	ASHEVILLE	
219	91408	KOROR	7.4N	134.5E	33	1		1		WT	ASHEVILLE	
220	91413	YAP	9.5N	138.1E	17	1				WT	TRC+ASHVL	TRAV. DATA PATCHY
221	91700	CANTON IS	2.8S	171.7W	3	1				WT	TRC+ASHVL	
222	98327	CLARK AFB	15.2N	120.6E	196	1				WT	TRC+ASHVL	
223	98645	CEBU	10.3N	123.9E	42	1				T	TRC TAPES	POOR COVERAGE
224	OWS K	WEATHER SHIP K	45.0N	16.0W	12	1				WT	TRC TAPES	
225	OWS T	WEATHER SHIP T	29.0N	135.0E	12	1				WT	TRC TAPES	TYPHOON SEASON ONLY
226	OWS D	WEATHER SHIP D	44.0N	41.0W	12	1				WT	TRC TAPES	
227	OWS E	WEATHER SHIP E	35.0N	48.0W	12	1				WT	TRC TAPES	
228	OWS N	WEATHER SHIP N	30.0N	140.0W	12	1				WT	TRC TAPES	
229	OWS V	WEATHER SHIP V	34.0N	164.0E	12	1				WT	TRC TAPES	
230	68032	MAUN	20.0S	23.0E	945		1			WT	FOREIGN	S. AFR. 7/57-12/60
231	68100	SWAKOPMUND	22.7S	14.0E	19		1			WT	FOREIGN	S. AFR. 7/57-12/58
232	68110	WINDHOEK	22.6S	17.1E	1725		1			WT	FOREIGN	S. AFR. 1/59-12/60
233	68262	PRAETORIA	25.8S	28.2E	1374		1			WT	FOREIGN	S. AFR. 7/57-11/60
234	61290	BAMAKO	12.6N	8.0W	330		1			W	FOREIGN	FR.
235	64501	PORT GENTIL	0.7S	8.8E	4		1			W	FOREIGN	FR. FEW T OBS ON MCDS
236	64870	N'GAOUNDERE	7.3N	13.3E	1115		1			W	FOREIGN	FR.
237	67001	MORONI	11.7S	43.2E	12					W	FOREIGN	FR. 1963 ONLY

KN	WMO	STATION NAME	LAT	LONG	HT	00	06	12	18	TYPE	SOURCE	COMMENTS
238	67009	DIEGO SUAREZ	12.4S	49.3E	114	1				W	FOREIGN	FR.
239	67197	FORT DAUPHIN	25.0S	47.0E	8	1				W	FOREIGN	FR.
240	81405	CAYENNE	4.8N	52.4W	8			1		WT	FOREIGN	FR.
241	64910	DOUALA	4.0N	9.7E	13		1			WT	FOREIGN	FR.
242	67085	TANANARIVE	18.9S	47.0E	1300	1				WT	FOREIGN	FR.
243	91938	TAHITI	17.6S	149.6W	2	1	1	1	1	WT	FOREIGN	BRT. 10/59- 12Z WINDS INTERP.
244	41350	GAN	0.7S	73.2E	2		1	1	1	WT	FOREIGN	BRT.
245	63741	NAIROBI	1.3S	36.0E	1792	1	1	1		WT	FOREIGN	BRT.
246	63894	DAR ES SALAAM	6.9S	39.3E	58		1	1		WT	FOREIGN	AUST.
247	91517	HONIARA	9.4S	160.0E	57	1				WT	FOREIGN	AUST.
248	94027	LAE	6.7S	147.0E	8	1				WT	FOREIGN	AUST.
249	94120	DARWIN	12.4S	130.9E	26	1				WT	FOREIGN	AUST.
250	94294	TOWNSVILLE	19.3S	146.8E	4	1				WT	FOREIGN	AUST.
251	94299	WILLIS IS.	16.3S	150.0E	8	1				WT	FOREIGN	AUST.
252	94300	CARNAVON	24.9S	113.7E	5	1				WT	FOREIGN	AUST.
253	94312	PT. HEDLAND	20.4S	118.6E	8	1				WT	FOREIGN	AUST.
254	94326	ALICE SPRINGS	23.8S	133.9E	546	1				WT	FOREIGN	AUST.
255	94335	CLONCURRY	20.7S	140.5E	188	1				WT	FOREIGN	AUST.
256	96996	COCOS IS.	12.1S	96.9E	3	1				WT	FOREIGN	AUST.
257	98836	ZAMBOANGA	6.9N	122.1E	6	1				T	ASHEVILLE	1958 MOSTLY
258	61900	ASCENSION IS.	7.9S	14.4W	6	1				WT	ASHEVILLE	
259	82400	FERNANDO NORONHA	3.9S	32.4W	45	1				WT	ASHEVILLE	
260	84129	GUAYAQUIL	2.2S	79.9W	4	1				WT	ASHEVILLE	
261	83781	SAO PAULO	23.6S	46.7W	795	1				WT	ASHEVILLE	
262	91643	FUNAFUTI	8.5S	179.2E	1	1		1		W	FOREIGN	N.Z.
263	91680	NANDI	17.8S	177.5E	16	1		1		WT	FOREIGN	N.Z. TEMPS 00Z ONLY
264	91843	RAROTONGA	21.2S	159.8W	3	1		1		W	FOREIGN	N.Z.
265	93997	RAOUL IS.	29.3S	177.9W	48	1				T	FOREIGN	N.Z.
266	82898	RECIFE	8.0S	34.9W	10	1				WT	ASHEVILLE	
267	84631	LIMA	12.1S	77.0W	155	1				WT	ASHEVILLE	
268	85442	ANTOFAGASTA	23.5S	70.4W	128	1				WT	MCDS+ASHVL	IGY PERIOD
269	91489	CHRISTMAS IS.	2.0N	157.4W	3	1				WT	FOREIGN	FR. 1/63-
270	91592	NOUMEA	22.3S	166.5E	75	1				WT	ASHEVILLE	7/57-11/57 ONLY
271	76225	CHIHUAHUA	28.6N	106.1W	1424	1				WT	MICROCARDS	3 MONTHS ONLY
272	41756	JIWANI	25.1N	61.8E	55	1				WT	MICROCARDS	
273	41917	DACCA	23.8N	90.4E	10	1				T	MCDS+ASHVL	ENDS 6/61
274	48097	RANGOON	16.8N	96.2E	22	1				WT	MICROCARDS	FEW WINDS
275	48802	CHAPA	22.4N	103.8E	1570	1				WT	MICROCARDS	
276	64005	COQUILHATVILLE	0.1N	18.3E	325			1		WT	MICROCARDS	
277	64210	LEOPOLDVILLE	4.3S	15.3E	309			1		WT	MICROCARDS	
278	64360	ELISABETHVILLE	11.6S	27.5E	1277			1		WT	MICROCARDS	
279	66160	LUANDA	8.9S	13.2E	70	1				WT	FRN. TAB.	
280	66422	MOCAMEDES	15.2S	12.2E	44	1				WT	FRN. TAB.	
281	67241	LUMBO	15.0S	40.7E	11	1				WT	FRN. TAB.	
282	67341	LOURENCO MARQUES	25.9S	32.6E	43	1				WT	FRN. TAB.	
283	67475	KASAMA	10.2S	31.1E	1395			1		WT	MICROCARDS	6 MONTHS ONLY
284	67587	LILONGWE	14.0S	33.7E	1135			1		WT	MICROCARDS	6 MONTHS ONLY
285	67663	BROKEN HILL	14.5S	28.5E	1207			1		WT	MICROCARDS	

KN	WMO	STATION NAME	LAT	LONG	HT	00	06	12	18	TYPE	SOURCE	COMMENTS
286	67774	SALISBURY	17.8S	31.0E	1473			1		WT	MICROCARDS	
287	80401	MARACAY	10.3N	67.7W	432	1				T	MICROCARDS	
288	87157	RESISTENCIA	27.5S	59.0W	52	1				WT	MCDS+ASHVL	
289	96745	DJAKARTA	6.2S	106.8E	8			1		T	MCDS+ASHVL	ENDS 2/61
290	96933	SURABAJA	7.2S	112.7E	3			1		WT	MCDS+ASHVL	
291	43466	COLOMBO	6.9N	79.9E	6	1		1		WT	ASHEVILLE	1960 MOSTLY
292	61967	DIEGO GARCIA	7.2S	72.4E	2	1				WT	FRN. TAB.	7/63-
293	61995	VACOAS	20.3S	57.5E	425	1				WT	FRN. TAB.	1/61-
294	61996	NOUVELLE AMSTERDAM	37.8S	77.6E	28	1				WT	ASHEVILLE	IGY PERIOD (FORMERLY 67198)
295	63450	ADDIS ABABA	9.0N	38.7E	2360			1		WT	ASHEVILLE	1/58-12/62 IRREGULAR
296	68406	ALEXANDER BAY	28.6S	16.5E	27	1				WT	ASHEVILLE	IGY PERIOD
297	68588	DURBAN	30.0S	31.0E	8	1				WT	ASHEVILLE	IGY PERIOD
298	68816	CAPE TOWN	34.0S	18.6E	46	1				WT	ASHEVILLE	IGY PERIOD
299	68906	GOUGH ISLAND	40.4S	9. W	54	1		1		WT	ASHEVILLE	IGY PERIOD Foreign, S. Africa, 7/57 - 4/66
300	85543	QUINTERO	32.8S	71.5W	2	1				WT	ASHEVILLE	
301	87576	EZEIZA	34.8S	58.5W	24	1				WT	ASHEVILLE	IGY PERIOD
302	94461	GILES	25.0S	128.3E	599	1				WT	FOREIGN	AUST.
303	94510	CHARLEVILLE	26.4S	146.3E	304	1				WT	FOREIGN	AUST.
304	94610	PERTH AP	31.9S	116.0E	18	1				WT	FOREIGN	AUST.
305	94995	LORD HOWE IS.	31.5S	159.1E	46	1				WT	FOREIGN	AUST.
306	94996	NORFOLK IS.	29.1S	167.9E	109	1				WT	FOREIGN	AUST.
	94646	Forrest	30.5S	128.1E	157	1				WT	FOREIGN	AUST. 6/57 - 12/69
	94659	Woomera	31.1S	136.5E	166	1				WT	FOREIGN	AUST. 2/57 - 12/69
	94672	Adelaide	34.6S	138.3E	11	1				WT	FOREIGN	AUST. 6/57 - 12/69
	94711	Cobar	31.4S	145.5E	265	1				WT	FOREIGN	AUST. 6/62 - 12/69
	94750	Nowra	34.6S	150.3E	117	1				WT	FOREIGN	AUST. 3/58 - 12/69
	94776	Williamtown	32.5S	151.5E	9	1				WT	FOREIGN	AUST. 3/57 - 12/69
	94865	Laverton	37.5S	144.5E	14	1				WT	FOREIGN	AUST. 5/57 - 12/69
	94910	Wagga	35.1S	147.3E	214	1				WT	FOREIGN	AUST. 12/65 - 12/69
	94975	Hobart Airport	42.5S	147.3E	7	1				WT	FOREIGN	AUST. 4/57 - 12/69
	94998	MacQuarie Isl.	54.3S	158.6E	6	1				WT	FOREIGN	AUST. 4/57 - 12/69
	93112	Whenuapai	36.5S	174.4E	28	1				WT	FOREIGN	N.Z. 7/57 - 4/66
	93780	Christchurch	43.3S	172.3E	34	1				WT	FOREIGN	N.Z. 7/57 - 4/66
	93844	Invercargill	46.3S	168.2E	1	1				WT	FOREIGN	N.Z. 7/57 - 4/66
	93944	Campbell Isl.	52.3S	169.1E	19	1				WT	FOREIGN	N.Z. 7/57 - 4/66
	78825	Clipperton	10.2N	109.1W	4	1				W	FOREIGN	FR. 4/67 - 7/67, 5/68 - 8/68
	85469	Easter Isl.	27.1S	109.3W	41	1				WT	FOREIGN	CHILE May, June, July 1970
	82193	Belem	01.2S	48.3W	16			1		WT	FOREIGN	BRAZIL 8/68 - 12/69
	82332	Manaus	03.1S	59.6W	84			1		WT	FOREIGN	BRAZIL 6/67 - 12/69
	82599	Natal	05.5S	35.1W	8			1		WT	FOREIGN	BRAZIL 1/67 - 12/69
	83208	Vilhena	12.4S	60.1W				1		WT	FOREIGN	BRAZIL 3/67 - 12/69
	83378	Brazilia	15.5S	47.6W	1061			1		WT	FOREIGN	BRAZIL 5/66 - 12/69
	83612	Campo Grande	20.3S	54.4W	563			1		WT	FOREIGN	BRAZIL 1/67 - 12/69
	83840	Curitiba	25.3S	49.1W	910			1		WT	FOREIGN	BRAZIL 1/67 - 12/69
	83971	Porto Alegre	30.0S	51.1W	4			1		WT	FOREIGN	BRAZIL 9/65 - 12/68
	83650	Trindade Isl.	20.3S	29.2W	21			1		T	FOREIGN	BRAZIL 2/60 - 12/63

Appendix II

The Objective Analysis Technique

The objective analysis technique used in this study grew out of the screening regression approach of Eddy (1967a, b), which is appropriate for the case of isotropic, stationary scalar fields. It produces grid-point values from station data and may be used to produce contour maps and zonally averaged quantities. Although rather complex and time-consuming, this regression technique has two main advantages over simpler schemes, for example that of Cressman (1959), which take an average of all station values around a grid point using a weighting function dependent on the distance from it. The main advantages are as follows: (1) Since the correlation between adjacent stations is relatively high, selection of one station from an area of high station density tends to prevent the selection of others, and the analyzed value is based on a more uniform distribution of stations about the grid point. Consequently better analyses are obtained where field gradients are sharper. (2) The sum of the regression coefficients or weights is not usually equal to one, and may be greater. For this reason, high and low values may be built up in areas surrounded by anomalies of the same sign. In the simpler schemes the analyzed value has to lie between the highest and lowest values of the nearby stations.

At first it was hoped that the Eddy scheme could be used without modification, and early results for temperature fields were encouraging. However, for quantities, such as the momentum flux, which are very susceptible to observational error and show less continuity between adjacent stations, the results were far from satisfactory. This was traced to a small zero lag correlation and a relatively short influence radius. The resulting analyses were lacking in continuity and gave a poor fit to apparently well-defined maxima and minima in the data fields. From a statistical viewpoint the Eddy scheme represents the ultimate analysis, and the low autocorrelations are simply a reflection of the uncertainty in the observations.

Examination of plotted maps showed that reasonable results could be obtained by hand analysis however, with some averaging of neighboring values being done by eye. With the belief that the hand analyses were a better representation of the true situation, and with a desire to reduce the computer time per map, several modifications of the Eddy scheme

247

were tried with the intention of reproducing the hand analyses as closely as possible. The scheme finally chosen proceeds as outlined below.

Step 1. Data checking and trend removal. Data are grouped into latitude bands, and within each band the mean and standard deviations are found. Data which differ from the mean by more than a prescribed number of standard deviations are eliminated. Following this, a mean latitudinal trend is computed by filtering the zonal means or by fitting a polynomial to them. The trend thus computed is the first approximation to the analysis, and it is removed from each of the station values.

Step 2. Regression analysis. A screening regression analysis is carried out at each grid point where the predictand is located. As predictors, a limited number of stations closest to the grid point are selected. Their "correlations" with each other and with the predictand are defined solely in terms of an arbitrary function of the separation distance. With the correlation (or covariance) matrix formed in this way, the screening procedure described by Efroymson (1962) is followed. The predictors which give the largest reduction in the predictand variance are selected in turn until no further significant reduction is obtained. The predictand value is obtained by multiplying the station residues by the regression coefficients, and it is added to the latitudinal trend. The percent reduction in predictand variance is retained and gives a rough indication of the reliability of the analysis. Typically it ranges from zero in isolated areas to about 90 percent where data coverage is good.

The arbitrary correlation function chosen was

$$r(d) = 0.9 \times \left(\frac{D^2 - d^2}{D^2 + d^2} \right),$$

where d is the separation distance and D the influence radius beyond which the correlation is taken as zero. The form of this function is identical to that frequently used for a weighting function in the simple Cressman-type schemes. The zero lag value of 0.9 was based on typical values for temperature fields which show good continuity and which are most suited to the Eddy approach. A suitable value for the influence radius D was obtained by experimenting with different parameters and levels. In general, shorter influence radii gave a better fit to station values, while longer radii afforded better continuity in regions of few observations. The difference in the resulting analyses, however, was not very great even when D was changed by a factor of 2, and this was taken as an indication that the analysis was not sensitive to the form of the autocorrelation curve. The initial set of analyses presented in Kidson, Vincent, and Newell (1969) were made with $D = 1900$ nm (3520 km) at the equator, while in the more recent analyses a value of 1600 nm (2965 km) was used. The resulting zonal means and energy integrals showed only slight changes.

The analyses were performed on a square grid on a mercator projection with the grid size corresponding to 10° longitude at the equator. The latitudes corresponding to the nine rows of the grid are 41.0°N, 33.0°N, 24.2°N, 14.8°N, 5.0°N, 5.0°S, 14.8°S, 24.2°S and 33.0°S. Zonal means were formed from the 36 grid-point values at each of these latitudes (without regard for surface elevation). The latitudinal trend removal and data checking were carried out using the original Eddy subroutine. A limit of 3 standard deviations was used to eliminate bad wind observations, and a limit of 4 standard deviations for the temperature and heat flux. Statistics calculated from fewer than 30 observations for means of single quantities, and from fewer than 90 observations for covariances, were rejected. The trend was obtained by fitting a 6th-degree polynomial to some 19 latitude band means with a spacing of one-half in grid length.

As well as converting the original Eddy program to work in grid coordinates instead of latitude and longitude, part of it was converted from Fortran to Assembler Language. The net result of all these changes was to reduce the execution time of the program from the order of a minute to 10 seconds per map. This made the analysis of a large number of maps economically feasible. The resulting analyses were printed out on the line printer using a contouring subroutine kindly provided by Dr. R. C. Gammill. This program fitted 3rd-degree polynomials to the grid-point values so that a smoothly varying field was obtained.

Although these maps are not included in this monograph in favor of subjectively analyzed charts, they are suitable for many purposes and provide a good check on the hand analyses. Zonally averaged values of the variables were found to be in reasonable agreement with those obtained from subjective analysis and have occasionally been used as noted in the text.

Appendix III

The Accuracy of [v̄] as Obtained from Objective Analyses

The uncertainty in the grid-point values used for determining $[\bar{v}]$ can be obtained directly from the original objective analysis scheme of Eddy (1967a, b) referred to in Chapter 2. Three examples are given to illustrate the probable error in the analysis at the 850, 500, and 200-mb levels for the period December–February.

The standard deviation of the possible error in the grid-point values may be obtained directly from the reduction in variance resulting from the screening procedure and the field variance remaining after the latitudinal trend has been removed. In isolated areas, where no reduction in the predicted variance is possible, maximum errors occur and the standard deviation of the grid point error is 1.75, 2.35, and 4.14 m sec^{-1} at the 850, 500, and 200-mb levels, respectively. In areas of high station density such as the continental United States the greatest reductions in variance are obtained and the standard deviations of the grid-point errors reduced to a minimum of 1.03 m sec^{-1} at the 850-mb and 500-mb levels. The analysis at the 200-mb level is unsatisfactory as only small reductions in variance are obtained. A maximum reduction in variance of 81 percent occurs at the 500-mb level, where the continuity of the data is apparently best. If we assume that the errors at the grid points are independent and that the standard deviation at each grid point is 1 m sec^{-1}, then the 95 percent confidence limits for the mean of 36 grid points is $2\sigma/\sqrt{N} = 33$ cm sec^{-1}. This implies that if the data coverage at all longitudes were as dense as over the continental United States, the uncertainty in the zonal average would still be of the same order as the magnitude of the zonal average at middle latitudes. In practice the confidence limits will be larger than this since, in areas of poor coverage, the errors at adjacent grid points are well correlated because they derive their values from the same station or stations. Furthermore, as can be seen from the maps and the zonal cross sections, the pattern of alternating north-south motion results in zonally averaged mean meridional velocities which are small compared to individual grid-point values, particularly at middle latitudes. This averaging of positive and negative values of almost equal magnitude results in zonal means subject to a large probable error.

With these difficulties in mind it will be appreciated that the middle-latitude values of $[\bar{v}]$ are not very well determined, but that the larger values in the tropics can be treated with some confidence.

References

Eddy, A. 1967a. The statistical objective analysis of scalar data fields. *J. Appl. Meteorol.* 9:597–609.

————. 1967b. *Two-Dimensional Statistical Objective Analysis of Isotropic Scalar Data Fields.* Austin: University of Texas Atmospheric Science Group, publication no. 5. 100 pp.

Cressman, G. P. 1959. An operational objective analysis system. *Mon. Weath. Rev.* 87:367–374.

Efroymson, M. A. 1962. Multiple regression analysis. In *Mathematical Methods for Digital Computers,* A. Ralston and H. S. Wilf, eds. New York: Wiley, vol. I, pp. 191–203.

Kidson, J. W., D. G. Vincent, and R. E. Newell. 1969. Observational studies of the general circulation of the tropics: long term mean values. *Quart. J. Roy. Meteorol. Soc.* 95:258–287.

Name Index

Subject Index

Stratosphere (continued)
  Northern Hemisphere middle, standing eddy fluxes in, 144
  polar, 37, 47
Stress
  frictional, 132
  surface, 47, 153, 154, 157
Subtropics
  cloudiness of, 217
  definition of, 1
  divergence in, 213
  hydrological regimes of, 219
  precipitation values in, 218
  relative humidity in, 202 (*see also* Humidity)
Summer, wind flow during, 31
Surface stress, 153, 154, 157
  calculation of, 47
Symbols, list of, 7

Taylor-Winston brightness values, 216
Technology, satellite, 221
"Teleconnections," 1
Temperature, 17
  daily station values for, 9
  equatorial, 26
  map of, 17
  mean, 26
  patterns of, 17
  potential, 3
  seasonal statistics for, 11
  seasonal values for, 14
  standard deviation of, 10, 26
  time-mean, 17
  zonally averaged, 19
  zonal mean, 50, 51
Topography, 193
Torques
  external, 132, 133, 151
  gravitational tidal, 131
  mountain, 132, 133, 151
  negative, 132
  surface, 155
Trace elements, atmospheric redistribution of, xiii
Trade-flow, average, 215
Trade winds
  northeast, 196, 206
  southeast, 197

Transequatorial flux
  seasonal variations in, 218
  values associated with, 219
Transient eddy flux, 6, 135, 207, 208. *See also* Eddy flux; Standing eddy flux
  vertically integrated values of, 207
Transient eddy momentum flux, 412
Transient eddy transfer, in low latitudes, 208
Travelers Research Center (TRC), xiii, 10, 196
TRC. *See* Travelers Research Center
Tropical atmosphere
  definition of, 1
  energy budget of, 4
  water balance of, 209 (*see also* Water balance)
Tropics
  convergence in, 213
  definition of, 196
  hydrologic cycle of, 214–219
  meridional motions in, 37 (*see also* Meridional motions)
  water balance of, 197
  water vapor content of, 201, 208
Troposphere
  Hadley cell in, 163 (*see also* Hadley cell
  relative angular momentum flux in, 135
  seasonal changes in, 47
  standing eddy fluxes in, 144
  temperature in, 21
  tropical, 4
  upper, 39
  vapor transport in, 196
Turbulent momentum flux, 133

Univac 1108, 196

Vapor transport
  easterly, 204
  easterly and westerly, 203
  vertical, 221
Vertical momentum flux, 150
Vertical velocities, zonal mean, 56–57

Water, precipitable, 199. *See also* Precipitable water

Water balance, of tropics, 194, 197, 211–214, 218
Water-balance equation, 218
Water-balance studies, 194
Water-budget components, computation of, 218
Water vapor
  general circulation statistics for, 194
  horizontal flux of, 203–211
  meridional flux of, 205, 210
    longitudinally averaged, 208
  meridional transfer of, 205
  transequatorial flow of, 216
  transequatorial transfer of, 210–211
  vertically integrated, 201
  zonal flux of, 203
Water-vapor content
  equatorial seasonal variations in, 216
  line of maximum atmospheric, 207
  mean monthly, 200
  zonally averaged values of, 215
Water vapor flux
  analysis of, 196
  mean meridional, 207
Water vapor transfer, 219
Wave motions, 4
Waves, 133
  baroclinic, 47
  standing, large-scale, 4
Weather Bureau, U.S., 197
Westerlies, 5, 28
  angular momentum in, 149
  peak, 30
  in Western Hemisphere, 31
  during winter, 37
Wind. *See also* Easterlies; Westerlies
  data on, 17
  geostrophic, 2
  mean meridional, 38, 54–55
  mean zonal, 52–53
  meridional component, standard deviation of, 60–61
  seasonal statistics for, 11
  surface, 28, 149
  tropical low-level, 2
  vertical gradients of, 5
  westerly, 132
  zonally averaged, maintenance of, 158
Wind components
  mean monthly, 195

meridional, 134
meridional, longitudinally averaged, 197
"typical" values of, 196
variance of, 47, 51, 63–129
zonal, 58–59
  longitudinally averaged, 197
Wind fields
  low-level, time-mean, 193
  mean temperature and, 50, 62–128
  tropical, low-level, 196–198
  zonal, 26–37
  zonal mean, 3
Wind stations, radiosonde, 10
Wind velocity
  associated with southern Hadley cell, 211
  daily station values for, 9
  in extratropical latitudes, 164
  seasonal distribution of, 39
  seasonal values for, 14
  and specific humidity, 209
  standard deviations of, 206
Winter
  meridional flow during, 38
  wind flow during, 31
World Atlas of Sea Surface Temperatures, 18
World Meteorological Organization, 10
World Weather Records, 17, 153

Zero zonal transport, line of, 203
Zonal flux of water vapor, 203–204. *See also* Water vapor
Zonal-momentum equation, 151
Zonal wind
  acceleration of, 164
  maintenance of, 157–163
  mean, 31
Zonal wind component, standard deviation of, 58–59
Zonal wind equation, 157–158, 163